Computational Medicine in Data Mining and Modeling

Goran Rakocevic • Tijana Djukic
Nenad Filipovic • Veljko Milutinović
Editors

Computational Medicine in Data Mining and Modeling

Editors
Goran Rakocevic
Mathematical Institute
Serbian Academy of Sciences
and Arts
Belgrade, Serbia

Nenad Filipovic
Faculty of Engineering
University of Kragujevac
Kragujevac, Serbia

Tijana Djukic
Faculty of Engineering
University of Kragujevac
Kragujevac, Serbia

Veljko Milutinović
School of Electrical Engineering
University of Belgrade
Belgrade, Serbia

ISBN 978-1-4939-4834-5 ISBN 978-1-4614-8785-2 (eBook)
DOI 10.1007/978-1-4614-8785-2
Springer New York Heidelberg Dordrecht London

Printed on acid-free paper

Springer is part of Springer Science+Business Media (www.springer.com)

Preface

Humans have been exploring the ways to heal wounds and sicknesses since times we evolved as a species and started to form social structures. The earliest of these efforts date back to prehistoric times and are, thus, older than literacy itself. Most of the information regarding the techniques that were used in those times comes from careful examinations of human remains and the artifacts that have been found. Evidence shows that men used three forms of medical treatment – herbs, surgery, and clay and earth – all used either externally with bandages for wounds or through oral ingestion. The effects of different substances and the proper ways of applying them had likely been found through trial and error. Furthermore, it is likely that any form of medical treatment was accompanied by a magical or spiritual interpretation.

The earliest written accounts of medical practice date back to around 3300 BC and have been created in ancient Egypt. Techniques that had been known at the time included setting of broken bones and several forms of open surgery; an elaborate set of different drugs was also known. Evidence also shows that the ancient Egyptians were in fact able to distinguish between different medical conditions and have introduced the basic approach to medicine, which includes a medical examination, diagnoses, and prognoses (much the same it is done to this day). Furthermore, there seems to be a sense of specialization among the medical practitioners, at least according to the ancient Greek historian Herodotus, who is quoted as saying that the practice of medicine is so specialized among them that each physician is a healer of one disease and no more. Medical institutions, referred to as Houses of Life, are known to have been established in ancient Egypt as early as the First Dynasty.

The ancient Egyptian medicine heavily influenced later medical practices in ancient Greece and Rome. The Greeks have left extensive written traces of their medical practices. A towering figure in the history of medicine was the Greek physician Hippocrates of Kos. He is widely considered to be the “father of modern medicine” and has invented the famous Oath of Hippocrates, which still serves as the fundamental ethical norm in medicine. Together with his students, Hippocrates began the practice of categorizing illnesses as acute, chronic, endemic, and epidemic. Two things can be observed from this: first, the approach to medicine was taking up a scholarly form, with groups of masters and students studying different medical conditions, and second, a systematic approach was taken. These observations lead to the conclusion that medicine had been established as a scientific field.

In parallel with the developments in ancient Greece and, later, Rome, the practice of medicine has also evolved in India and China. According to the sacred text of Charaka, based on the Hindu beliefs, health and disease are not predetermined and life may be influenced by human effort. Medicine was divided into eight branches: internal medicine, surgery and anatomy, pediatrics, toxicology, spirit medicine, aphrodisiacs, science of rejuvenation, and eye, ear, nose, and throat diseases. The healthcare system involved an elaborate education structure, in which the process of training a physician took seven years. Chinese medicine, in addition to herbal treatments and surgical operations, also introduced the practices of acupuncture and massages.

During the Islamic Golden Age, spanning from the eighth to the fifteenth century, scientific developments had been centered in the Middle East and driven by Islamic scholars. Central to the medical developments at that time was the Islamic belief that Allah had sent a cure for every ailment and that it was the duty of Muslims to take care of the body and spirit. In essence, this meant that the cures had been made accessible to men, allowing for an active and relatively secular development of medical science. Islamic scholars also gathered as much of the already acquired knowledge as they could, both from the Greek and Roman sources, as well as the East. A sophisticated healthcare system was established, built around public hospitals. Furthermore, physicians kept detailed records of their practices. These data were used both for spreading and developing knowledge, as well as could be provided for peer review in case a physician was accused of malpractice. During the Islamic Golden Age, medical research went beyond looking at the symptoms of an illness and finding the means to alleviate them, to establishing the very cause of the disease.

The sixteenth century brought the Renaissance to Europe and with it a revival of interest in science and knowledge. One of the central focuses of that age was the "man" and the human body, leading to large leaps in the understanding of anatomy and the human functions. Much of the research that was done was descriptive in nature and relied heavily on postmortem examinations and autopsies. The development of modern neurology began at this time, as well as the efforts to understand and describe the pulmonary and circulatory systems. Pharmacological foundations were adopted from the Islamic medicine, and significantly expanded, with the use of minerals and chemicals as remedies, which included drugs like opium and quinine. Major centers of medical science were situated in Italy, in Padua and Bologna.

During the nineteenth century, the practice of medicine underwent significant changes with rapid advances in science, as well as new approaches by physicians, and gave rise to modern medicine. Medical practitioners began to perform much more systematic analyses of patients' symptoms in diagnosis. Anesthesia and aseptic operating theaters were introduced for surgeries. Theory regarding microorganisms being the cause of different diseases was introduced and later accepted. As for the means of medical research, these times saw major advances in chemical and laboratory equipment and techniques. Another big breakthrough was brought on by the development of statistical methods in epidemiology. Finally, psychiatry had been established as a separate field. This rate of progress continued well into the twentieth century, when it was also influenced by the two World Wars and the needs they had brought forward.

The twenty-first century has witnessed the sequencing of the entire human genome in 2003, and the subsequent developments in the genetic and proteomic sequencing technologies, following which we can study medical conditions and biological processes down to a very fine grain level. The body of information is further reinforced by precise imaging and laboratory analyses. On the other hand, following Moore's law for more than 40 years has yielded immensely powerful computing systems. Putting the two together points to an opportunity to study and treat illnesses with the support of highly accurate computational models and an opportunity to explore, in silico, how a certain patient may respond to a certain treatment. At the same time, the introduction of digital medical records paved the way for large-scale epidemiological analyses. Such information could lead to the discovery of complex and well-hidden rules in the functions and interactions of biological systems.

This book aims to deliver a high-level overview of different mathematical and computational techniques that are currently being employed in order to further the body of knowledge in the medical domain. The book chooses to go wide rather than deep in the sense that the readers will only be presented the flavors, ideas, and potentials of different techniques that are or can be used, rather than giving them a definitive tutorial on any of these techniques. The authors hope that with such an approach, the book might serve as an inspiration for future multidisciplinary research and help to establish a better understanding of the opportunities that lie ahead.

Belgrade, Serbia Goran Rakocevic

Contents

Chapter 1
Mining Clinical Data

Argyris Kalogeratos, V. Chasanis, G. Rakocevic, A. Likas, Z. Babovic, and M. Novakovic

1.1 Data Mining Methodology

The prerequisite of any machine learning or data mining application is to have a clear target variable that the system will try to learn [27]. In a supervised setting, we also need to know the value of this target variable for a set of training examples (i.e., patient records). In the case study presented in this chapter, the value of the considered target variable that can be used for training is the ground truth characterizations of the coronary artery disease severity or, as a different scenario, the progression of the patients. We either set as target variable the disease severity, or disease progression, and then we consider a two-class problem in which we aim to discriminate a group of patients that are characterized as "severely diseased" or "severely progressed," from a second group containing "mildly diseased" or "mildly progressed" patients, respectively. This latter mild/severe characterization is the actual value of the target variable for each patient.

In many cases, neither the target variable nor its ground truth characterization is strictly specified by medical experts, which is a fact that introduces high complexity

A. Kalogeratos (✉) • V. Chasanis • A. Likas
Department of Computer Science, University of Ioannina, GR-45110 Ioannina, Greece
e-mail: argyriskalogeratos@gmail.com; akaloger@cs.uci.gr

G. Rakocevic
Mathematical Institute, Serbian Academy of Sciences and Arts, Belgrade 11000, Serbia

Z. Babovic • M. Novakovic
Innovation Center of the School of Electrical Engineering, University of Belgrade, Belgrade 11000, Serbia

G. Rakocevic et al. (eds.), *Computational Medicine in Data Mining and Modeling*,
DOI 10.1007/978-1-4614-8785-2_1,

and difficulty to the data mining process. The general data mining methodology we applied is a procedure divided into six stages:

Stage 1: Data mining problem specification

- Specify the objective of the analysis (the target variable).
- Define the ground truth for each training patient example (the specific value of the target variable for each patient).

Stage 2: Data preparation, where some preprocessing of the raw data takes place

- Deal with data inconsistencies, different feature types (numeric and nominal), and missing values.

Stage 4: Data subset selection

- Selection of a feature subset and/or a subgroup of patient records

Stage 5: Training of classifiers

- Build proper classifiers using the selected data subset.

Stage 6: Validate the resulting models

- Using techniques such as v-fold cross-validation.
- Compare the performance of different classifiers.
- Evaluate the overall quality of the results.
- Understand whether the specification of the data mining problem and/or the definition of the ground truth values are appropriate in terms of what can be extracted as knowledge from the available data.

A popular methodology to solve these classification problems is to use a decision tree (DT) [28]. DTs are popular tools for classification that are relatively fast to both train and make predictions, while they also have several other additional advantages [10]. First, they naturally handle missing data; when a decision is made on a missing value, both subbranches are traversed and a prediction is made using a weighted vote. Second, they naturally handle nominal attributes. For instance, a number of splits can be made equal to the number of the different nominal values. Alternatively, a binary split can be made by grouping the nominal values into subsets. Most important of all, a DT is an interpretable model that represents a set of rules. This is a very desirable property when applying classification models to medical problems since medical experts can assess the quality of the rules that the DTs provide.

There are several algorithms to train DT models, among the most popular of them are ID3 and its extension C4.5 [2]. The main idea of these algorithms is to start building a tree from its root, and at each tree node, a split of the data in two subsets is determined using the attribute that will result in the minimum entropy (maximum information gain).

DTs are mainly used herein because they are interpretable models and have achieved good classification accuracy in many of the considered problems.

However, other state-of-the-art methods such as the support vector machine (SVM) [3] may provide better accuracy at the cost of not being interpretable. Another powerful algorithm that builds non-interpretable models is the random forest (RF) [18]. An RF consists of a set of random DTs, each of them trained using a small random subset of features. The final decision for a data instance is taken using strategies such as weighted voting on the prediction of the individual random DTs. This also implied that a decision can be made using voting on contradicting rules and explains why these models are not interpretable. In order to assess the quality of the DT models that we build, we compare the classification performance of DTs to other non-interpretable classifiers such as the abovementioned SVM and RF.

Another property of DTs is that they automatically provide a measure of the significance of the features since the most significant features are used near the root of the DT. However, other feature selection methods can also be used to identify which features are significant for the classification tasks that we study [7]. Most feature selection methods search over subsets of the available features to find the subset that maximizes some criterion [4]. Common criteria measure the correlation between features and the target category, such as the information gain (IG) or chi-squared measures. Among the state-of-the-art feature selection techniques are the RFE-SVM [6], mRMR [22], and MDR [13] techniques. They differ to the previous approaches in that they do not use single-feature evaluation criteria. Instead, they try to eliminate redundant features that do not contain much information. In this way, a feature that is highly correlated with other features is more probable to be eliminated than a feature that may have less IG (as single-feature evaluation measure) comparing to the IG of the first but at the same time carries information that is not highly correlated with other features [11].

1.2 Data Mining Algorithms

In this section we briefly describe the various algorithms used in our study for classifier construction and feature evaluation/selection, as well as the measures we used to assess the generalization performance of the obtained models.

1.2.1 Classification Methods

1.2.1.1 Decision Trees

A decision tree (DT) is a decision support tool that uses a treelike graph representation to illustrate the sequence of decisions made in order to assign an input instance to one of the classes. The internal node of a decision tree corresponds to an attribute test. The branches between the nodes tell us the possible values that these attributes can have in the observed samples, while the terminal (leaf) nodes provide the final value (classification label) of the dependent variable.

A popular solution is the J48 algorithm for building DTs that has been implemented in the very popular Weka software for DM [2]. It is actually an implementation of the well-known and widely studied C4.5 algorithm for building decision trees [15]. The tree is built in a top-down fashion, and at each step, the algorithm splits a leaf node by identifying the attribute that best discriminates the subset of instances that correspond to that node. A typical criterion that is commonly used to quantify the splitting quality is the information gain. If a node of high-class purity is encountered, then this node is considered as a terminal node and is assigned the label of the major class. Several post-processing pruning operations also take place using a validation in order obtain relatively short trees that are expected to have better generalization.

It is obvious that the great advantage of DTs as classification models is their interpretability, i.e., their ability to provide the sequence of decisions made in order to get the final classification result. Another related advantage is that the learned knowledge is stored in a comprehensible way, since each decision tree can be easily transformed to a set of rules. Those advantages make the decision trees very strong choices for data mining problems especially in the medical domain, where interpretability is a critical issue.

1.2.1.2 Random Forests

A random forest (RF) is an ensemble of decision trees (DTs), i.e., it combines the prediction made by multiple DTs, each one generated using a different randomly selected subset of the attributes [18]. The output combination can be done using either simple voting or weighted voting. The RF approach is considered to provide superior results to a single DT and is considered as a very effective classification method competitive to support vector machines. However, its disadvantage compared to DTs is that model interpretability is lost since a decision could be made using voting on contradicting rules.

1.2.1.3 Support Vector Machines

The support vector machine classifier (SVM) [6, 16] is a supervised learning technique applicable to both classification and regression. It provides state-of-the-art performance and scales well even with large dimension of the feature vector. More specifically, suppose we are given a training set of l vector with d dimensions, $x_i \in R^d$, $i = 1, \ldots, n$, and a vector $y \in R_l$ with $y_i \in \{1, -1\}$ denoting the class of vector x_i. The classical SVM classifier finds an optimal hyperplane which separates data points of two classes in such way that the margin of separation between the two classes is maximized. The margin is the minimal distance from the separating hyperplane to the closest data points of the two classes. Any hyperplane can be written as the set of points x satisfying $w^T x + b = 0$. The vector w is a normal vector and is perpendicular to the hyperplane. A mapping function $\varphi(x)$ is assumed that maps each training vector to a higher dimensional space, and the corresponding kernel function defined as the inner product $K(x,y) = \varphi^T(x) \cdot \varphi(y)$.

Then the SVM classifier is obtained by solving the following primal optimization problem:

$$\min_{w,b,\xi} \frac{1}{2} w^T w + C_i \sum_{i=1}^{l} \xi_i \tag{1.1}$$

$$\begin{aligned} &\textit{subject to } y_i(w^T \varphi(x_i) + b) \geq 1 - \xi_i, \\ &\xi_i \geq 0, i = 1, \ldots, l \end{aligned} \tag{1.2}$$

where ξ_i is called slack variable and measures the extent to which the example x_i violates the margin condition and C a tuning parameter which controls the balance between training error and the margin. The decision function is thus given from the following equation:

$$sqn\left(\sum_{i=1}^{l} w_i K(x_i, x) + b\right), \quad where K(x_i, x_j) = \phi^T(x_i)\phi(x_j) \tag{1.3}$$

A notable characteristic of SVMs is that, after training, usually most of the training instances x_i have $w_i = 0$ in the above equation [17]. In other words, they do not contribute to the decision function. Those x_i for which $w_i = 0$ are retained in the SVM model and called support vectors (SVs). In our approach we tested the linear SVM (i.e., with linear kernel function $K(x_i,x_j) = x_i^T \cdot x_j$) and the SVM with RBF kernel function with no significant performance difference. For this reason we have adopted the linear SVM approach. The optimal value of the parameter C for each classification problem was determined through cross-validation.

1.2.1.4 Naïve Bayes Classifier

The naïve Bayes (NB) [19] is a probabilistic classifier that builds a model $p(x|C_k)$ for the probability density of each class C_k. These models are used to classify a new instance x as follows: First the posterior probability $P(C_k|x)$ is computed for each class C_k using the Bayes theorem:

$$P(C_k|x) = \frac{P(x|C_k)P(C_k)}{P(x)} \tag{1.4}$$

where P(x) and $P(C_k)$ represent the a priori probabilities. Then the input x is assigned to the class with maximum $P(C_k|x)$.

In the NB approach, we made the assumption that the attributes x^i of x are independent to each other. Thus, $P(x|C_k)$ can be computed as the product of the one-dimensional densities $p(x^i|C_k)$. The assumption of variable independence drastically simplifies model generation since the probabilities $p(x^i|C_k)$ can be easily estimated, especially in the case of the discrete attributes where they can be computed using histograms (frequencies). The NB approach has been proved successful in the analysis of the genetic data.

1.2.1.5 Bayesian Neural Networks

A new methodology has been recently proposed for training feed-forward neural networks and more specifically the multilayer perceptron (MLP) [29]. This Bayesian methodology provides a viable solution to the well-studied problem of estimating the number of hidden units in MLPs. The method is based on treating the MLP as a linear model, whose basis functions are the hidden units. Then, a sparse Bayesian prior is imposed on the weights of the linear model that enforces irrelevant basis functions (equivalently unnecessary hidden units) to be pruned from the model. In order to train the model, an incremental training algorithm is used which, in each iteration, attempts to add a hidden unit to the network and to adjust its parameters assuming a sparse Bayesian learning framework. The method has been tested on several classification problems with performance comparable to SVMs. However, its execution time was much higher compared to SVM.

1.2.1.6 Logistic Regression

Logistic regression (LR) is the most popular traditional method used for statistical modeling [20] of binary response variables, which is the case in most problems of our study. LR has been used extensively in the medical and social sciences. It is actually a linear model in which the logistic function is included in the linear model output to constraint its value in the range from zero to one. In this way, the output can be interpreted as the probability of the input belonging to one of the two classes. Since the underlying model is linear, it is easy to train using various techniques.

1.2.2 Generalization Measures

In order to validate the performance of the classification models and evaluate their generalization ability, a number of typical cross-validation techniques and two performance evaluation measures were used. In this section we will cover two of them: classification accuracy and the kappa statistic.

In k-fold cross-validation [1], we partition the available data into k-folds. Then, iteratively, each of these folds is used as a test set, while the remaining

Table 1.1 Interpretation of the kappa statistic value

Kappa value	<0	0.0–0.2	0.2–0.4	0.4–0.6	0.6–0.8	0.81–1
Interpretation	No agreement	Slight agreement	Fair agreement	Moderate agreement	Substantial agreement	Almost perfect agreement

folds are used to train a classification model, which is evaluated on the test set. The average classifier performance on all test sets provides a unique measure of the classifier's performance on the discrimination problem. Leave-one-out validation technique is a special case of cross validation, where the test set contains only a single data instance each time that is left out of the training set, i.e., leave-one-out is actual N-fold cross validation where N is the number of data objects.

The accuracy performance evaluation measure is very simple and provides the percentage of correctly classified instances. It must be emphasized that its absolute value is not important in the case of unbalanced problems, i.e., an accuracy of 90 % may not be considered important when the percentage of data instances belonging to the major class is 90 %. For this reason we always report the accuracy gain as well, which is the difference between the accuracy of the classifier and the percentage of the major class instances.

The kappa statistic is another reported evaluation measure calculated as

$$Kappa = \frac{P(A) - P(E)}{1 - P(E)} \tag{1.5}$$

where P(A) is the percentage of observed agreement between the predictions and actual values and P(E) the percentage of chance agreement between the predictions and actual values. A typical interpretation of the values of the kappa statistic is provided in Table 1.1.

1.2.2.1 Feature Selection and Ranking

A wide variety of feature (or attribute) selection methods have been proposed to identify which features are significant for a classification task [4]. Identification of significant feature subsets is important for two main reasons. First, the complexity of solving the classification problem is reduced, and data quality is improved by ignoring the irrelevant features. Second, in several domains such as medical domain, the identification of discriminative features is actually new knowledge for the problem domain (e.g., discovery of new gene markers using bioinformatics datasets or SNPs in our study using the genetic dataset).

1.2.2.2 Single-Feature Evaluation

Simple feature selection methods rank the features using various criteria that measure the discriminative power of each feature when used alone. Typical criteria compute the correlation between the feature and the target category, such as the information gain and chi-squared measure, which we have used in our study.

Information Gain

Information gain (IG) of a feature X with respect to class Y(I(Y;X)) is the reduction in uncertainty about the value of Y when the value of X is known. The uncertainty of a variable X is measured by its entropy H(X), and the uncertainty about the value of Y, when the value of X is known, is given by its conditional entropy H(Y|X). Thus, information gain I(Y;X) can be defined as

$$I(Y;X) = H(Y) - H(Y|X) \tag{1.6}$$

For discrete features, the entropies are calculated as

$$H(Y) = -\sum_{j=1}^{l} P\left(Y = y_j\right) log_2\left(P\left(Y = y_j\right)\right) \tag{1.7}$$

$$H(Y|X) = -\sum_{j=1}^{l} P(X = x_j) H\left(Y|X = x_j\right) \tag{1.8}$$

Alternatively, IG can be calculated as

$$I(Y;X) = H(X) + H(Y) - H(Y,X) \tag{1.9}$$

For continuous features, discretization is necessary.

Chi-Square

The chi-square (also denoted as chi-squared or χ^2) is another popular criterion for feature selection. Features are individually evaluated by measuring their chi-squared statistic with respect to the classes [21].

1.2.2.3 Feature Subset Selection

The techniques described below are more powerful but computationally expensive. They differ from previous approaches in that they do not use single-feature evaluation criteria and result in the selection of feature subsets. They aim to eliminate features that are highly correlated to other already-selected features. The following methods have been used:

Recursive Feature Elimination SVM (RFE-SVM)

Recursive feature elimination SVM (RFE-SVM) [6] is a method that recursively trains an SVM classifier in order to determine which features are the most redundant, non-informative, or noisy for a discrimination problem. Based on the ranking produced at each step, the method eliminates the feature of the lower ranking (or more than one feature). More specifically, the trained SVM uses the linear kernel, and its decision function for a data vector x_i of class $y_i = \{-1 \text{ or } +1\}$ is

$$D(x) = w \cdot x_1 + b, \tag{1.10}$$

where b the bias and w the weight vector computed as a linear combination of the N data vectors:

$$w = \sum_{i=1}^{N} a_i y_i x_i, \tag{1.11}$$

$$b = \frac{1}{N} \sum_{i=1}^{N} (y_i - w \cdot x_i). \tag{1.12}$$

Most of a_i weights are zero, while the weights that correspond to the marginal support vectors (SVs) are greater than zero and sum to the cost parameter C. These parameters are the output of the trained SVM of a step, and then the algorithm computes the w feature weight vector that describes how useful each feature is based on the derived SVs. The ranking criterion used by the RFE-SVM is the w_i^2, and the feature that is eliminated is given by $r = \text{argmin}(w_i^2)$.

Minimum Redundancy, Maximum Relevance (mRMR)

Minimum redundancy, maximum relevance (mRMR) [22] is an efficient incremental feature subset selection method that adds features to the subset based on the trade-off between feature relevance (discriminative power) and feature redundancy (correlation with the already-selected features).

Feature redundancy is computed through minimizing the mutual information (information gain of one feature with respect to the others) of the selected features:

$$W_I = \frac{1}{|S|^2} \sum_{i,j \in S} I(i,j), \tag{1.13}$$

where S is the subset of the selected features. Relevance is computed as the total information gain of all features in S:

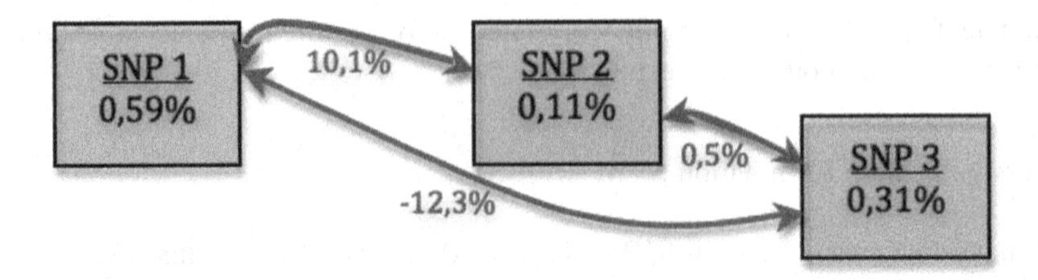

Fig. 1.1 Example of feature interaction graphs. Features (in this example SNPs) are represented as graph nodes and a selection of the three-way interactions as edges. Numbers in nodes represent individual information gains, and the numbers on edges represent the two-way interaction information between the connected attributes, all with respect to the class attribute

$$V_I = \frac{1}{|S|} \sum_{i \in S} I(h, i), \tag{1.14}$$

Optimization with respect to both criteria requires to combine them into a single criterion function: $\max(V_I - W_I)$ or $\max(V_I/W_I)$.

K-Way Interaction Information/Interaction Graphs

K-way interaction information (KWII) [30] is a multivariate measure of information gain, taking into the account the information that cannot be obtained without observing all k features at the same time [25]. Feature interaction can be visualized by use of interaction graphs [31]. In such a graph, individual attributes are represented as graph nodes and a selection of the 3-way interactions as edges (Fig. 1.1).

Multifactor Dimensionality Reduction (MDR)

Multifactor dimensionality reduction (MDR) [13] is an approach for detecting and characterizing combinations of attributes that interact to influence a class variable. Features are pooled together into groups taking a certain value of the class label (original target of MDR were genetic datasets, thus most commonly, multilocus genotypes are pulled together into low-risk and high-risk groups). This process is referred to as constructive induction. For low orders of interactions and numbers of attributes, an exhaustive search is possible to be conducted. However, for higher numbers, exhaustive search becomes intractable, and other approaches are necessary (preselecting the attributes, random searches, etc.). The MDR approach has been used for SNP selection in the genetic dataset (Fig. 1.2).

AMBIENCE Algorithm

AMBIENCE [12] is an information theoretic search method for selecting combinations of interacting attributes based around KWII. Rather than calculating

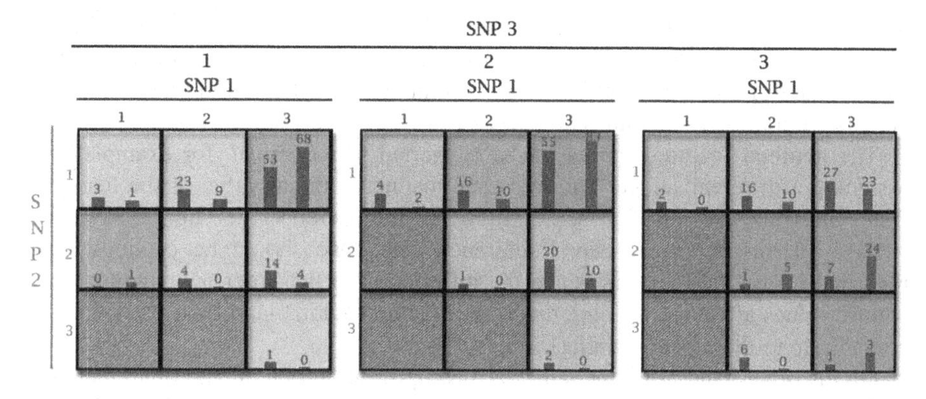

Fig. 1.2 MDR example. Combinations of attribute values are divided into "buckets." Each bucket is marked as low or high risk, according to a majority vote

KWII in each step (a procedure which requires the computations of super-sets, thus growing exponentially), AMBIENCE employs the total correlation information (TCI) defined as

$$TCI(X_1, X_2, \cdots X_k) = \sum_{i=1}^{k} H(X_i) - H(X_1 X_2 \cdots X_k) \tag{1.15}$$

where H denotes the entropy.

A metric called phenotype-associated information (PAI) is constructed as

$$PAI(X_1, X_2, \ldots, X_k, Y) = TCI(X_1, X_2, \ldots, X_k, Y) - TCI(X_1, X_2, \ldots, X_k) \tag{1.16}$$

The algorithm starts from n subsets of attributes, each containing one of the n attributes with the highest individual information gain with respect to the class label. In each step, n new subsets containing combinations with highest PAI are greedily selected, from all of the combinations created by adding each attribute to each subset from the previous step. The procedure is repeated t times. After t iterations KWII is calculated for the resulting n subsets. The AMBIENCE algorithm has been successfully employed in the analysis of the genetic dataset.

1.2.3 *Treating Missing Values and Nominal Features*

Missing values problem is a major preprocessing issue in all kinds of data mining applications. The primary reason is that not all classification algorithms are able to handle data with missing values. Another reason is that when a feature has values that are missing for some patients, then the algorithm may under-/overestimate its importance for the discrimination problem. A second preprocessing issue of less importance is the existence of nominal features in the dataset, e.g., features that take string values or date features. There are several methods that require numeric data vectors without missing values (e.g., SVM).

The nominal features can easily be converted to numerical, for example, by assigning a different integer value to each distinct nominal value of the feature. Dates are often converted to some kind of time difference (i.e., hours, days, or years) with respect to a second reference date. One should be cautious and renormalize the data vectors, since the differences in the order of magnitude of feature values affect the training procedure (features taking larger values will play crucial role to the model training).

On the other hand, missing values is a complicated problem, and often there is not much space for sophisticated things to do. Among the simple and straightforward approaches to treat missing values are:

- The complete elimination of features that have missing values. Obviously, if a feature is important for a classification problem, this may be not acceptable.
- The replacement with specific computed or default values
  - Such values may be the average or median value of the existing numeric values and, for a nominal feature, the nominal value with higher frequency. This latter can also be used when the numeric values are discrete and generally small in number. In some cases it is convenient to put zero values in the place of missing values, but this can also be catastrophic in other cases.
  - Another approach is to use the K-nearest neighborhood for the data objects that have missing values and then try to fill them with values that are more frequent in the neighborhood objects. If an object is similar to another, based on all the data features, then it is highly probable that the missing value would be similar to the respective value of its neighbor.
  - In some cases, it is possible to take advantage of the special properties of a feature and its correlation to other features in order to figure out good estimations for the missing values. We describe such a special procedure in the case study at end of the chapter.
- The conversion of a nominal feature to a single binary when the existing values are quite rare in terms of frequency and have similar meaning. In this way, the binary feature takes a "false" value only in the cases where the initial feature had a missing value.
- The conversion of a nominal feature to multiple binary features. This approach is called feature extension, or binarization, or 1-out-of-k encoding (for k nominal values). More specifically, a binary feature is created for each unique nominal value, and the value of the initial nominal feature for a data object is indicated by a "true" value at the respective created binary feature. Conversely, a missing value is encoded with "false" values to all the binary extensions of the initial feature.

1.3 Case Study: Coronary Artery Disease

This section presents a case study based on the mining on medical data carried out as a part of ARTreat project, funded by the European Commission under the umbrella of the Seventh Framework Program for Research and Technological Development, in the period 2008–2013 [32]. The project was a large, multinational collaborative effort to advance the knowledge and technological resources related to treatment of coronary artery disease. The specific work used as the background for the following text was carried out in a cooperation of Foundation for Research and Technology Hellas (Ioannina, Greece), University of Kragujevac (Serbia), and Consiglio Nazionale delle Ricerche (Pisa, Italy). Moreover, the patient databases used in our analysis were collected and provided by the Consiglio Nazionale delle Ricerche.

1.3.1 Coronary Artery Disease

Coronary artery disease (CAD) is the leading cause of death in both men and women in developed countries. CAD, specifically coronary atherosclerosis (ATS), occurs in about 5–9 % of people aged 20 and older (depending on sex and race). The death rate increases with age and overall is higher for men than for women, particularly between the ages of 35 and 55. After the age of 55, the death rate for men declines, and the rate for women continues to climb. After age 70–75, the death rate for women exceeds that for men who are the same age.

Coronary artery stenosis is almost always due to the gradual, lasting even years, buildup of cholesterol and other fatty materials (called atheromas or atherosclerotic plaques) in the wall of a coronary artery [24]. As an atheroma grows, it may bulge into the artery, narrowing the interior of the artery (lumen) and partially blocking blood flow. As an atheroma blocks more and more of a coronary artery, the supply of oxygen-rich blood to the heart muscle (myocardium) becomes more inadequate. An inadequate blood supply to the heart muscle, by any cause, is called myocardial ischemia. If the heart does not receive enough blood, it can no longer contract and pump blood normally. An atheroma, even one that is not blocking much the blood flow, may rupture suddenly. The rupture of an atheroma often triggers the formation of a blood clot (thrombus) which further narrows, or completely blocks, the artery, causing acute myocardial ischemia (AMI).

The ATS disease can be medically treated using pharmaceutical drugs, but this cannot decrease the existing stenoses but rather delay their development. A different treatment approach applies an interventional therapeutic procedure to a stenosed coronary artery, such as percutaneous coronary artery angioplasty (PTCA, balloon angioplasty) and coronary artery bypass graft surgery (CABG). PTCA is one way to widen a coronary artery. Some patients who undergo PTCA have restenosis (i.e., renarrowing) of the widened segment within about 6 months after the procedure. It is believed that the mechanism of this phenomenon, called "restenosis," is not related with the progression of ATS disease but rather with the body's immune system response to the injury of the angioplasty. Restenosis that is caused by neointimal hyperplasia is a slow process, and it was suggested that the local administration of a drug would be helpful in preventing the phenomenon. Stent-based local drug delivery provides sustained drug release with the use of stents that have special features for drug release, such as a polymer coating. However, cell-culture experiments indicate that even brief contact between vascular smooth-muscle cells and lipophilic taxane compounds can inhibit the proliferation of such cells for a long period. Restenosed arteries may have to undergo another angioplasty. CABG is more invasive than PTCA as a procedure. Instead of reducing the stenosis of an artery, it bypasses the stenosed artery using vessel grafts.

Coronary angiography, or coronography, (CANGIO) is an X-ray examination of the artery of the heart. A very small tube (catheter) is inserted into an artery. The tip of the tube is positioned either in the heart or at the beginning of the arteries supplying the heart, and a special fluid (called a contrast medium or dye) is injected. This fluid is visible by X-ray and hence pictures are obtained. The severity, or degree, of stenosis is measured in the cardiac cath lab by comparing the area of narrowing to an adjacent normal segment. The most severe narrowing is determined based on the percentage reduction and calculated in the projection. Many experienced cardiologists are able to visually determine the severity of stenosis and semiquantitatively measure the vessel diameter. However, for greatest accuracy, digital cath labs have the capability of making these measurements and calculations with computer processing of a still image. The computer can provide a measurement of the vessel diameter, the minimal luminal diameter at the lesion site, and the severity of the stenosis as a percentage of the normal vessel. It uses the catheter as a reference for size.

The left coronary artery, also called left main artery (TC), usually divides into two branches (Fig. 1.3), known as the left anterior descending (LAD) and the circumflex (CX) coronary arteries. In some patients, a third branch arises in between the LAD and the CX known as the ramus intermediate (I). The LAD travels in the anterior interventricular groove that separates the right and the left ventricle, in the front of the heart. The diagonal (D) branch comes off the LAD and runs diagonally across the anterior wall towards its outer or lateral portion. Thus, D artery supplies blood to the anterolateral portion of the left ventricle. A patient may have one or several D branches. The LAD gives rise to septal branches (S). The CX travels in the left atrioventricular groove that separates the left atrium from the left ventricle. The CX moves away from the LAD and wraps around to the back of the heart. The major branches that it gives off in the proximal or initial portion are known as obtuse, or oblique, marginal coronary arteries (MO). As it makes its way to the posterior portion of the heart, it gives off one or more left posterolateral (PL) branches. In 85 % of cases, the CX terminates at this point and is known as a nondominant left coronary artery system.

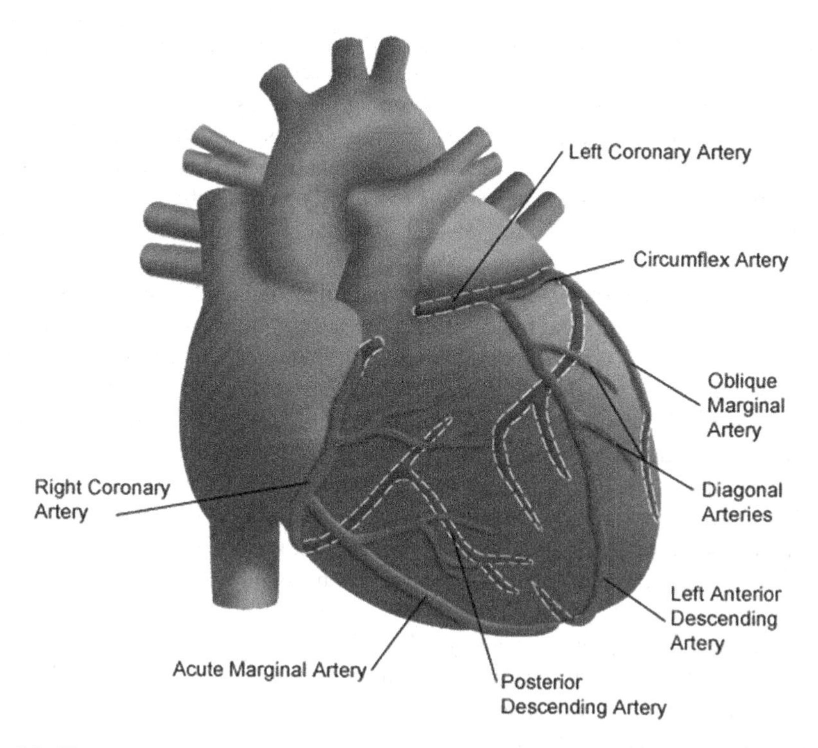

Fig. 1.3 The coronary arteries structure of the heart

The right coronary artery (RC) travels in the right atrioventricular (RAV) groove, between the right atrium and the right ventricle. The right coronary artery then gives rise to the acute marginal branch that travels along the anterior portion of the right ventricle. The RC then continues to travel in the RAV groove. In 85 % of cases, the RC is a dominant vessel and supplies the posterior descending (DP) branch that travels in the PIV groove. The RC then supplies one or more posterolateral (PL) branches. The dominant RC system also supplies a branch to the right atrioventricular node just as it leaves the right AV groove, and the PD branch supplies septal perforators to the inferior portion of the septum. In the remaining 15 % of the general population, the CX is "dominant" and supplies the branch that travels in the posterior interventricular (PIV) groove. Selective coronary angiography offers the only means of establishing the seriousness, extent, and site of coronary sclerosis.

Extensive clinical and statistical studies have identified several factors that increase the risk of coronary heart disease and heart attack [9]. Note that coronary heart disease usually implies CAD where the stenoses are caused by atherosclerosis; however there can be also causes other than that. Important risk factors are those that research has shown to significantly increase the risk of heart and blood vessel (cardiovascular) disease [8]. Other factors are associated with increased risk of cardiovascular disease, called contributing risk factors, but their significance and prevalence have not yet been precisely specified. The more risk factors you have, the greater your chance of developing the disease. However, the disease may develop without the presence of any classic risk factor. Researchers are studying other possible factors, including C-reactive protein and omocistein. On the other way, researchers are moving to identify in risk subgroups of subjects, a decisive factor for the selection of high-risk patients to be submitted to most aggressive treatment.

Genetic studies of coronary heart disease and infarction are lagging behind other cardiovascular disorders. The major reason for the limited success in this field of genetics is that it is a complex disease which is believed to be caused by many genetic factors, environmental factors, as well as interactions among these factors. Indeed, many risk factors have been identified, and, among these factors, family history is one of the most significant independent risk factor for the disease. Unlike single-gene disorders, complex genetic disorders arise from the simultaneous contribution of many genes. Genetic variants or single-nucleotide polymorphisms (SNPs) are identified in the literature, and many candidate genes with physiologic relevance to coronary artery disease have been found to be associated with increased or decreased risks for coronary heart disease [23, 26]. The frequencies of SNP alleles or genotypes are analyzed and an allele or genotype is associated with the disease if its occurrence is significantly different from that reported in the control [14]. The identification of the key complement of genes that contribute to cardiovascular diseases, in particular CAD, will lead to new types of genetic tests that can assess an individual's risk for disease development. Subsequently, the latter may also lead to more effective treatment strategies for the delay or even prevention of the disease altogether.

1.3.2 The Main Database (M-DB)

We have considered two databases: the main database (M-DB) concerning 3,000 patients on which most of the data mining work was focused and a second database with about 676 patient records with detailed scintigraphy results.

M-DB contains detailed information for 3,000 patients who suffer from some kind of symptoms related to the ATS disease that were presented to them and made them go to the hospital. For most of the patients, these symptoms correctly indicate that they have stenosed arteries in a sensible extend, while for not quite a small number of other patients, their symptoms are a false-positive indication of important stenoses in critical arteries for the heart function. Patient's history describes the profile of a patient when hospitalized and includes the following:

- Age when hospitalized, sex
- Family history related to the ATS disease

- History of interventional treatment (bypass or angioplasty operations)
- Acute myocardial infarction (AMI) and history of previous myocardial infarction (PMI)
- Angina on effort/at rest
- Ischemia on effort/at rest
- Arrhythmias, cardiomyopathy, diabetes, cholesterol, and akinesia
- The presence of risk factors such as obesity and smoking

A series of medical examinations is provided:

- Blood tests
- Functional examinations
- Electrocardiogram (ECG) during exercise stress test
- ECG during rest
- Imaging examinations
- A first coronary angiography (CANGIO) examination
- A second CANGIO examination available only for 430 patients
- Medical treatment after the entrance of patient to the hospital include,
- Pharmaceutical treatment
- Interventional procedures (bypass or PTCA operations)

Follow-up information reports events such as:

- Death events and a diagnosed reason for it
- Events of acute myocardial infarctions
- Interventional treatment procedures (also mentioned in the medical treatment category)
- Other cardiac events (pacemaker implantation, etc.)

Genetic information that includes the expressions of 57 genes is available only for 450 patients.

Particularly for the CANGIO examination, the database reports the stenosis level on the four major coronary arteries TC, LAD, CX, and RC if that level is at least 50 %. For each of the major arteries it is also available, for many but not all cases, the exact site of the artery where the narrowing is located, namely, proximal, medial, and distal. A stenosis is more severe when sited at the proximal part of the artery and less severe at distal, since the blood flow at the early part of the artery affects the flow in larger part of the heart (Fig. 1.3). Moreover, the CANGIO also provides the degree of stenosis for a number of secondary arteries, such as D, I, and MO. Table 1.2 presents some examples of CANGIO examinations, the extent of stenosis for the major and secondary vessels (luminal diameter reduction). The Max columns indicate the maximum stenosis in the length of the respective artery. For some cases the medical expert was not in position to specify the site of a stenosis, whereas he identified the extent of the functional problem, i.e., the percentage of the stenosis.

Table 1.2 Some examples of CANGIO examinations for patients included in the M-DB

CANGIO	Stenosed	Main coronary arteries													Secondary coronary arteries							
	vessels	LAD				CX				RC												
date	(≥50 %)	Max	P	M	D	Max	P	M	D	Max	P	M	D	TC	I	S	D	MO	DP	RAV	MA	DP
19/11/1982	2	75								75												
19/01/1983	3	75			75	75	75			50		50						50				50
13/12/1978	2	90								100												
04/02/1983	1	100	100																			
14/01/1983	1	90	90																			
09/05/1979	2	90								75							90					
03/11/2003	3	100				100	100			100	100											
16/02/1983	1									90	75	90										
16/02/1983	0																					
22/04/1983	2	90	90							90	90											
02/03/1983	3	90	90	50		75				50	50		50					75				
06/03/2003	4	50		50		90	90			100		100		50				75				
30/03/1983	3	90		90		90	75		90	90		90						90				75
20/04/1983	2	100				50		50								90						
15/05/1979	3	100				90				75	75							100				
14/11/2007	3	90			90	100			100	50	50		50			50	75					
27/04/1983	3	90	90			75	75			75		75	75									
09/03/1983	4													100								

1.3.3 The Database with Scintigraphies (S-DB)

The scintigraphic dataset (S-DB) is a dataset containing records for about 440 patients with laboratory tests, 12-lead electrocardiography (ECG), stress/rest gated SPECT, clinical evaluation, and the results of CANGIO. More specifically:

- Clinical Examinations
  The available clinical variables include patient age, sex, and history of angina (at rest, on effort, or mixed), previous MI, and cardiovascular risk factors: family history of premature IHD, presence of diabetes mellitus, arterial hypertension, hypercholesterolemia, hypertriglyceridemia, obesity, and being a current or former smoker.
- Laboratory Examinations
  The laboratory data available include erythrocyte sedimentation rate, fasting glucose, serum creatinine, total cholesterol, HDL and LDL levels, triglycerides, lipoprotein, thyrotropin, free triiodothyronine, free thyroxine, C-reactive protein, and fibrinogen.
- Electrocardiographic Data
  The ECG data include 12-lead ECG results (normal/abnormal), exercise stress test results, and maximal workload on effort.
- Echocardiographic Data
  Two-dimensional echocardiographic data include left ventricular ejection fraction (LVEF), left ventricular end-diastolic diameter, wall motion score index, and end-diastolic thickness of the interventricular septum and posterior wall.
- Scintigraphic Data
  The detailed scintigraphic data available include the values of SRS, SSS, SDS, EDV on effort, ESV on effort, SMS on effort, and STS on effort.

The objectives of the analysis are the same as with the main database, i.e., to build classification models predicting the severity of ATS using the other features and mainly the scintigraphic information.

1.3.4 Defining Disease Severity

As mentioned before, the target variable needed for the present learning problem is the "correct" ground truth class, namely, severe or mild-normal, of each patient instance and this must be set in advance of any supervised model training. Next, the classification algorithms try to learn how to discriminate the patients of each category. Generally, the characteristics of the real-world problem under investigation and the quality/quantity of the provided examples affect directly the level of difficulty of the learning problem.

Apart from any data quality issues, the real problem of predicting the severity of a patient's ATS condition presents additional difficulties regarding the very fundamental definition of the disease severity categories for the known training dataset. To define the target variable of the classification problems, we used the information of the CANGIO examinations which can express the atherosclerotic burden of a patient at the time being examined. The CANGIO indicates which arteries are stenosed, when the narrowing percentage is at least 50 %, and the stenosis is characterized by that percentage. In particular, five different percentage values are reported in the database: 0 %, 50 %, 75 %, 90 %, and 100 %.

The first issue that arises is that we need to define a way to utilize all these measurements to a single indication about disease severity. The second issue is that these indications about stenotic vessels are provided by the doctor that did the CANGIO, and the diagnosis may depend on the personal opinion of the expert (may vary for different doctors) and the technology of the hardware and the procedures used for the examination (e.g., the CANGIO back in 1970 cannot be as good as a modern diagnosis). In the following paragraphs of this section, we describe the different severity definitions we considered and how a two-class classification problem was set up.

1.3.4.1 The Number of Diseased Vessels

The number of the diseased major vessels (TC, LAD, CX, RC) and the extent of stenosis on each of them can be used to quantify the ATS disease severity. Thus, patients can be categorized by the following simple rule:

- Severely diseased having $>=$ A diseased vessels with $>=$ T stenosis
- Mild, otherwise

The values of the two parameters vary:

- $\mathrm{A} = \{1, 2, 3\}$
- $\mathrm{T} = \{50\%, 75\%, 100\%\}$

This disease severity definition is denoted as DefA.

1.3.4.2 Angiographic Score17

The more detailed special angiographic score proposed in [5] can be utilized for quantifying the severity of the disease. This score, herein denoted as Score17, assigns a severity level to a patient in the range of [0,. . .,17] with 17 being the most severe condition, while zero correspond to a normal patient. More specifically, this metric examines all the sites of the 4 major coronary arteries (e.g., the proximal, medial, and distal site of LAD) for lesions exceeding a predefined stenosis threshold. The exact computation of Score17 is presented in Fig. 1.4.

```
if stenosis is found in TC then
    Score17 = 12 points
    Ignore stenosis in LAD and CX
    if there is a stenosis in RC then
        Score17 = Score17 + the most severe case from RC
                        (5 for proximal and medial, or 3 for distal)
    end
else
    Score17 = the most severe stenosis from LAD
                        (7 points for proximal, 5 medial, or 3 for distal)
    Score17 = Score17 + the most severe stenosis from CX
                        (5 for proximal and medial, or 3 for distal)
    Score17 = Score17 + the most severe stenosis from RC
                        (5 for proximal and medial, or 3 for distal)
end
```

Fig. 1.4 The algorithm to compute Score17

Based on this score, four medically meaningful categories are defined:

a. Score17 = 0: Normal vessels
b. Score17 less or equal to 7: Mild ATS condition
c. Score17 between 7 and 10: Moderate ATS condition
d. Score17 between 10 and 17: Severe ATS condition

These can be used to directly set up a four-class problem denoted as S-vs-M-vs-M-vs-N. Furthermore, we defined a series of cases by grouping together the above subgroups, e.g., SM-vs-MN is the problem where the "Severe" class contains patients with severe ATS (case (a)) or moderate ATS severity (case (b)), while the mild and normal ATS diseased patients (cases (c) and (d)) constitute the "Mild" class. This definition is denoted as DefB.

1.3.4.3 HybridScore: A Hybrid Angiographic Score

Undoubtedly, Score17 gives more freedom to the specification of the target value of the problem. However, the need to define the threshold leads again in a large set of problem variants. To tackle this situation, we have developed an extension of this score that does not depend on a stenosis threshold. The basic idea is the use of a set of weights, each of them corresponding to different ranges of stenosis degree. These weights are incorporated to the score computation in order to add fuzziness to patient characterization. An example would explain the behavior of the modified Score17 denoted as HybridScore (Table 1.3).

Examples:

a. Supposing that a patient has 50 % stenosis at TC, 50 % at RC proximal, 90 % at RC distal, and the rest of his vessels are normal, then the classic Score17, with a threshold at 75 % stenosis, assigns a disease severity level 3 for the DX distal stenosis.

Table 1.3 The weights used by the HybridScore

Stenosis range	<50 %	50–75 %	75–90 %	90–100 %
Weight value	0	1/2	2/3	1

The developed HybridScore17 assigns 12*1/2(for TC) + max{5*1/2, 3*1} = 9. Note that for multiple stenoses at the same vessel, this score takes into account the most serious with respect to the combined weighted severity.

b. Let us examine another patient with exactly the same TC and RC findings, but having as well 90 % stenosis at LAD proximal and 90 % at CX medial. The traditional Score17 ignores these latter two, because they belong to the left coronary tree where TC is the most important part and exceeds the elementary threshold of 50 % stenosis (over which a vessel is generally considered as occluded). On the other hand, HybridScore17 would assign a severity value by computing the max {9(the previous result), 7 * 1(for LAD proximal) + 5 * 1(for CX medial)} = 12.

Table 1.4 provides the values for the different CANGIO scores. For the Score17 the table provides the values with different stenosis thresholds: 50 % (T50), 75 % (T75), and 90 % (T90). Note also that the site of the stenosis might not reported by the medical expert during the examination. In these cases we assume that the stenosis is located at the proximal site (the most serious scenario). It is worth mentioning that the threshold of Score17 plays a crucial role in evaluating the ATS burden of a patient. In the eleventh line of Table 1.4, we observe that using a threshold of 50 % stenosis, the score gives a value equal to 17 and with 75 % threshold the score is 12, while for 90 % threshold this value becomes 7. On the other hand, HybridScore is a single measurement with a value equal to 12.

To illustrate the way the presented scores work, we provide the following graph that presents the cumulative density function (cdf) for the range of values 0–17, for the original Score17 using three different thresholds and the HybridScore. The scores have been computed for the 3,000 patient records of M-DB dataset. The value at ATS score = 0 corresponds to the number of patients that have a score value in [0,1], for ATS score = 1 a computed score in [0,1] or in [1,2], and so on. For example, looking at the Score17-T90 line, over 40 % of the 3,000 patients database are assigned with a score value equal to 0 and very few patients exist with score values larger than zero and less than or equal to 3. Apparently, there is a large group of patients (about 20 % of the total patients) that have a score over 3 and at most 4 (Figs. 1.5 and 1.6).

Next, we present the respective figures, Figs. 1.7 and 1.8, for the M-DB after excluding a subset of patients with a recorded history of PMI or AMI. These patients are generally cases of more serious ATS burden. This is depicted by the increased frequencies of the lower ATS scores in the cdf of Fig. 1.7 compared with the cdf in Fig. 1.5 of the full database of 3,000 patients.

To define a classification problem based on this angiographic score, a proper threshold needs to be specified. A value of HybridScore over that threshold would imply that a patient is severely diseased, and is mildly diseased, or even in normal condition, if his score is below threshold. This definition of ATS disease severity is denoted as DefC.

Table 1.4 The angiographic scores, Score17 for various stenosis thresholds and the proposed HybridScore, computed based on the CANGIO examinations of some patients of M-DB. When the site of a stenosis is not reported (e.g., for LAD and RC of the first patient), we assume that it is located at the proximal region of the artery

CANGIO	Stenosed	Main coronary arteries																
	vessels	LAD				CX				RC					Score17	Score17	Score17	Hybrid
date	(≥50 %)	Max	P	M	D	Max	P	M	D	Max	P	M	D	TC	T50	T75	T90	Score
19/11/1982	2	75								75					12	12	0	8.00
19/01/1983	3	75			75	75	75			50		50			13	8	0	7.83
13/12/1978	2	90								100					12	12	12	12.00
04/02/1983	1	100	100												7	7	7	7.00
14/01/1983	1	90	90												7	7	7	7.00
09/05/1979	2	90								75					12	12	7	10.33
03/11/2003	3	100				100	100			100	100				17	17	17	17.00
16/02/1983	1									90	75	90			5	5	5	5.00
16/02/1983	0														0	0	0	0.00
22/04/1983	2	90	90							90	90				12	12	12	12.00
02/03/1983	3	90	90	50		75				50	50		50		17	12	7	12.S3
06/03/2003	4	50		50		90	90			100		100		50	17	10	10	12.50
30/03/1983	3	90		90		90	75		90	90		90			15	15	13	13.33
20/04/1983	2	100				50		50							12	7	7	9.50
15/05/1979	3	100				90				75	75				17	17	12	15.33
14/11/2007	3	90			90	100			100	50	50		50		11	6	6	S.50
27/04/1983	3	90	90			75	75			75		75	75		17	17	7	13.67
09/03/1983	4													100	12	12	12	12.00

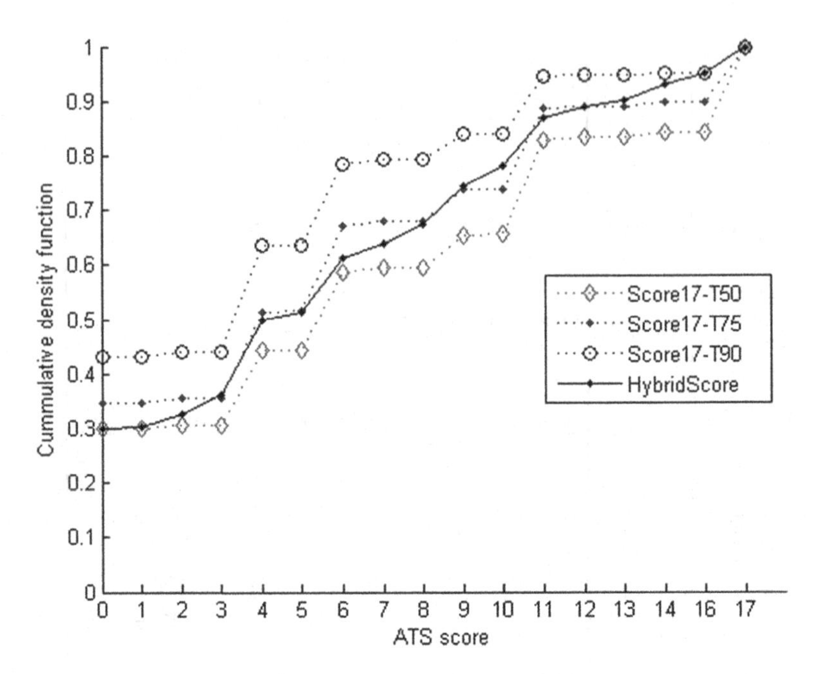

Fig. 1.5 The cumulative density functions of the Score17 and HybridScore for M-DB

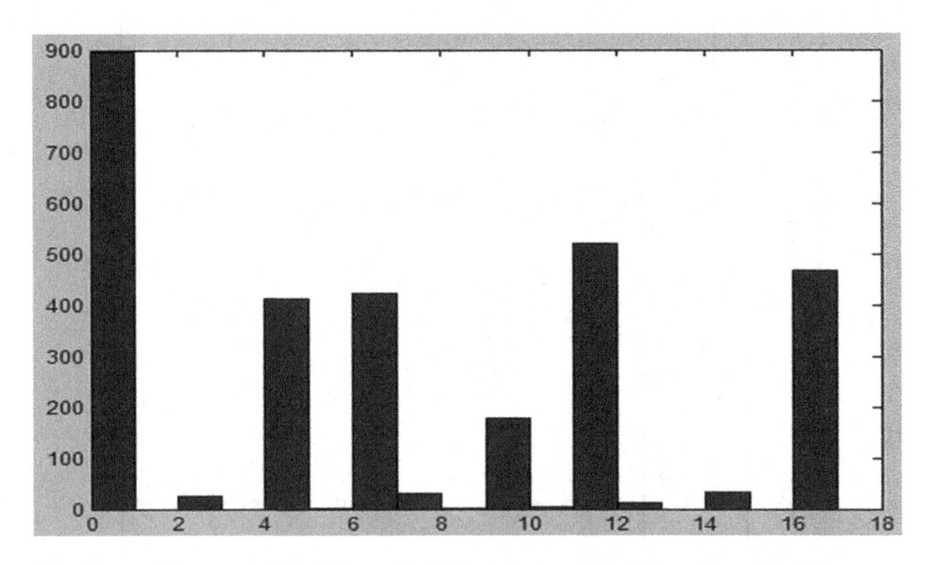

Fig. 1.6 The histogram of the different HybridScore values for M-DB patients (*x-axis*)

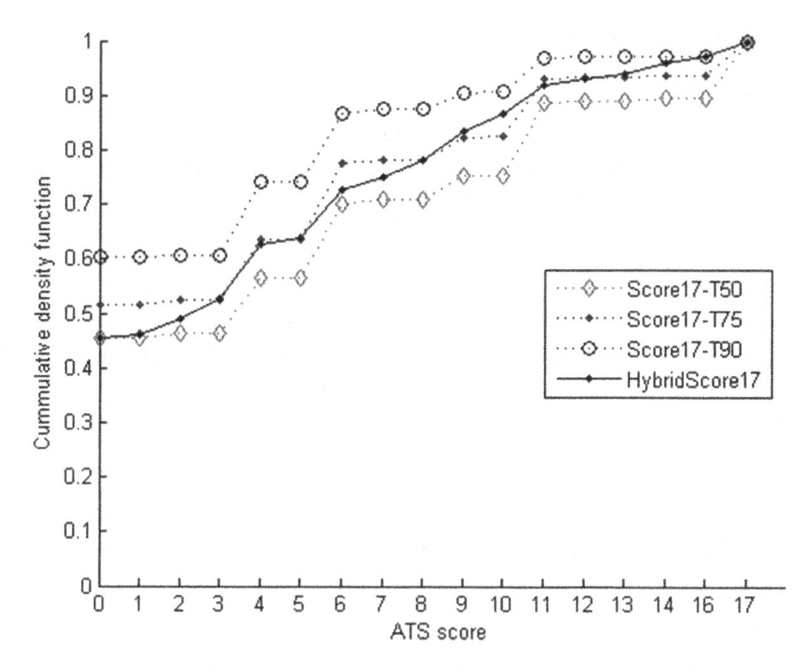

Fig. 1.7 The cumulative density functions of the Score17 and HybridScore for M-DB, excluding the patients with PMI/AMI history

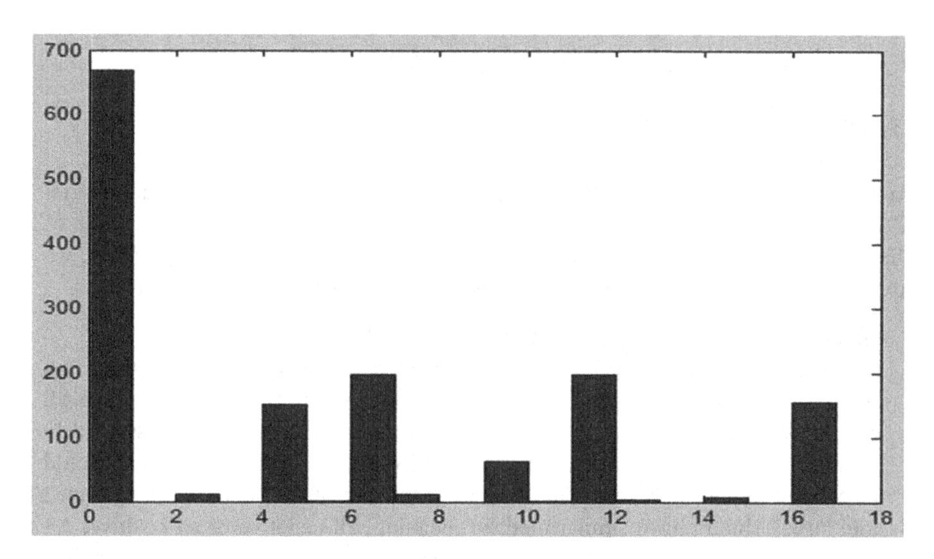

Fig. 1.8 The histogram of the different HybridScore values for M-DB, excluding the patients with PMI/AMI history (*x-axis*)

1.3.4.4 Discussion on Angiographic Scores

The introduction of the HybridScore proposed in this study has been proved very beneficial since it allows the complete characterization of the CANGIO examination using a single numeric value, while the existing characterization is using two numeric values (namely, Score17 and a stenosis threshold to compute the score). In this way it is straightforward to define the various classification problems that emerge by setting a threshold value (th) to this HybridScore (Mild class, hybrid score $<$ th and Severe class, hybrid score $> =$ th). As the threshold value (th) increases from 0 to 17, we obtain a sequence of meaningful classification problems.

The proposed HybridScore definition allows for the direct computation of the difference between two coronary examinations. To our knowledge, this is the first time such a difference is quantified in literature with a convenient measure which is also applicable for the quantification of ATS progression.

1.3.5 Results for Data Mining Tasks

This section will illustrate some of the results obtained during the analyses of the data in the described tasks. It should be noted that the presented results are provided as examples of the results that can be achieved by mining clinical data, and not as a facts that should be considered, or accepted, as having medical validity.

1.3.5.1 Correlating Patients' Profile and ATS Burden

Data Preprocessing

In this task we used the information about the patients' history and the first CANGIO examination. Initially, each patient record contains 70 features, some of them having missing values. For the nominal features that have missing values, we apply some of the feature transformations presented earlier in the chapter.

- Binarize a Feature by Merging Rare Feature Values
  For nominal features that take several values each of them having a very low frequency while at the same time having many missing values, we merge all existing different values to a "true" value, and the "false" value was assigned to the missing value cases. To be this transformation appropriate, the values that would be merged should express similar findings for the patient, i.e., all the values grouped into "true" should have similar medical meaning, all negative or all positive. An example is the feature describing the diagnosis for akinesia that takes values such as API (1.70 %), SET (0.97 %), INF (6.70 %), POS (0.23 %), LAT (0.13 %), ANT (0.03 %), and combinations of these values (14.96 %), while the rest 75.30 % are missing values. Apparently, all the reported values have the same negative medical meaning about negative findings diagnosed to the patients. In this case, the new

(1) Compute the average and standard deviation for the EF values of each ventricular dysfunction category (Normal, Regional, Global).

(2) Fill the missing EF values for patients without a dysfunction (Normal) cases with the average EF value measured for the Normal patients. The same for the other dysfunctions (Regional and Global).

(3) Using the probability p(type of dysfunction | EF), computed assuming a Gaussian distribution to model the values of each dysfunction type, fill the missing dysfunction characterization based on the available EF value.

(4) Apply feature extension to Echo left ventricular dysfunction.

Fig. 1.9 The procedure of filling EF and ECHO left ventricular dysfunction missing values

binary feature has 75.30 % the "true" value and 24.70 % the "false" value. Other similar cases are dyskinesia, hypokinesia, and AMI complications.

- Feature Extension for Nominal Features with Missing Values
  Only one of the new binary features can be "true," while a missing value is encoded as an instance where all these new features are "false." This transformation is used for features such as AMISTEMI and PMISTEMI.

Missing values are present for numeric features as well. To deal with these cases, we apply the following transformations:

- Firstly, we eliminated all such features that have a frequency of missing values over 11 %. These features were hdl (missing, 25.53 %), ldl (missing, 27.73 %), rpp (missing, 57.83 %), watt (missing, 57.87 %), septup (missing, 16.97 %), and posteriorwall (missing, 17.90 %).
- For the features that have less that 11 % missing values percentage, we filled them with the average feature value. This category of features includes hr (missing, 7.23 %), pressmin (missing, 1.20 %), pressmax (missing, 1.20 %), creatinine (missing, 9.70 %), cholesterol (missing, 6.60 %), triglic (missing, 8.63 %), and glicemia (missing, 10.53 %).

Special cases of features with missing values are the ejection fraction of the left ventricular of the heart (EF) and the diagnosis of a dysfunction of that ventricular (ECHO left ventricular dysfunction). These two findings are commonly measured by an electrocardiogram and are closely correlated since, usually, a dysfunction of the ventricle results in a low ejection fraction. The more serious a problem is diagnosed to the ventricle, the less fraction of the blood in the ventricle in end-diastole state is pumped out of the ventricle. In the M-DB, there are patient records where (a) both measurements are provided and (b) only one of the measurements is reported. We developed the heuristic procedure of Fig. 1.9.

The final step of the above procedure applies feature expansion to the dysfunction of the ventricular. This is done in order to prepare the data for classification algorithms such as SVM, where the different nominal values cannot be handled. After the preprocessing we described, each patient record of the M-DB contains 92 features. This is the full set feature that we finally used.

The AMI date was converted to an integer feature expressing the positive time difference in days between that date and the hospitalization date of the patient, similarly for the PMI date. The missing values of these features are filled with zeros.

The results regarding the feature evaluation did not indicate that the elimination of certain features could lead to better predictions. In fact, there are some features that do not have much information associated with the value of the target variable that is predicted (the class of each patient) and are ranked in low positions, but, at the same time, when eliminated the performance of the models does not improve at all. Thus, we did not aim further on feature selection by means of computational feature evaluation. Instead, we considered a second version of each database for these two tasks where we discarded a number of features that are known to be medically high correlated with the ATS disease. This approach would force the training algorithms to use the remaining features and may reveal nontrivial connections between patient characteristics and the disease. The exact features discarded are ischemia at rest, AMI, AMI date, AMI STEMI (all the binary expansions), AMI NSTEMI, AMI complications, PMI, PMI date, PMI STEMI, PMI NSTEMI, history of CABG, history of PTCA, and ischemia on effort before (hospitalization). For AMI STEMI, AMI NSTEMI, PMI STEMI, and PMI NSTEMI, all the features of the feature expansion were eliminated. The set of features is then called "reduced."

Evaluating the Trained Classification Models

In this task we aimed to build efficient systems that can discriminate the patients into two classes regarding the severity of their ATS disease condition that can be characterized as normal-mild or severe. In the previous section, we discussed how we can quantify the CANGIO examination into one single value using the proposed HybridScore. Based on that, we have defined the target variable for training classifiers. From the machine learning standpoint, we also need the value of the target variable for each patient, i.e., the indication about the class each patient belongs. Unfortunately, this requires medical knowledge about specific values of the ATS scores that could be used as thresholds. This cannot be provided since there are not any related studies in the literature that propose such a cutoff value. In fact one could make reasonable choices but there is no gold standard to use.

As a result, we should test all possible settings of ATS score and build classifiers for all these cases. For example, we choose to use all integer values of the HybridScore in [0,17]. Then we need to evaluate the classifiers produced for a fixed classification problem, with a specific cutoff threshold. An evaluation of the produced classifiers is also needed in a second level: to understand which classification is medically more meaningful or easier to solve based on the available data. In other words, the objectives of the analysis are both to find the interesting discrimination problems as well as to find interesting solutions for them. In fact, this is a complex procedure where the final classifiers are somehow evaluated by both supervised and unsupervised way. And this is the most challenging issue we had to deal with in this study.

Supposing we have produced all classifiers for all thresholds of HybridScore, we evaluate the produced system using multiple quality indicators. The first category of indicators is the classification accuracy and indices such as kappa statistic. Different classifiers trained on the same target values (same threshold) can be directly compared in terms of their classification accuracy measured using cross-validation. On the other hand, if two classifiers have been trained on different values of the target variable, then it is not trivial to compare them in a strict way.

Thus a different level in which we examine how interesting each specific classifier is compared to other classifiers produced for different HybridScore thresholds is to measure the gain in classification accuracy they present with respect to an "empty-box" decision system that decides always for the largest class of patients in every case. For instance, let us consider a discrimination problem with 60 % seriously diseased patients and 40 % normal-mild cases for which a classifier gives 75 % prediction accuracy. Let us consider the second problem with 80–20 % distribution of patients and a respective classifier achieving 82 % accuracy. We can conclude that the first system with 15 % gain in accuracy retrieves a greater amount of information from the available classes compared to the 2 % of the second one.

The class distribution is also called "class balance" and is an important determinant for most of training algorithms. When one of the classes is overrepresented in a training set, then the classification algorithm will eventually focus on the larger data class and probably will lose the fine-detail information in the smaller class. To this end, we adopted an additional evaluation strategy for the classification problems. In particular, we selected all the patients from the smaller class and an equal number of randomly selected patients from the larger class to train a classifier. This is repeated five times and the reported accuracy is the average accuracy of the five classifiers. This approach is denoted as "Balanced." Secondly, we select at most 200 patients from the two classes and follow the previous workflow. This strategy is called "Balanced200." The second strategy may reveal how a classifier scales to the size of database, the number of patients provided for training, in a problem with a fixed HybridScore threshold. If the accuracy does not drop dramatically when fewer patients are used for training, then this is an indication of getting stable results. Note that this is only an evaluation methodology since the final classifiers we created were trained on the full dataset at each time, for the selected class definition.

Classification Results

Defining ATS Disease Severity Using DefA

According to the ATS severity definition DefA, which combines the number of diseased vessels and the stenosis level, we trained classifiers for all possible discrimination problems that could be set. In Fig. 1.10 we used the SVM classifiers to evaluate the different discriminating problems. The last one considers the normal or mildly diseased patients to be those with at most two arteries with at most 75 % stenosis. The green line indicates the size of the largest class in each definition of the mild-severe classes. The brown line is the SVM accuracy on all the data of the M-DB, and the blue is the gain in accuracy, i.e., the difference between the SVM accuracy and the largest class (green line). The large gain values indicate the settings under which the classifier managed to retrieve much more information from the data than that of the empty-box classifier.

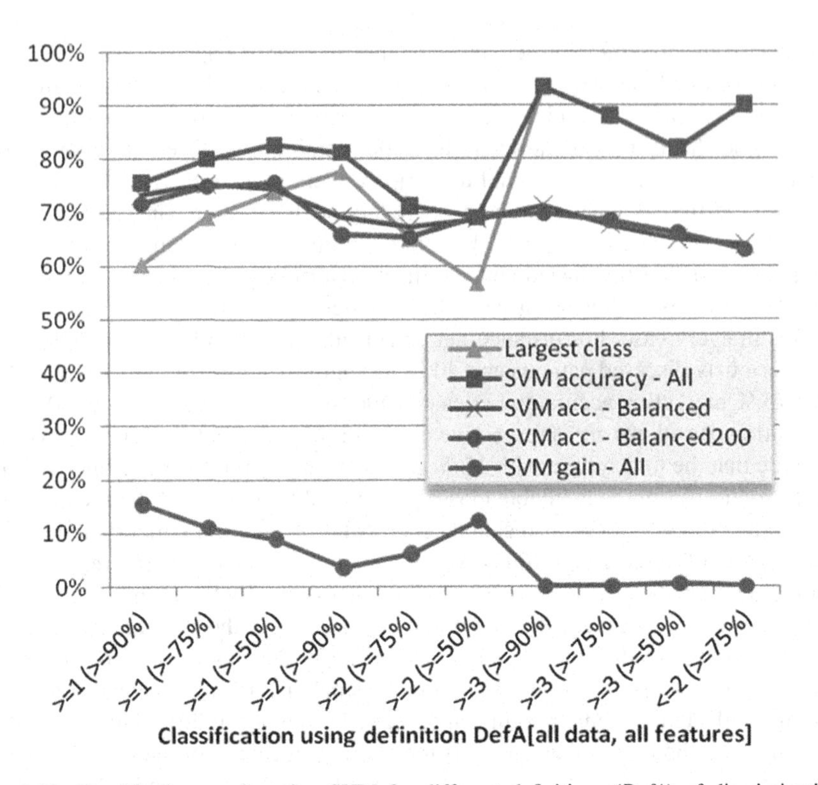

Fig. 1.10 Classification results using SVM for different definitions (*DefA*) of discrimination problems on M-DB

The gain might be increased for mainly two possible reasons. The first is the fact that the problem setting we considered is more separable comparing to the other settings. Thus the data properties (the characteristics of patients) are better described by this specific class definition, and this also indicates that it is interesting to understand the properties that led to this classification grouping "preference." The second reason might be the class balance. When classes are balanced in size, the classifier may achieve lower accuracy value, but still with remarkable gain. This is the role of the experiments we do on balanced subsets of the M-DB. The deep blue line shows the average accuracy of the SVM for the 5 balanced data subsets, and the red is the balanced case of 200 patients per class. If these two lines are close enough in performance, it is evident that the classification performance is not heavily dependent on the amount of available data (similarly one could state that the classifier retrieves as much information as the data let under the specific class definition).

Having all these in mind, we can look back in Fig. 1.10 to observe that the balanced class definitions clearly indicate the first three cases as the best out of all. In particular, the first one, defining the severe class as having all patients with at least one artery with at least 90 % stenosis level, is the overall best since the class sizes of the full DB are more balanced. Table 1.5 summarizes the classification results for all the patients and features of M-DB. The largest class is indicated as

Table 1.5 Classification results using definition DefA for ATS disease severity. All patients and features are used

Problem description			Accuracy						Kappa	Gain				
Diseased arteries	Largest class		DT-J48		Random	Linear SVM			J48		RF	Linear SVM		
number (stenosis)	size (%)	class	All	(size)	Forest	All	Bal	Bal200	All	All	All	All	Bal	Bal200
> = 1 (> = 90 %)	60.23	(S)	72.50	(06/11)	74.23	75.57	73.21	71.55	0.4102	12.27	14.00	15.34	23.21	21.55
> = 1 (> = 75 %)	69.00	(S)	76.73	(10/19)	78.73	80.07	75.34	74.90	0.4054	08.60	09.73	11.07	25.34	24.90
> = 1 (> = 50 %)	73.80	(S)	79.00	(09/17)	81.53	82.70	74.55	75.55	0.3910	05.90	07.73	08.90	24.55	25.55
> = 2 (> = 90 %)	77.63	(M)	81.13	(02/03)	79.93	81.23	69.21	65.80	0.2834	02.40	02.30	03.60	19.21	15.80
> = 2 (> = 75 %)	65.08	(M)	70.83	(09/17)	70.60	71.23	67.26	65.30	0.2560	05.45	05.52	06.15	17.26	15.30
> = 2 (> = 50 %)	56.70	(M)	66.03	(07/13)	66.90	69.00	68.75	69.00	0.2781	10.06	10.20	12.30	18.75	19.00
> = 3 (> = 90 %)	93.50	(M)	93.5	(01/01)	93.50	93.50	71.28	69.74	0.0000	00.00	00.00	0.000	21.28	19.74
> = 3 (> = 75 %)	88.10	(M)	87.8	(05/09)	88.23	88.10	67.56	68.40	0.1163	−00.30	00.13	00.00	17.56	18.40
> = 3 (> = 50 %)	81.60	(M)	81.57	(16/31)	82.23	82.10	64.93	66.15	0.1271	−00.03	00.63	00.50	14.93	16.15
<=2 (> = 75 %)	90.23	(S)	90.23	(01/01)	90.20	90.23	63.96	63.00	0.1763	00.00	−00.03	00.00	13.96	13.00

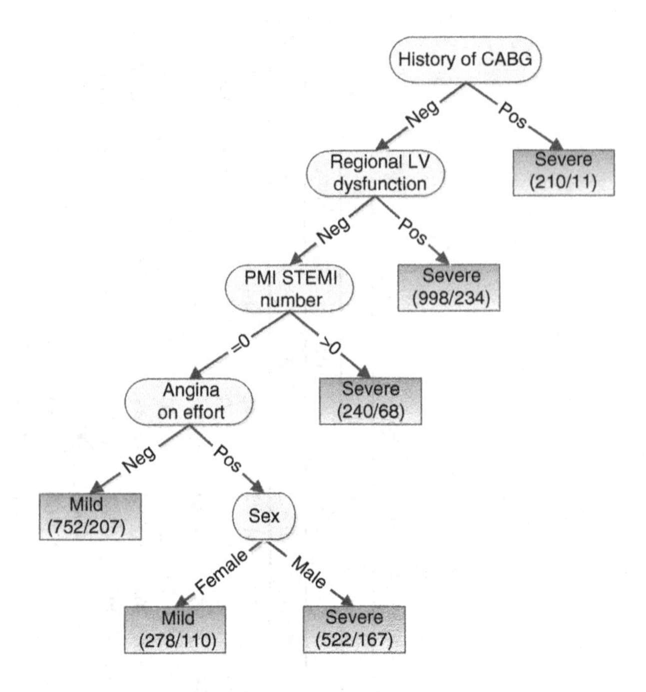

Fig. 1.11 The DT for the classification problem that considers the ATS condition of a patient as severe when he has at least 1 artery with at least 90 % level of stenosis

Severe (S) or Mild (M); the size of a decision tree (DT) is denoted as (number of leaves/number of total tree nodes). The performance of random forest (RF) and SVM is also reported. The three different evaluation indices are denoted in the first line: accuracy, kappa statistic, and accuracy gain. "Bal" indicates the subset of the M-DB with balanced data classes, and "Bal200" are subsets that contain balanced classes with at most 200 patients each. The k-statistic also indicates that in the first three cases, the DT J48 retrieves the "real structure" of the data defined by the considered class labels for the patients.

The DT corresponding to the first line of Table 1.5 is presented in Fig. 1.11. The quality of a rule is indicated by the two numbers inside the rectangle leaf, the first is the number of patients that this leaf decides for and – after the "/" character – the number of patients that were incorrectly classified in the class of the leaf.

References

1. M.W. Browne, "Cross-validation methods", Journal of Mathematical Psychologyvol. 44, Issue 1, pp. 108–132, March 2000.
2. I.H. Witten, Eibe Frank, "Data Mining: Practical Machine Learning Tools and Techniques", Morgan Kaufmann, June 2005.
3. R. Herbrich, Learning Kernel Classifiers, MIT Press, Cambridge, MA, 2002.
4. I. Guyon and A. Elisseeff, "Variable and feature selection", Journal of Machine Learning Research, vol. 3, March 2003.

5. A. Gimelli, G. Rossi, P. Landi, P. Marzullo, G. Iervasi, A. L'Abbate, and Daniele Rovai, "Stress/Rest Myocardial Perfusion Abnormalities by Gated SPECT: Still the Best Predictor of Cardiac Events in Stable Ischemic Heart Disease", Journal of Nuclear Medicine, vol. 50, Issue 4, April 2009.
6. I. Guyon, J. Weston, S. Barnhill, V. Vapnik, "Gene Selection for Cancer Classification using Support Vector Machines", Machine Learning, vol. 46, Issue 1–3, pp. 389–422, 2002.
7. University of California – Irvine (UCI) Machine Learning Repository: http://archive.ics.uci.edu/ml.
8. R. Das, I. Turkoglu and A. Sengur, "Effective Diagnosis of Heart Disease through Neural Network Ensembles", Expert Systems with Applications, vol. 36, pp. 7675–7680, 2009.
9. M.G. Tsipouras, T.P. Exarchos, D.I. Fotiadis, A.P. Kotsia, K.V. Vakalis, K.K. Naka, L.K. Michalis, "Automated Diagnosis of Coronary Artery Disease Based on Data Mining and Fuzzy Modeling", IEEE Transactions on Biomedical Engineering, vol. 12, Issue 4, pp. 447–458, 2008.
10. C. Ordonez, "Comparing Association Rules and Decision Trees for Disease Prediction", Proceedings of the ACM HIKM'06, Arlington, 2006.
11. C. Ordonez, N. Ezquerra and C. Santana, "Constraining and Summarizing Association Rules in Medical Data", Knowledge and Information Systems, vol. 9, Issue 3, pp. 259–283, 2006.
12. P. Chanda, L. Sucheston, A. Zhang, D. Brazeau, J.L. Freudenheim, C. Ambrosone and M. Ramanathan, "AMBIENCE: A Novel Approach and Efficient Algorithm for Identifying Informative Genetic and Environmental Associations with Complex Phenotypes", Genetics, vol. 180, pp. 1191–1210, October 2008.
13. J.H. Moore, J.C. Gilbert, C.T. Tsai, F.T. Chiang, T. Holden, N. Barney and B.C. White, "A flexible computational framework for detecting, characterizing, and interpreting statistical patterns of epistasis in genetic studies of human disease susceptibility", Journal of Theoretical Biology, vol. 241, pp. 252–261, 2006.
14. International HapMap Project: http://hapmap.ncbi.nlm.nih.gov/abouthapmap.html
15. J. R. Quinlan and J. R. C4.5, "Programs for machine learning", Morgan Kaufmann Publishers, 1993.
16. C. Cortes and V. Vapnik, "Support-vector network", Machine Learning, vol. 20, Issue 3, pp. 273–297, 1995.
17. N. Cristiannini and J. Shawe-Taylor, "An Introduction to Support Vector Machines and Other Kernel-Based Learning Models", Cambridge University Press, 2000.
18. L. Breiman, "Random Forests", Machine Learning, vol. 45, Issue 1, pp. 5–32, 2001.
19. P. Tan, M. Steinbach and V. Kumar, "Introduction to Data Mining", Addison-Wesley, 2005.
20. T. Hastie, R. Tibshirani and J. Friedman, "The Elements of Statistical Learning", Springer-Verlag, 2008.
21. H. Liu and R. Setiono, "Chi2: Feature selection and discretization of numeric attributes", Proceedings of the IEEE 7th International Conference on Tools with Artificial Intelligence, pp. 338–391, 1995.
22. H. Peng, F. Long, and C. Ding, "Feature selection based on mutual information: criteria of max-dependency, max-relevance, and min-redundancy", IEEE Transactions on Pattern Analysis and Machine Intelligence, vol. 27, Issue 8, pp. 1226–1238, 2005.
23. A. Reiner, C. Carlson, B. Thyagarajan, M. Rieder, J. Polak, D. Siscovick, D. Nickerson, D. Jacobs Jr, and M. Gross. "Soluble P-Selectin, SELP Polymorphisms, and Atherosclerotic Risk in European-American and African-African Young Adults", Arteriosclerosis, Thrombosis and Vascular Biology, August 2008.
24. A. Timinskas, Z. Kucinskiene, and V. Kucinskas. "Atherosclerosis: alterations in cell communication", in ACTA MEDICA LITUANICA, vol. 14, Issue 1. P. 24–29, 2007
25. S. Szymczak, B.W. Igl, and A. Ziegler. "Detecting SNP-expression associations: A comparison of mutual information and median test with standard statistical approaches", Statistics in Medicine, vol. 28, pp. 3581–3596, 2009.

26. J. Stangard, S. Kardia, S. Hmon, R. Schmidt, A. Tybjaerg-Hansen, V. Salomaa, E. Boerwinkle, and C. Sing. “Contribution of regulatory and structural variations in APOE to predicting dyslipidemia”, The Journal of Lipid Research, vol. 47, pp. 318–328, 2006.
27. N. Yosef, J. Gramm, Q. Wang, W. Noble, R. Karp, and R. Sharan.“Prediction Of Phenotype Information From Genotype Data”, Communications In Information And Systems, vol. 10, Issue 2, pp. 99–114, 2010.
28. F. Pan, L. McMilan, F. Pardo-Manuel De Villena, D. Threadgill, and W. Wang.“TreeQA: Quantitative Genome Wide Association Mapping Using Local Perfect Phylogeny Trees”, Pac Symposium of Biocomputing, pp. 415–426, 2009.
29. D. Tzikas and A. Likas, “An Incremental Bayesian Approach for Training Multilayer Perceptrons”, Proceedings of the International Conference on Artificial Neural Networks (ICANN’10), Thessaloniki, Greece, Springer, 2010.
30. X. Wu, D.l Barbar, L. Zhang, and Y. Ye, “Gene Interaction Analysis Using k-way Interaction Loglinear Model: A Case Study on Yeast Data”, ICML Workshop, Machine Learning in Bioinformatics, 2003.
31. A. Jakulin, I. Bratko, “Testing the Significance of Attribute Interactions”,Proceedings of the 21st International Conference on Machine Learning (ICML-2004), Eds. R. Greiner and D. Schuurmans, pp. 409–416, Banff, Canada, 2004.
32. The ARTreat Project, site: http://www.artreat.org

Chapter 2
Applications of Probabilistic and Related Logics to Decision Support in Medicine

Aleksandar Perović, Dragan Doder, and Zoran Ognjanović

2.1 Introduction

Since the late 60s, probability theory has found application in development of various medical expert systems. Bayesian analysis, which is essentially an optimal path finding through a graph called Bayesian network, has been (and still is) successfully applied in so-called sequential diagnostics, when the large amount of reliable relevant data is available. The graph (network) represents our knowledge about connections between studied medical entities (symptoms, signs, diseases); the Bayes formula is applied in order to find the path (connection) with maximal conditional probability. Moreover, a priori and conditional probabilities were used to define a number of measures designed specifically to handle uncertainty, vague notions, and imprecise knowledge. Some of those measures were implemented in MYCIN in the early 70s [96]. The success of MYCIN has initiated construction of rule-based expert systems in various fields.

However, expert systems with the large number of rules (some of them like CADIAG-2 have more than 10,000) are designed without any proper knowledge of mathematical logic. As an unfortunate consequence, most of them are turned to be inconsistent. On the other hand, the emergence of theoretical computer science as a

A. Perović (✉)
Faculty of Transportation and Traffic Engineering, University of Belgrade,
Vojvode Stepe 305, 11000 Belgrade, Serbia
e-mail: pera@sf.bg.ac.rs

D. Doder
Faculty of Mechanical Engineering, University of Belgrade, Kraljice Marije 16,
11000 Belgrade, Serbia
e-mail: ddoder@mas.bg.ac.rs

Z. Ognjanović
Mathematical Institute of Serbian Academy of Sciences and Arts,
Kneza Mihaila 36, 11000 Belgrade, Serbia
e-mail: zorano@mi.sanu.ac.rs

G. Rakocevic et al. (eds.), *Computational Medicine in Data Mining and Modeling*,
DOI 10.1007/978-1-4614-8785-2_2,

new scientific discipline has lead to discovery that the completeness techniques form mathematical logic are the only known methods for proving correctness of hardware and software. Consequently, mathematical logic has become a theoretical foundation of artificial intelligence.

The last two decades has brought a rapid development of various formal logics that can describe plethora of AI settings. Arguably, the most significant among those are fuzzy logics [18, 28, 29, 35], possibilistic logics [16–18], and probability logics. Our focus will be solely on the probability logics, since it is our field of expertise.

Besides the introduction, this chapter is divided into six sections: probability logic, nonmonotonic reasoning and conditional probabilities, probabilistic approach to measuring inconsistency, reasoning about evidence, MYCIN, and CADIAG-2.

In the second section we give a short overview of the meaning and scope of probability logic. The emphasis is on decidability (existence of automated satisfiability checking of probability assertions), complexity of decision procedure (assessment of the running time and hardware resources needed for the execution of decision procedure), heuristical SAT (SAT means satisfiability) solving based on genetic algorithms, and the novel application of probability in classification problems (e.g., handling queries in fuzzy relational databases).

The third section brings a short overview of the theory of nonmonotonic inference and its deep connections with probability logic. A number of inference systems are presented. Though this part is a rather technical, we believe that it should be useful to any physician who wishes to identify the atomic steps in the personal diagnostic practice – a necessary step for the construction of any successful medical expert system.

The fourth section presents recently developed probabilistic approach to managing inconsistency in knowledge bases. The fifth section is a brief overview of recent results in the logical treatment of evidence and probabilistic spatiotemporal reasoning.

The last two sections present a brief overview of the two significant medical support systems: MYCIN, developed at the Stanford University in the early 70s by Edward Shortliffe and others, and CADIAG-2, an ongoing project of the Institute for medical expert and knowledge-based systems and the Medical University of Vienna that is initiated and carried out primarily by K.-P. Adlassing.

2.2 Probability Logic

2.2.1 Overview

In broader sense, probability (or probabilistic) logic can be understood as a tool for reasoning with incomplete or imprecise information (knowledge), where the uncertainty of the premises is expressed by qualitative or quantitative probability statements. The typical forms of the qualitative statements are "α is the probable cause for β," like in "the seasonal flue is the probable cause for the patient's fever," and "α is more probable than β," like in "the HPV is more probable cause for the observed cervical cancer than exposure to gamma radiation." The typical form of the quantitative probability statements is "the probability of α is approximately equal to s," like in "the incidence of monozygotic twinning is about 3/1,000," or in "around 30 % cases of the coronary thrombosis are caused by smoking." Generally, the quantitative statements can express more complex conditions on the probability values (e.g., linear or polynomial inequalities).

Mathematical representation of probabilistic reasoning extends basic logical language (that involves propositional connectives and universal and existential quantifiers) with probabilistic operators and probabilistic quantifiers. Though the roots of probability logic can be traced at least to Leibnitz, the modern era of probability logic has started with the work of Jerome Keisler [47–49] throughout the 70s and the mid-80s of the twentieth century. It is worth mentioning that Bayesian analysis (application of Bayes formula in determination of optimal diagnostic/therapy strategies) has been successfully applied in early clinical decision support systems specialized in sequential diagnostics in the late 60s (see for instance [30]).

The modal representation of probability, i.e., introduction of modal-like probability operators in classical reasoning, deeply motivated by intensive application in various expert systems, was initiated by Nils Nilsson [59, 60] in the mid-80s and early 90s. A major breakthrough along these lines was made by Ronald Fagin, Joseph Halpern, and Nimrod Megiddo [21], especially in terms of decidability and computational complexity (so-called small model theorems). Though the introduced syntax was not modal per se, it was very similar to it and the developed modal probability semantics has become the standard one.

The first probability logic with unary modal probability operator was introduced by Miodrag Rašković in the early 90s [86]. Soon after, a rather rapid development of the subject has followed. We shall track a selection of the existing research in the field of probability logic: primarily our own contributions, "seasoned" with certain papers (books) that are closely related to our work.

To begin with, an extensive study of finitely additive probability measures was given in [93]. Historical development, various boundaries of probability functions, and many other important concepts regarding sentential probability logics are given in [34]. An extensive study of uncertainty and its connection with probability was given in [36]. Various formalizations of probability with variety of scopes – simple probabilities, higher-order (nesting of probability operators) probabilities, conditional probabilities, representation of default reasoning, etc. – are presented in [1, 3, 4, 10–15, 20, 21, 28, 29, 42–46, 55, 57–73, 75–78, 82–92, 104]. Authors of those papers usually did not consider assumptions about the corresponding probabilistic distributions. Then, as probabilities are generally not truth-functional, the best one can do is to calculate bounds on probabilities of conclusions starting from probabilities of assumptions [34].

One of the main proof-theoretical problems is providing an axiom system that would be strongly complete in the sense that every consistent theory has a model. This problem originates from the inherent non-compactness of the so-called nonrestricted real-valued probability logics. Namely, in such formalisms it is possible to define an inconsistent infinite set of formulas, every finite subset of which is consistent. For example, one such theory is given by

$$\{P_{>0}\alpha\} \cup \left\{P_{<\frac{1}{n}}\alpha : n \text{ is a positive integer}\right\}.$$

As it was pointed in [63, 104], there is an unpleasant consequence of finitary axiomatization in that case: there exist unsatisfiable sets of formulas that are consistent with respect to the assumed finite axiomatic system (since all finite subsets are consistent and deductions are finite sequences). Another important theoretical problem is related to the decidability issue.

2.2.2 LPP$_2$ *Logic*

We shall briefly describe the LPP_2 probability logic. Detailed exposition can be found in [71]. It is an extension of the classical propositional logic with probability operators of the form $P_{\geq s}$, where s can be any rational number between 0 and 1 (including both of them).

The initial syntactical layer is formed of classical propositional formulas; they will be denoted by α, β and γ, indexed or primed if necessary. Basic probability formulas are expressions of the form

$$P_{\geq s}\alpha.$$

The intended meaning of $P_{\geq s}\alpha$ is rather obvious: the probability of α is at least s. Finally, complex probability formulas are formed from the basic ones by application of logical connectives: negation (denoted by $\neg$) and implication (denoted by $\rightarrow$). Probability formulas will be denoted by A, B and C, indexed or primed if necessary. Some standard abbreviations (e.g., formal introduction of conjunction, disjunction and equivalence) are defined in the usual way:

- $A \wedge B =_{def} \neg(A \rightarrow \neg B)$.
- $A \vee B =_{def} \neg A \rightarrow B$.
- $A \leftrightarrow B =_{def} (A \rightarrow B) \wedge (B \rightarrow A)$.
- $P_{\leq s}\alpha =_{def} P_{\geq s}\neg\alpha$.
- $P_{>s}\alpha =_{def} P_{\geq s}\alpha \wedge \neg P_{\leq s}\alpha$.
- $P_{<s}\alpha =_{def} P_{\leq s}\alpha \wedge \neg P_{\geq s}\alpha$.
- $P_{=s}\alpha =_{def} P_{\geq s}\alpha \wedge P_{\leq s}\alpha$.

Due to the modal nature of probability operators, the standard probabilistic semantics is defined on the so-called probability Kripke structures. A probability Kripke structure is any triple (W,H,μ) with the following properties:

1. W is a nonempty subset of the set of all classical evaluations $\{0,1\}^{Var}$.
2. H is an algebra of sets (it is nonempty and closed under intersection, union and complement) that contains all sets of the form $[\alpha]$. Here by $[\alpha]$ we have denoted the set of all evaluations satisfying α.
3. $\mu : H \rightarrow [0,1]$ is a finitely additive probability measure.

The satisfiability relation $\models$ is defined in the following way:

- $(W,H,\mu) \models \alpha$ iff $[\alpha] = W$.
- $(W,H,\mu) \models P_{\geq s}\alpha$ iff $\mu([\alpha]) \geq s$.
- $(W,H,\mu) \models \neg A$ iff $(W,H,\mu) \not\models A$.
- $(W,H,\mu) \models A \rightarrow B$ iff either $(W,H,\mu) \not\models A$, or $(W,H,\mu) \models A$ and $(W,H,\mu) \models B$.

The axioms of the LPP_2 logic are the following ten schemata:

- A × 1: $\alpha \rightarrow (\beta \rightarrow \alpha)$.
- A × 2: $(\alpha \rightarrow (\beta \rightarrow \gamma)) \rightarrow ((\alpha \rightarrow \beta) \rightarrow (\alpha \rightarrow \gamma))$.
- A × 3: $(\neg\beta \rightarrow \neg\alpha) \rightarrow (\alpha \rightarrow \beta)$.
- A × 4: $A \rightarrow (B \rightarrow A)$.
- A × 5: $(A \rightarrow (B \rightarrow C)) \rightarrow ((A \rightarrow B) \rightarrow (A \rightarrow C))$.
- A × 6: $(\neg B \rightarrow \neg A) \rightarrow (A \rightarrow B)$.
- A × 7: $P_{\geq 0}\alpha$.
- A × 8: $P_{\geq s}\alpha \rightarrow P_{> r}\alpha$ for all $r < s$.
- A × 9: $P_{\geq s}\alpha \wedge P_{\geq r}\beta \wedge P_{= 0}(\alpha \wedge \beta) \rightarrow P_{\geq \min(r+s,1)}(\alpha \vee \beta)$.
- A × 10: $P_{\leq s}\alpha \wedge P_{\leq r}\beta \rightarrow P_{\leq \min(r+s,1)}(\alpha \vee \beta)$.

The inference rules of the LPP_2 logic are the following one:

- Modus ponens for classical formulas: from α and $\alpha \rightarrow \beta$ infer β.
- Modus ponens for probability formulas: from A and $A \rightarrow B$ infer B.
- Necessitation: from α infer $P_{= 1}\alpha$.
- Archimedean rule: from the set of premises $\{A \rightarrow P_{\geq r}\alpha : r < s\}$ infer $A \rightarrow P_{\geq s}\alpha$.

The notion of deduction differs from the classical one only in the length of the inference: since Archimedean rule has countably many premises, the length of the inference can be any countable successor ordinal.

Intuitively, Archimedean rule should be understood in the following way: if the probability of α is infinitely close to the rational number s, then it must be equal to s. As a consequence, "problematic" finitely satisfiable but unsatisfiable theories such as previously mentioned theory $\{P_{>0}\alpha\} \cup \left\{P_{<\frac{1}{n}}\alpha : n \text{ is a positive integer}\right\}$ become inconsistent in LPP_2. The proof of the strong completeness theorem for LPP_2 logic can be found in [71].

2.2.3 Decidability and Complexity

Any potential or actual application of weighted logics in artificial intelligence is closely related to the satisfiability problem and related computational complexity estimation. Here we shall outline the satisfiability procedure for the LPP_2-formulas and give its exact complexity. So, let $A \in For_P$. Recall that an atom a of A is a formula of the form $\pm p_1 \wedge \cdots \wedge \pm p_n$, where $\pm p_i$ is either p_i or $\neg p_i$, and $p_1, \ldots, p_n$ are all primitive propositions appearing in A. For example, if A is the formula $P_{\geq 0.9}(p \vee q)$, then its atoms are $p \wedge q, p \wedge \neg q, \neg p \wedge q$ and $\neg p \wedge \neg q$.

Note that atoms are pairwise disjoint. Hence, for any probability measure μ and any pair of atoms a_i and $a_j (a_i \neq a_j)$, we have that

$$\mu(a_i \vee a_j) = \mu(a_i) + \mu(a_j).$$

As a next step we can equivalently transform the given formula A into its complete disjunctive normal form

$$DNF(A) = \bigvee_{i=1}^{m} \bigwedge_{j=1}^{k_i} X^{i,j}(p_1, \ldots, p_n),$$

where:

- $X^{i,j}$ is one of probability operators $P_{\geq s_{i,j}}$ and $P_{< s_{i,j}}$.
- $X^{i,j}(p_1, \ldots, p_n)$ denotes the fact that the propositional formula which is in the complete disjunctive normal form, i.e., the propositional formula is a disjunction of the atoms of A.

Example. The complete disjoint normal form of the formula A defined by

$$P_{<0.1}p \vee P_{\geq 0.8}q$$

is disjunction of the following four formulas:

- $P_{<0.1}((p \wedge q) \vee (p \wedge \neg q)) \wedge P_{\geq 0.8}((p \wedge q) \vee (\neg p \wedge q))$.
- $P_{<0.1}((p \wedge q) \vee (p \wedge \neg q)) \wedge P_{<0.8}((p \wedge q) \vee (\neg p \wedge q))$.
- $P_{\geq 0.1}((p \wedge q) \vee (p \wedge \neg q)) \wedge P_{\geq 0.8}((p \wedge q) \vee (\neg p \wedge q))$.
- $P_{\geq 0.1}((p \wedge q) \vee (p \wedge \neg q)) \wedge P_{<0.8}((p \wedge q) \vee (\neg p \wedge q))$.

The logic LPP_2 is decidable, i.e., the satisfiability and validity of LPP_2-formulas is algorithmically solvable. Firstly we will outline satisfiability algorithm in general case, and then we will apply it on the previous example.

The first step is to transform the given For_P-formula A to its complete disjunctive normal form $DNF(A) = \bigvee_{i=1}^{m} \bigwedge_{j=1}^{k_i} X^{i,j}(p_1, \ldots, p_n)$. So, satisfiability of A is equivalently reduced to the satisfiability of $DNF(A)$. Thus, A is satisfiable iff at least one disjunct from $DNF(A)$ is satisfiable. Let the measure of the atom a_i be denoted by y_i. We use an expression of the form $a_t \in X(p_1, \ldots, p_n)$ to denote that the atom a_t appears in the propositional part of $X(p_1, \ldots, p_n)$. Furthermore, a disjunct $D = \wedge_{j=1}^{k} X^j(p_1, \ldots, p_n)$ from $DNF(A)$ is satisfiable iff the following system of linear equalities and inequalities is satisfiable:

$$
\begin{gathered}
y_1 + \cdots + y_{2^n} = 1 \\
y_1 \geq 0 \\
\vdots \\
y_{2^n} \geq 0 \\
\sum_{a_t \in X^1(p_1,\ldots,p_n) \in D} y_t \geq^{X^1} s_1 \\
\vdots \\
\sum_{a_t \in X^k(p_1,\ldots,p_n) \in D} y_t \geq^{X^k} s_k,
\end{gathered}
$$

where $\geq^{X^i} = \geq$ if $X^i = P_{\geq s_i}$, otherwise $\geq^{X^i} = <$.

Since the satisfiability of A is reduced to the linear systems solving problem, the satisfiability problem for LPP_2-logic is decidable. Finally, since A is valid iff $\neg A$ is unsatisfiable, the validity problem is also decidable.

Back to the previous example: the atoms of the given probability formula $A = P_{<0.1}p \vee P_{\geq 0.8}q$ are $p \wedge q$, $p \wedge \neg q$, $\neg p \wedge q$ and $\neg p \wedge \neg q$. By $y_1, \ldots, y_4$ we will denote their unknown probabilities. The first disjunct in $DNF(A)$ generates the following system:

$$
\begin{gathered}
y_1 + y_2 + y_3 + y_4 = 1 \\
y_1 \geq 0 \\
y_2 \geq 0 \\
y_3 \geq 0 \\
y_4 \geq 0 \\
y_1 + y_2 < 0.1 \\
y_1 + y_3 \geq 0.8.
\end{gathered}
$$

One solution of this system is given by $y_1 = y_2 = 0$, $y_3 = 0.8$ and $y_4 = 0.2$, so the formula A is satisfiable.

Concerning complexity estimation of the decision procedure, we shall show that it is NP-complete. Indeed, the lower bound follows from the complexity of the same problem for classical propositional logic. The upper bound is a consequence of the NP-complexity of the satisfiability problem for linear weight formulas from [21].

2.2.4 A Heuristical Approach to Satisfiability Problem

Since the LPP_2-satisfiability problem is NP-complete, it is natural to try to solve its instances using certain heuristics. Here we shall describe such an approach based on the so-called genetic algorithms.

As the name suggests, genetic algorithms (GA) imitate the basic genetic postulates (e.g., selection, mutation) in the automated search of the solution within the specified data set. In order to endow the method as fully as possible, the technical vocabulary was not changed in this new setting. As a consequence, in the application of genetic algorithms, one will refer to various potentially inanimate objects (e.g., quadruple of real numbers from the unit real interval [0,1]) as chromosomes.

Each individual (also called chromosome) is seen as a possible solution in the search space for the particular problem. Thus, a GA can be seen as a searching procedure for the global optima of the corresponding problem. Individuals are represented by genetic code over a finite alphabet.

An evaluation function assigning fitness values to individuals has to be defined. Fitness values indicate quality of the corresponding individuals, while average fitness of entire populations may be good measures of obtained quality of the procedures. GAs consist of applications of the genetic operators to populations that must ensure that average fitness values are continually improved from each generation to subsequent. Basic genetic operators are selection, crossover, and mutation, but some additional operators such as inversion or local search may be used as well.

Selection mechanism favorizes highly fitted individuals (as well as parts of genetic code of individuals, i.e., genes) to have better chances for reproduction into next generations. On the other hand, chances for reproduction for less fitted members are reduced, and they are gradually wiped out from populations.

Crossover operator partitions a population into a set of pairs of individuals named parents. For each pair a recombination of their genetic material is performed with some probability. In that way nondeterministic exchange of genetic material in populations is obtained.

Multiple applications of selection and crossover operators may produce that the variety of genetic materials is lost. It means that some areas of search spaces become not reachable. This usually causes the convergence in local optimums far from the global optimal values. Mutation operator can help to avoid this shortcoming. Parts of individuals (genes) can be changed with some small probability to increase diversibility of genetic material. An initial population is usually generated by random, although sometimes it may be fully or partially produced by an initial heuristic.

A general description of GAs is given in Fig. 2.1, where N_{pop} and p_i denote the number of individuals and their objective values, respectively. The objective value of an individual corresponds to the value which the individual owns in the case of the considered problem. The for-loop is repeated until a finishing criterion

Fig. 2.1 A general description of GAs (Reproduced from [71])

```
InputData();
PopulationInit();
while ( not FinishedGA() ) {
    for ( i = 0 ; i < N_pop ; i++) p_i = ObjectiveFunction();
    HeuristicImprovement();
    ComputeFitnesses();
    Selection();
    Crossover();
    Mutation();
}
OutputResults();
```

(the global optima is found, the maximal number of iterations is reached,...) is satisfied. Since the procedure is not complete, if the maximal number of iterations is reached, we do not know whether the considered problem is solvable. Heuristic-Improvement() can be optionally included to improve efficiency of GA and/or to help the procedure to escape from local optima.

Genetic algorithm will be applied on linear weight formulas, i.e., formulas of the form

$$a_1 w(\alpha_1) + \cdots + a_n w(\alpha_n) \geq c,$$

where a_i's and c are rational numbers and α_i's are classical propositional formulas containing primitive propositions from A. The intended meaning of $w(\alpha)$ is "the probability of α." In the wider context $w(\alpha)$ refers to the weight of α, which is a mathematical model for agent's (expert's) estimation of importance (or agent's confidence) of α.

A weight literal is an expression of the form $\Sigma_i a_i w(\alpha_i) \geq c$ or $\Sigma_i a_i w(\alpha_i) < c$.

The logic that allows such kind of formulas is still NP-complete; so by using this logic, we just add some expressiveness to our language.

Since For_P-formulas can be equivalently translated into their disjunctive normal forms and a disjunction is satisfiable if at least one disjunct is satisfiable, from now on we will only consider formulas of the following form:

$$a_{j,1} w(DNF(\alpha_{j,1})) + \cdots + a_{j,n_j}(DNF(\alpha_{j,n_j}))\rho_j c_j,$$

where $\rho_j \in \{\geq, <\}$, $a_{j,i}$'s and c_j are rational numbers, and $DNF(\alpha)$ denotes the complete disjunctive normal form of α. We say that such a formula is in the weight conjunctive form (wfc-form). Also, we will use $at \in DNF(\alpha)$ to denote the fact that the atom at appears in $DNF(\alpha)$.

The input for the LPP_2-satisfiability checker based on genetic algorithms is a weight formula f in the wfc-form with L weight literals. Without loss of generality, we demand that classical formulas appearing in weight terms are in complete disjunctive normal form. Let $\phi(f) = \{p_1, \ldots, p_N\}$ denote the set of all primitive propositions from f, and $|\phi(f)| = N$.

An individual M consists of L pairs of the form (atom, probability) that describe a probabilistic model. The first coordinate is given as a bit string of length N, where 1 at the position i denotes $\neg p_i$, while 0 denotes p_i. Probabilities are represented by floating point numbers.

For an individual $M = ((at_1, \mu(at_1)), \ldots,(at_N, \mu(at_N)))$, the linear system is equivalent to $\vee_{i=1}^{L}(\sum_{j=1}^{L} a_{i,j}\mu(at_j)\rho_i c_i)$. Note that it is possible that some $a_{i,j} = 0$ though $[a_{i,j}]$ matrix is usually not sparse. The individuals are evaluated using function $d(M)$, which measures a degree of unsatisfiability of an individual M. Function $d(M)$ is defined as the distance between the left- and right-hand side values of the weight literals not satisfied in the model described by M:

$$d(M) = \sqrt{\sum_{M \not\models t_i\rho_i c_i} \left(\left(a_{i,1} \cdot \sum_{at \in DNF(\alpha_{i,1})} \mu(at) \right) + \cdots + \left(a_{i,n_i} \cdot \sum_{at \in DNF(n_i)} \mu(at) \right) - c_i \right)^2}$$

If $d(M) = 0$, all the inequalities in the linear system are satisfied, hence the individual M is a solution.

Some features of GA have been set for all tests:

- The population consists of ten individuals.
- One set of tests has been performed with a population of 20 individuals.
- Selection is performed using the rank-based roulette operator (with the rank from 2.5 for the best individual to 1.6 for the worst individual – the step is 0.1).
- The crossover operator is one point, with the probability 0.85.
- The elitist strategy with one elite individual is used in the generation replacement scheme.
- Multiple occurrences of an individual are removed from the population.

Two problem-specific (*two-part*) mutation operators were used. The first operator (*TP1*) features two different probabilities of mutation for the two parts (*atoms*, *probabilities*) of an individual; after mutation, the real numbers in *probabilities* part of an individual have to be scaled since their sum must be equal to 1. The second operator (*TP2*) is a combination of ordinary mutation on *atoms* part and a special mutation on *probabilities* part of an individual.

Instead of performing mutation on two bits in the representation of *probabilities* part, two members p_{i_1}, p_{i_2} of *probabilities* part are chosen randomly and then replaced with the random p'_{i_1}, p'_{i_2}, such that $p_{i_1} + p_{i_2} = p'_{i_1} + p'_{i_2}$ and $0 \leq p'_{i_1}, p'_{i_2} \leq 1$. The sum of probabilities does not change and no scaling is needed.

We have experimented with the following choices in the local search procedure:

LS1 (LS denotes "local search"): For an individual M all the weight literals are divided into two sets: the first set (B) contains all satisfied literals, while the second one (W) contains all the remaining literals. The literal $t_B\rho_B c_B \in B$ (called the best one) with the biggest difference $|\mu(t_B) - c_B|$ between the left and the right side and the literal

$t_W \rho_W c_W \in W$ (the worst one) with the biggest difference $|\mu(t_W) - c_W|$ are found. Two sets of atoms are determined: the first set $B_{At(f)}$ contains all the atoms from M satisfying at least one classical formula α_i^B from $t_B = a_1^B w(\alpha_1^B) + \cdots + a_{k_B}^B w\left(\alpha_{k_B}^B\right)$, while the second one $W_{At(f)}$ contains all the atoms from M satisfying at least one classical formula α_i^W from $t_W = a_1^W w(\alpha_1^W) + \cdots + a_{k_W}^W w\left(\alpha_{k_W}^W\right)$. The probabilities of a randomly selected atom from $B_{At(f)} \setminus W_{At(f)}$ and a randomly selected atom from $W_{At(f)} \setminus B_{At(f)}$ are changed so that $t_B \rho_B c_B$ remains satisfied, while the distance $|\mu(t_W) - c_W|$ is decreased or $t_W \rho_W c_W$ is satisfied.

LS2: For an individual M, the *worst* weight literal $t_W \rho_W c_W$ from W (the set of unsatisfied literals) with the biggest difference $|\mu(t_W) - c_W|$ is found. The literal can be represented as $\Sigma_{j=1}^{L} a_{W_j} \mu(at_j) \rho_W c_W$. We try to change the vector of probabilities $[\mu(at_j)]$, so that the linear equation $\Sigma_{j=1}^{L} a_{W_j} \mu(at_j) = c_W$ is satisfied. The equation $\Sigma_{j=1}^{L} a_{W_j} \mu(at_j) = c_W$ represents a hyperplane in $\mathbb{R}^n$ while $\left[a_{W_j}\right]$ denotes a vector normal to the hyperplane. The projection of $[\mu(at_j)]$ to the hyperplane that satisfies the given equation is $\left[\mu'(at_j)\right] = \left[\mu(at_j)\right] + k_W \left[a_{W_j}\right]$. The calculation of k and the projection vector is simple and straightforward and gives

$$k = \frac{c_W - \sum_{j=1}^{L} \mu(at_j) a_{W_j}}{\sum_{j=1}^{L} a_{W_j}^2}.$$

We set the new vector of probabilities to be

$$\left[\mu''(at_j)\right] = \frac{[\max(\mu'(at_k), 0)]}{\sum_{k=1}^{L} \max(\mu'(at_k), 0)}.$$

Negative coordinates are replaced with 0, and the vector is scaled so that the sum of its coordinates $\Sigma_{j=1}^{L} \mu''_{at_j}$ equals 1.

LS3 is similar to LS2, with the difference being made when choosing the weight literal $t_W \rho_W c_W$ from W (the set of unsatisfied literals). The chosen literal is the one with the smallest difference $|\mu(t_W) - c_W|$; it is the *best bad literal*.

LS4 is similar to LS2 and LS3. Instead of calculating the projection $\left[\mu'(at_j)\right] = \left[\mu(at_j)\right] + k_W \left[a_{W_j}\right]$ for one chosen weight literal $t_W \rho_W c_W$ from W, we calculate $k_{W_i} \left[a_{W_{ij}}\right]$ for each literal $t_{W_i} \rho_{W_i} c_{W_i}$ from W (the set of unsatisfied literals) and calculate the *intermediate* vector $[\mu'(at_j)]$, by adding the linear combination to the original vector: $\left[\mu'(at_j)\right] = \left[\mu(at_j)\right] + \Sigma_{W_i} k_{W_i} \left[a_{W_{ij}}\right]$. The new vector of probabilities $[\mu''(at_j)]$ is then calculated in same fashion as in LS2.

In our methodology, introduced in [64], the performance of the system is evaluated on a set of PSAT-instances, i.e., on a set of randomly generated formulas in the wfc-form (with classical formulas in disjunctive normal form). The advantage of this approach is that a formula can be randomly generated according to the following parameters: N, the number of propositional letters; L, the number of weight literals; S, the maximal number of summands in weight terms; and D, the maximal number of disjuncts in DNFs of classical formulas. The considered set of test problems contains 27 satisfiable formulas. Three PSAT-instances were generated for each of nine pairs of (N,L), where $N \in \{50,100,200\}$, and $L \in \{N, 2N, 5N\}$. For every instance $S = D = 5$.

Having the above parameters, L atoms and their probabilities (with the constraint that the sum of probabilities must be equal to 1) are chosen. Next, a formula f containing L basic weight formulas is generated. It contains primitive propositions from the set $\{p_1, \ldots, p_N\}$ only. Every weight literal contains at most S summands in its weight term. Every classical formula is in disjunctive normal form with at most D disjuncts, while every disjunct is a conjunction of at most N literals. For every weight term t coefficients are chosen and the value of t is computed. Next, the sum $sp(t)$ of positive coefficients and the sum $sn(t)$ of negative coefficients are computed. Finally, the right-side value of the weight literals between $sp(t)$ and $sn(t)$ and the relation sign are chosen such that f is satisfiable.

We prefer to test more problem instances of different sizes (even very large scale instances) rather than making more trials on a smaller set of instances (of smaller or average size). Since the tests are of large sizes, the necessity to perform them in a reasonable time imposed to set the maximal number of generations to be 10,000 for $N = 50$, 7,000 for $N = 100$ and 5,000 for $N = 200$.

As an illustration of the corresponding results, we give Table 2.1 containing the average running time of successful tests as measured on our test computer (a Pentium P4 2.4 GHz, 512 MB-based Linux station). The table shows running times only for selected tests. Columns 2 and 3 show times for tests without LSs, with different population sizes (10 individuals vs. 20 individuals). Increased population size does result in smaller number of iterations needed to find the solution, but the computational cost for each iteration is increased and the overall computational cost is greater than with smaller population size. In columns 4–7 and 8–11, we can compare the efficiency of various LSs. It is clear that LS2 and LS3 are more efficient than LS1 and LS4 when used for large problem instances; however, it is not clear which of them is the most efficient. The running times in columns 8–11 (LSs applied in each third generation) are on average smaller than times in columns 4–7 (LSs applied in each generation). However, this does not mean that the principle of reducing application of LSs to each third generation is always more efficient. Finally, columns 12–14 show execution times for tests using combination of LSs. Combined usage of LSs is not justified in terms of time efficiency, but it is justified in terms of increased success rate. Higher mutation rate in this setup leads to better time efficiency and higher success rate, except for a few less complex problem instances.

Table 2.1 Average time (rounded to seconds) used by the test computer to execute successful tests for some selected parameters (Note: value 0 means that the average time was less than 0.5 s.) (Reproduced from [71])

	Table 1		Table 2				Table 3				Table 4		
	TP2(12, 4)		TP2(12, 4)				TP2(12, 4)				TP2(12, 4)	TP2(24, 8)	TP2(48, 16)
	10 ind.	20 ind.	10 individuals				10 individuals				10 individuals		
			LS's applied in each generation				LS's applied in each third generation						
L, N, inst. no.	No LS		LS1	LS2	LS3	LS4	LS1	LS2	LS3	LS4	Combination of LS's applied in each generation		
50, 50, 1	0	1	0	0	0	0	0	0	0	0	0	0	0
50, 50, 2	0	1	0	0	0	0	0	0	0	0	0	0	0
50, 50, 3	0	1	0	1	1	0	0	0	0	0	0	0	0
50, 100, 1	1	1	1	0	0	2	1	0	0	1	1	0	1
50, 100, 2	1	2	1	1	2	2	1	1	1	2	2	2	3
50, 100, 3	3	3	1	2	7	10	1	2	3	4	1	3	3
50, 250, 1	16	20	28	16	16	30	22	14	11	21	40	35	42
50, 250, 2	51	56	24	38	34	97	26	35	30	50	68	70	132
50, 250, 3	18	20	18	9	17	25	10	8	13	14	15	16	19
100, 100, 1	0	1	0	0	0	0	0	0	0	0	0	0	0
100, 100, 2	0	1	0	0	0	0	0	0	0	0	0	0	0
100, 100, 3	0	1	0	0	0	1	1	0	0	0	1	1	1
100, 200, 1	8	12	10	3	8	9	6	3	8	7	5	5	7
100, 200, 2	2	3	1	3	2	4	1	3	1	4	4	1	2
100, 200, 3	1	3	4	1	2	26	2	1	1	2	2	2	2
100, 500, 1	187	236	170	130	149	384	94	145	244	228	269	294	271
100, 500, 2	295	309	242	241	298	333	169	306	151	228	236	260	480
100, 500, 3	484	575	326	509	416	775	296	390	355	461	1019	777	671
200, 200, 1	58	91	71	108	56	134	34	78	66	3471	146	270	202
200, 200, 2	5	6	11	7	7	14	11	7	10	9	13	11	9
200, 200, 3	2	3	4	1	2	2	4	1	1	4	4	2	3
200, 400, 1	12	11	4	6	5	25	6	7	5	14	8	11	7
200, 400, 2	238	286	N/A	195	163	484	N/A	171	161	296	479	686	1128
200, 400, 3	205	230	N/A	174	205	247	N/A	153	201	208	419	334	374
200, 1000, 1	1593	2173	3064	888	1347	2972	2307	811	1271	1865	2363	2087	2032
200, 1000, 2	N/A	N/A	N/A	N/A	N/A	N/A	N/A	N/A	N/A	N/A	N/A	19582	19977
200, 1000, 3	1489	1861	3298	1364	792	3548	2456	1135	1080	2023	2818	2770	2778

2.2.5 *Probabilistic Classification*

A rather important and widely neglected feature of any probability on propositional formulas is the fact that it can lead to effective methods for extending of any fuzzy evaluation $e : Var \to [0,1]$ of the set of propositional letters Var to the evaluation $e^* : For \to [0,1]$, of the set of all propositional formulas.

Consequently, probability logic can be applied in problems that are traditionally handled by fuzzy logic or possibilistic logic. A typical example of this kind is any classification of the given objects $O_1, \ldots, O_n$ according to certain criteria.

Due to finite additivity, any probability measure on propositional formulas is uniquely determined by its values on finite conjunctions of pairwise distinct propositional letters. As it was shown in [76], any fuzzy evaluation $e : Var \to [0,1]$ can be uniquely extended to measures e^{Π} and e^G, where

$$e^{\Pi}(p_1 \wedge \cdots \wedge p_n) =_{def} e(p_1) \cdots e(p_n).$$

and

$$e^G =_{def} \min(e(p_1), \ldots, e(p_n))$$

($p_1, \ldots, p_n$ are pairwise distinct).

Product measures e^{Π} correspond to one extreme situation: stochastic or probability independence of propositional letters. Gödel's measures e^G correspond to another kind of extreme situation: logical dependence of propositional letters. While stochastic independence is a measure-theoretic property and cannot be forced by some nontrivial logical conditions (see [34]), logical dependence is expressible in classical propositional calculus. For instance, logical condition

$$p \to q$$

clearly entails that $e^*(p \wedge q) = \min((e(p), e(q)))$.

Linear convex combinations of finitely additive probability measures are finitely additive probability measures as well, so using e^G and e^{Π} we can construct an infinite scale of probability measures $e^{(s)} = se^{\Pi} + (1 - s)e^G$.

From the uncertainty point of view, measures e^s correspond to various degrees of dependence between propositional letters. From the fuzziness point of view, measures $e^{(s)}$ provide countably many ways to extend the initial evaluation e of propositional letters: a fact that enables probability evaluations of fuzzy quantities. We will try to illustrate our intended meaning with the following simple example:

Example. Suppose that we have to classify compounds C_1, C_2, and C_3 of the substances p and q according to the criteria of minimal harmfulness of a compound. It is known that both p and q are harmful, but they neutralize each other. Concentrations of substances p and q in compounds C_1, C_2, and C_3 are given in the Table 2.2 below:

Table 2.2 Reproduced from [76]

Compound	Concentration of p	Concentration of q
C_1	0.95	0.05
C_2	0.15	0.85
C_3	0.65	0.35

Table 2.3 Reproduced from [76]

Evaluation	$e(p)$	$e(q)$	$e^G(p \leftrightarrow q)$	$e^\Pi(p \leftrightarrow q)$	$e^{(0.25)}(p \leftrightarrow q)$
C_1	0.95	0.05	0.1	0.095	0.099
C_2	0.15	0.85	0.3	0.255	0.289
C_3	0.65	0.35	0.7	0.455	0.639

Syntactically, we will consider both p and q as propositional letters. Since the substances p and q neutralize each other, the minimal harmfulness criteria is adequately represented by the formula $p \leftrightarrow q$. Moreover, for any [0,1]-evaluation e is any of p and q we can easily calculate $e^\Pi(p \leftrightarrow q)$, $e^G(p \leftrightarrow q)$ and say $e^{(0.25)}(p \leftrightarrow q)$. For example, the calculation for e^Π goes as follows:

$$\begin{aligned} e^\Pi(p \leftrightarrow q) &= e^\Pi((p \wedge q) \vee (\neg p \wedge \neg q)) = e^\Pi(p \wedge q) + e^\Pi(\neg p \wedge \neg q) \\ &= e(p)e(q) + e^\Pi(\neg p) - e^\Pi(\neg p \wedge q) \\ &= e(p)e(q) + 1 - e(p) - e(q) + e(p)e(q) \\ &= 1 - e(p) - e(q) + 2e(p)e(q). \end{aligned}$$

It is usual to interpret C_1, C_2 and C_3 as [0,1]-evaluations of p and q. Evaluation results are displayed in the Table 2.3 above:

All three columns $e^G(p \leftrightarrow q)$, $e^\Pi(p \leftrightarrow q)$ and $e^{(0.25)}(p \leftrightarrow q)$ induce the obviously correct classification: the least harmful compound is C_3, then follows the compound C_2, and the most harmful compound is C_1.

In practice, we can apply measures $e^{(s)}$ in any classification problem where at least one part of the computation of the criterion function f involves computation of the truth value of certain formula $\alpha(p_1, \ldots, p_n)$.

For example, suppose that we want to develop a fuzzy relational database for automated trade of furniture, where database entries are evaluations of predefined quality attributes. User's queries should be stated in the form of propositional formulas over the quality attributes. The resolution process will be illustrated on the example of the query "find me a sturdy but light wooden chair that is not too expensive":

- Prompt the user to chose rational number $s \in [0,1]$. Here s represents the user's estimation of dependence between quality attributes in the query. 1 represents stochastic independence, while 0 represents the logical dependence.
- Compute $e^{(s)}(p_1 \wedge p_2 \wedge p_3 \wedge \neg p_4)$ for all relevant database entries e. Here p_1 denotes sturdiness, p_2 lightness, and so on.
- Return all relevant database entries e with maximal $e^{(s)}$-values.

Table 2.4 Reproduced from [76]

Object	$e^{\Pi}(\alpha)$	$e^{G}(\alpha)$	$e^{(0.25)}$
A	0.3	0.4	0.35
B	0.3	0.0	0.45
C	0.6	0.6	0.55

Table 2.5 Reproduced from [76]

Object	p_1	p_2	p_3
A	1.75	0.9	0.3
B	0.75	0.8	0.4
C	0.3	0.65	0.1
D	0.3	0.55	0.2

The intermediate measures $e^{(s)}$ are particularly useful in the cases where both e^G and e^{Π} do not classify observed objects. Namely, it is easy to construct an example with the following measurement results (Table 2.4):

As a consequence, neither Gödel's nor product measure provides classification of objects A, B and C according to the classification criteria α, while their arithmetic mean $e^{(0.5)}$ provides a classification – linear ordering $A < B < C$ that is sound with both partial orderings induced by e^{Π} and e^G. Note that e^{Π} induces partial ordering $A < C$ and $B < C$ (A and B are incomparable), while e^G induces partial ordering $A < B$ and $A < C$ (B and C are incomparable).

What we want to say is that, in cases where we disregard independence issues and only evaluate formulas with both e^{Π} and e^G, the intermediate measures $e^{(s)}$ can offer an additional information that is sound with partial classifications given by e^{Π} and e^G, thus provide a finer classification.

Example. This is a well-known example of the classification problem unsolvable by the standard aggregation via discrete Choquet integral proposed by Michel Grabisch. As it is shown in [76], instead of using bicapacities in the process of aggregation, we can obtain the obviously correct classification using the product measure e^{Π} in the aggregation process.

Let us start with the formulation of the problem. Objects A, B, C and D are described by quality attributes p_1, p_2 and p_3, whose values are given in the following Table 2.5:

Objects A, B, C and D should be classified according to the following criteria:

- ϕ_1: The arithmetic mean of the values of quality attributes.
- ϕ_2: If the analyzed object is good with respect to p_1, then p_3 is more important than p_2. Otherwise, p_2 is more important than p_3.

Clearly, given objects are ordered increasingly with respect to the given classification criteria in the following way: $D < C < A < B$. As a first step towards the automated decision mechanism, we need to formalize ϕ_1 and ϕ_2. Obviously, ϕ_1 must be evaluated as the arithmetic mean of p_1, p_2 and p_3. The formalization of the second criterion is not so obvious. One of possible propositional representations of ϕ_2 is given by

Table 2.6 Reproduced from [76]

Object	p_1	p_2	p_3	Arithmetic mean
A	0.75	0.9	0.3	0.65
B	0.75	0.8	0.4	0.65
C	0.3	0.65	0.1	0.65
D	0.3	0.55	0.2	0.35

Table 2.7 Reproduced from [76]

Object	p_1	p_2	p_3	$e^{\Pi}((p_1 \wedge p_3) \vee (\neg p_1 \wedge p_2))$
A	0.75	0.9	0.3	0.45
B	0.75	0.8	0.4	0.5
C	0.3	0.65	0.1	0.458
D	0.3	0.55	0.2	0.445

Table 2.8 Reproduced from [76]

Object	ϕ_1	ϕ_2	Aggregation
A	0.65	0.45	0.55
B	0.65	0.5	0.575
G	0.35	0.458	0.4175
D	0.35	0.445	0.3725

$$(p_1 \wedge p_3) \vee (\neg p_1 \wedge p_2).$$

We will separately evaluate ϕ_1 and ϕ_2 and then aggregate obtained evaluations. Since there is no information about significance of ϕ_1 and ϕ_2, we will assume that they are equally important, so the aggregation coefficient would be equal to 0.5. In other words, the final evaluation would be equal to the arithmetic mean of evaluations of ϕ_1 and ϕ_2. The arithmetic means of values of p_1, p_2 and p_3 for objects A, B, C and D are displayed in the Table 2.6 above:

Finally, $e^{\Pi}((p_1 \wedge p_3) \vee (\neg p_1 \wedge p_2))$-values and the final aggregation are displayed in the following two Tables 2.7 and 2.8:

As we can see, the last column gives us the obviously right classification $D < C < A < B$.

2.3 Nonmonotonic Reasoning and Conditional Probabilities

2.3.1 Nonmonotonic Reasoning

Nonmonotonic reasoning is a field of artificial intelligence that studies behavior of the so-called common sense reasoning from available, but incomplete data. Often, an expert posses incomplete knowledge and use it to infer further information in order to make decisions and plan actions.

Thus, nonmonotonic logics deal with principled reasoning about normal or typical situations. Let us consider the following example, a variant of well-known Penguin triangle.

Example. Carcinomas are malign tumors that usually produce metastasis. Basal cell carcinoma is a malign tumor. However, basal cell carcinoma almost never produces metastasis, so the monotonicity fails.

This example explains the term "nonmonotonic": in classical logic, if a conclusion is derived from a set of premises A, it can be derived from any larger set $B \supseteq A$. On the other hand, in common sense reasoning the conclusion might be retracted after new information has been added to A. The example also suggests the following desirable property of nonmonotonic logic: consequences of more specific information are more reliable then consequences of the more general information.

In this approach, knowledge base is always given abstractly, as a set of formulas of propositional logic. Given formulas α and β the default rule $\alpha \sim \beta$ has a meaning: if α, then normally β (or if α is true, then I plausibly (but not necessarily infallible) jump to the conclusion that β is also true). Here we want to point out the difference between the assertion "if α, then normally β," and "normally, if α is true, then β is true" (see [53]). The former assertion is represented by the formula $\alpha \sim \beta$, while the latter is represented by the formula $T \sim \alpha \sim \beta$, and they may have very different meanings.

In this notation, nonmonotonicity of the corresponding binary consequence relation $\sim$ on formulas means that from $\alpha \sim \gamma$ we cannot infer $\alpha \wedge \beta \sim \gamma$. In other words, the rule

$$\mathrm{M} : \frac{\alpha \sim \gamma \; \alpha \wedge \beta \sim \gamma}{\alpha \wedge \beta \sim \gamma}$$

called monotonicity does not hold for incomplete information. This means that classical logic is not adequate for the formal capture and possible automatization of the common sense reasoning, since we rarely have the complete knowledge (description, exact model) about the underlying problem.

There are two more important properties of classical logic which are not desirable: transitivity

$$\mathrm{TR} : \frac{\alpha \sim \beta \; \beta \sim \gamma}{\alpha \sim \gamma}$$

and contraposition

$$\mathrm{CP} : \frac{\alpha \sim \beta}{\neg\beta \sim \neg\gamma}$$

The following examples are designed in order to illustrate inadequacy of transitivity and contraposition in nonmonotonic reasoning.

Example [Transitivity]. Infective diseases are usually caused by microorganisms. On the other hand, microorganisms are usually useful (e.g., planktons). However, infective diseases are not useful, so transitivity fails.

Example [Contraposition]. Anaphylactic shock is not the usual manifestation of allergic reaction. However, anaphylactic shock is a type of allergic reaction, so contraposition fails.

In [94], Reiter gives several reasons for formal approach to nonmonotonic reasoning. One reason is that the inferences the default rules sanction can be complicated. He illustrates that two default assumptions can conflict in the following example called the Nixon diamond.

Example [Nixon Diamond]. Quakers are normally pacifists. Republicans are normally not pacifists. If Nixon is both a Quaker and a Republican, he inherits contradictory default assumptions: he is simultaneously both pacifist and non-pacifist.

In [25], Gabbay suggested that the study of nonmonotonic reasoning should be focused on the corresponding consequence relations. Natural question is what properties should a nonmonotonic consequence relation $\sim$ satisfy? Soon after Gabbay's paper, Kraus, Lehmann, and Magidor proposed in [53] a set of properties, named System P, that every nonmonotonic consequence relation should satisfy. Those postulates are commonly regarded as the minimal core of the nonmonotonic reasoning (see, e.g., [24]).

2.3.2 Preferential and Rational Relations

We assume that knowledge is represented by propositional formulas $For_{\mathcal{P}}$ built over an at most countable set of propositional letters $\mathcal{P}$.

A preferential relation (see [53]) is a binary relation $\sim$ on $For_{\mathcal{P}}$ that satisfies the properties reflexivity (REF); left logical equivalence (LLE); right weakening (RW); AND, OR; and cautious monotonicity (CM) that will be described below. As we have said earlier, those properties are usually called System P, where P stands for preferential.

Reflexivity is a common property for all consequence relations:

$$\mathrm{R} : \alpha \sim \alpha.$$

Left logical equivalence states that classically equivalent formulas α and β have the same nonmonotonic consequences:

$$\mathrm{LLE} : \frac{\alpha \Leftrightarrow \beta \; \alpha \sim \gamma}{\beta \sim \gamma}.$$

Right weakening postulates that defeasible consequences are closed under strict (classical) logical consequences:

$$\text{RW}: \frac{\beta \Rightarrow \gamma \; \alpha \sim \beta}{\alpha \sim \gamma}.$$

The rule

$$\text{AND}: \frac{\alpha \sim \beta \; \alpha \sim \gamma}{\alpha \sim \beta \wedge \gamma},$$

the conjunction of two plausible consequences, is also a plausible consequence.

The OR rule that is given by

$$\text{OR}: \frac{\alpha \sim \gamma \; \beta \sim \gamma}{\alpha \vee \beta \sim \gamma}$$

is the only rule of System P that doesn't belong to the weaker system C presented in [53]. Thus, the status of OR can be seen as more problematic than the status of other rules. The following example supports OR:

Example. The patient with obstipation will usually respond well to the laxative medicament. The same conclusion will hold if the patient drinks the plum juice. Hence, laxative medicament or plum juice (as a natural laxative) should improve condition of the patient with obstipation.

Finally, the cautious monotonicity rule is a very restricted form of monotonicity, allowing as adding the new information that are expected to be true (according to previous knowledge), without retracting the previous conclusions:

$$\text{CM}: \frac{\alpha \sim \beta \; \alpha \sim \gamma}{\alpha \wedge \beta \sim \gamma}.$$

In [53], CM is supported by the following example:

Example. Let as assume that we expect it will be raining tonight. Also, the horse named Fireball should win the race tomorrow. Then we can plausibly conclude that if it rains tonight, Fireball should still win the race tomorrow.

An arbitrary preferential relation is non-monotonic, in the sense that it doesn't need to satisfy monotonicity rule M. It turns out that the system P takes into account all the examples from the previous subsection and it takes care of the most specific information, when inferring the conclusion from knowledge base. Also, the popularity of the system P is, arguably, a consequence of existence of very natural, preferential semantics. Roughly speaking, $\alpha \sim \beta$ holds if β is satisfied in all "most normal" evaluations (with respect to a given preference or partial ordering of evaluations) that satisfy α.

Note that all the properties of the system P are so-called Horn sentences: if such and such pairs are in the relation, then such another pair is also in the relation.

After the work of Kraus, Lehmann, and Magidor, many researchers studied the subclasses of preferential relations obtained by adding a number of non-Horn rules to the system P; see [8, 9, 22, 23, 100, 105]. The most popular non-Horn rule is the so-called rational monotonicity.

A preferential relation $\sim$ is called a rational relation if it additionally satisfies the rational monotonicity rule

$$\text{RM} : \frac{\alpha \sim \gamma \; \alpha \neg \beta}{\alpha \wedge \beta \sim \gamma}.$$

While some researchers object that RM is too strong [56, 100], Lehmann and Magidor claim that any nonmonotonic relation should satisfy RM. In [54], they present a restricted family of preferential models, called the ranked models, that is shown to be a proper semantic for the rational relations. Additional evidence in support of their claim is the nonstandard probabilistic semantics.

2.3.3 *Probabilistic Semantics*

In [3], Adams studied (followed by Pearl [74]) a nonstandard probabilistic approach to nonmonotonic reasoning and the properties of the corresponding consequence relation. The notion "nonstandard" refers to the fact that that probabilities may have infinitesimally small values, where an infinitesimal may be zero, or any object strictly lesser than every positive real number and strictly greater than every negative real number. Furthermore, $a \approx b$ means that $a - b$ is an infinitesimal. For details, see [101].

In Adams' approach, default information $\alpha \sim \beta$ expresses that conditional probability $\mu(\beta|\alpha)$ is infinitely close to 1, i.e., there is an infinitesimal ε such that $\mu(\beta|\alpha) = 1 - \varepsilon$. It is easy to show that any nonstandard probability μ on $For_{\mathcal{P}}$ defines the rational relation $\sim_{\mu}$ as follows:

$$\alpha {\sim}_{\mu} \beta \text{ iff } \mu\left(\beta \middle| \alpha\right) \approx 1 \, \mu(\alpha) = 0.$$

Lehmann and Magidor have proved the converse in [54]: each rational relation is generated by some neat probability measure, i.e., for each rational relation $\sim$ there is a finitely additive hyeprreal valued probability measure μ on $For_{\mathcal{P}}$ such that $\sim = \sim_{\mu}$.

On the other hand, if we consider standard probability measures and define $\alpha \sim \beta$ as $\mu(\beta|\alpha) = 1 - \varepsilon$ for some positive real number ε, the obtained relation does not satisfy the rules of the system P. For example, if we choose $\varepsilon = \frac{1}{n}$, for some positive integer n (we assume that n is large), and define $\alpha \sim {}_n\beta$ as $\mu(\beta|\alpha) = 1 - \frac{1}{n}$, then we obtain (see [11]):

- REF_n: $\alpha \sim_n \alpha$.
- $\text{LLE}_n : \dfrac{\alpha \Leftrightarrow \beta \; \alpha \sim_n \gamma}{\beta \sim_n \gamma}$.
- $\text{RW}_n : \dfrac{\beta \Rightarrow \gamma \; \alpha \sim_n \beta}{\alpha \sim_n \gamma}$.
- $\text{AND}_n : \dfrac{\alpha \sim_{2n} \beta \; \alpha \sim_{2n} \gamma}{\alpha \sim_n \beta \wedge \gamma}$.
- $\text{OR}_n : \dfrac{\alpha \sim_{2n} \gamma \; \beta \sim_{2n} \gamma}{\alpha \vee \beta \sim_n \gamma}$.
- $\text{CM}_n : \dfrac{\alpha \sim_n \beta \; \alpha \sim_n \gamma}{\alpha \wedge \beta \sim_{n-1} \gamma}$.

In our opinion, this result supports the system P, since if the premises are highly reliable, conclusion is reliable as well. On the other hand, it warns us that when we increase length of the inference, we also decrease reliability of the conclusion.

In [6], Benferhat, Dubois, and Prade proposed a standard semantics for the system P, using a special subclass of the standard (real-valued) probability measures, the so-called big-stepped probabilities. They assume that the set of propositional letters P is finite, i.e., $\mathcal{P} = \{p_1, \ldots, p_n\}$. Big-stepped probabilities are the neat probability measures that satisfy the following conditions:

1. $\mu(at) > \Sigma_{at' \in At_{\mathcal{P}}, \mu(at') < \mu(at')} \mu(at')$ for all atoms $at \in At_{\mathcal{P}}$.
2. $\mu(at) = \mu(at')$ iff the atoms at and at' coincide.

A literal is either a propositional letter or negated propositional letter. An atom over $\mathcal{P} = \{p_1, \ldots, p_n\}$ is any conjunction $l_1 \wedge \cdots \wedge l_n$ of literals such that each propositional letter p_k appears in exactly one literal l_j. For example, if $\mathcal{P} = \{p, q\}$, then the corresponding atoms are $p \wedge q$, $p \wedge \neg q$, $\neg p \wedge q$ and $\neg p \wedge \neg q$.

Probabilistic semantics for the system P is given by the condition

$$\alpha \sim \beta \,\text{iff}\, \mu(\beta|\alpha) > \mu(\neg\beta|\alpha).$$

An important reformulation of the previous definition is given by

$$\alpha \sim \beta \,\text{iff}\, \mu(\beta|\alpha) > \frac{1}{2}.$$

The same paper also offers possibilistic semantics for the system P.

This research motivated development of probabilistic logics, appropriate for modeling nonmonotonic inference relation. Nonstandard approach is modeled in [92], and logic with big-stepped probabilities is presented in [12].

2.3.4 Some Non-Horn Rules and Nonstandard Probabilities

As it is the case with the rule of rational monotonicity, many other non-Horn rules are considered as desirable and natural properties of nonmonotonic consequence relations. We will compare the strength of those rules, always assuming presence of the rules of the system P. In papers [22, 23] the rules named negation rationality and disjunction rationality are introduced as follows:

- NR : $\frac{\alpha \sim \beta \alpha \wedge \gamma \nsim \beta}{\alpha \wedge \neg \gamma \sim \beta}$.
- DR : $\frac{\alpha \vee \beta \sim \gamma \alpha \nsim \gamma}{\beta \sim \gamma}$.

The rule NR is strictly weaker than DR and that DR is strictly weaker than RM. Another inference rule, the so-called weak rational monotonicity, that is strictly weaker than RM and incomparable with both NR and DR, is introduced in [105] as follows:

$$\text{WRM} : \frac{\alpha \sim \gamma \alpha \wedge \beta \nsim \gamma \alpha \nsim \neg \beta}{\sim \neg \alpha}.$$

The determinacy preservation rule

$$DP : \frac{\alpha \sim \beta \alpha \wedge \gamma \nsim \neg \beta}{\alpha \wedge \gamma \sim \beta},$$

introduced by Makinson in [56], is the only restricted form of monotonicity stronger than RM. On the other hand, rational contraposition

$$\text{RC} : \frac{\alpha \sim \beta \neg \beta \nsim \alpha}{\neg \beta \nsim \neg \alpha}$$

and weak determinacy

$$\text{WD} : \frac{\sim \neg \alpha \alpha \nsim \beta}{\alpha \sim \neg \beta}$$

are incomparable with RM but weaker than DP [8, 9]. It is also known that RC implies WD and that FD $=$ WD + RM.

Fragmented disjunction

$$\text{FD} : \frac{\alpha \sim \beta \vee \gamma \alpha \nsim \beta \alpha \nsim \gamma}{\neg \beta \sim \gamma}$$

and conditional excluding middle

$$\text{CEM} : \frac{\alpha \nsim \beta}{\alpha \sim \neg \beta}$$

are strictly above DP; see [8]. Furthermore, CEM is strictly above FD. Those two rules are not restricted form of monotonicity – they are incomparable with M.

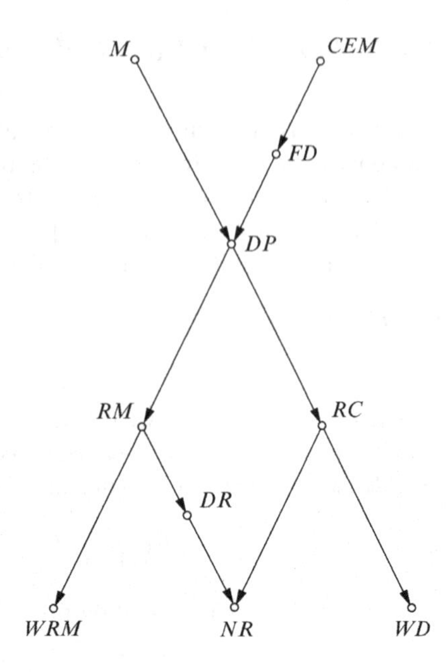

Fig. 2.2 The relationship between nonmonotonic consequence relations (Reproduced from [13])

From now on, we will call a preferential relation which satisfies some of the additional rules by that rule (e.g., a DP-relation is a preferential relation which satisfies the rule determinacy preservation). The following diagram summarizes the relationships between the mentioned rules explained above (Fig. 2.2).

The position of rules DP, FD, and CEM is below RM, which allowed a search for a probabilistic representation of those rules. We say that a neat nonstandard probability measure μ induces a rational relation $\sim$ iff

$$\alpha \sim \beta \text{ iff } \mu\big(\beta\big|\alpha\big) \approx 1 \, \text{or}\, \mu(\alpha) = 0.$$

The following classes of the hyperreal probability measures are proposed as their semantics in [13]:

- A probability measure μ is a CEM-measure iff $\frac{\mu(\alpha)}{\mu(\beta)} \approx 0$ or $\frac{\mu(\beta)}{\mu(\alpha)} \approx 0$ for all $\alpha, \beta \in For_{\mathcal{P}}$ so that $\alpha \wedge \beta$ is a contradiction and either $\mu(\alpha) > 0$ or $\mu(\beta) > 0$.
- μ is a DP-measure iff $\frac{\mu(\alpha)}{\mu(\beta)} \approx 0$ or $\frac{\mu(\beta)}{\mu(\alpha)} \approx 0$ for all $\alpha, \beta \in For_{\mathcal{P}}$ so that $\alpha \wedge \beta$ is a contradiction (we also say that α and β are disjoint), either $\mu(\alpha) > 0$ or $\mu(\beta) > 0$, and $\mu(\alpha) \approx 0$ and $\mu(\beta) \approx 0$.
- μ is an FD-measure iff μ is a DP-measure such that for any pairwise disjoint formulas α, β and γ we have that at least one of $\mu(\alpha)$, $\mu(\beta)$ and $\mu(\gamma)$ is an infinitesimal.

It is proved in [13] that a preferential relation $\sim$ satisfies CEM (DP, FD) if and only if there is a CEM (DP, FD)-measure μ that induces $\sim$.

2.4 Probabilistic Approach to Measuring Inconsistency

2.4.1 Measures of Inconsistency

Although it is the overruling opinion that conflicting knowledge bases are undesirable, they easily appear in practice when experts share their knowledge in order to build knowledge of the group. Hunter and Konieczny [39] give the example of a group of clinicians advising on a patient. Although every set of formulas T_i which represents the opinion of i-th clinician is consistent, it is possible that their union $\cup T_i$ is inconsistent.

As inconsistent information is often the only available information, we have to deal with it. Many logical formalisms are proposed for reasoning under inconsistency, like paraconsistent logics, default reasoning, possibility theory, belief revision and formal argumentation [5–7, 19, 26, 27, 38, 40, 41, 53, 54, 56, 81, 95, 97–99]. Unlike classical logic, they enable drawing nontrivial conclusions from inconsistent knowledge bases, where two different inconsistent sets can lead to different sets of conclusions.

Development of those techniques points out to the need for analyzing and comparing inconsistent sets [39]. It is definitely not true that all inconsistent sets are equally bad. For example, in the previously mentioned example, it makes the difference whether each two clinicians are in conflict or only several of them taken together are in conflict. Similarly, some inconsistent information are less significant than the other (two conflicting data about weight of a patient in a database are often irrelevant for making a diagnosis from the other information from database).

There are various applications of measures of conflicts. If we have to choose between several inconsistent databases, measures of inconsistency may be used for answering the question which one is the most tolerable. Similarly, in diagnosis, if assumptions are in conflict with observed symptoms, a measure of contradiction may be useful for detecting the wrong assumption. All that indicates that approaches to measure the degree of inconsistency need to be context sensitive (and different approaches lead to measures incompatible with one another [33]).

In one approach, the measure of inconsistency depends on the proportion of the language that is affected by the inconsistency in a theory [31, 51]. The second approach considers the number of formulas needed to produce a contradiction. This idea implies that the set of formulas $\phi_1, \ldots, \phi_n$ is not equivalent to the singleton $\{\phi_1 \wedge \cdots \wedge \phi_n\}$ (this property is valid in the so-called non-adjunctive logics [52]). The idea to consider the distribution of contradiction among the formulas approach turns to be closely related to probabilistic measure on formulas. In [11, 50], this idea was compared with semantic notion of existence of a probability measure which assigns a high probability to each formula of the theory.

Besides existence of probability-based measures of inconsistency of propositional theories, some researchers investigate degrees of inconsistency of sets of probabilistic formulas. In [79, 102], the level of inconsistency of a set of probabilistic formulas is proportional to distance from the function on formulas (given by the conditions of theory) to the closest probabilistic measure on formulas. This approach has a potential in analyzing CADIAG-2 expert system with applications in medicine.

2.4.2 Probabilistic Approach for Measuring Propositional Theories

According to Sorensen, the significance of any inconsistency is correlated to the minimal number of formulas that are needed for its derivation. The first formalization of this idea was presented in [50] by Knight. He proposed a real number $\eta \in [0,1]$ as a degree of consistency of a given set of formulas T, based on the existence of the appropriate probability measure on formulas, and called it η-consistency:

- A knowledge base T is η-consistent iff there exists a probability measure μ on formulas such that $\mu(\phi) \geq \eta$ for all $\phi \in T$.
- An η-consistent knowledge base T is maximally η-consistent iff it is not ζ-consistent for all $\zeta < \eta$.

In particular, consistency degree of any consistent theory (knowledge base) is equal to 1. On the other hand, the worst theories (0-consistent) are those containing a contradictory formula. It is proved in [50] that every theory T is maximally η-consistent for some rational number η.

Example. For a patient with a sniffle, we may consider the following two hypotheses as possible causes:

- A: The patient has a flu.
- B: The patient is allergic to pollen.

In addition, let us assume that A and B have not occurred simultaneously in our patient. Denote the last assumption by C. Now our knowledge base $T = \{A,B,C\}$ is inconsistent, since C is equivalent to $\neg A \vee \neg B$. It can be shown that T is $\frac{2}{3}$-consistent. T is also minimally inconsistent in the sense that any proper subset of T is consistent.

Knight proved the following characterization of minimally inconsistent knowledge bases:

- If T is a minimally inconsistent set of formulas, then T is also maximally $1 - \frac{1}{|T|}$ consistent.

For arbitrary inconsistent sets, the following generalization holds:

- If T' is the smallest minimally inconsistent subset of T, then T is $\frac{|T'|-1}{|T|}$-consistent.

Note that maximality of η is not stated in the previous result. Precise relationship between syntactic notion of cardinality of the smallest minimal inconsistent subset of a knowledge base and semantic approach based on probability measures is studied in [11]. For that purpose, the following notion is introduced:

- A theory T is n-consistent iff each subset T' of T with n elements is consistent.

The notion of n-consistency is a syntactical notion and it is not comparable with the Knight's η-consistency. The corresponding semantical notion is the notion of n-probability:

- A theory T is n-probable if there exists a probability measure μ such that $\mu(\phi) > 1 - \frac{1}{n}$ for all $\phi \in T$.

The notion of n-probability is similar to a special case of Knight's η-consistency, for $\eta = 1 - \frac{1}{n}$. It turns out that the n-consistency is a weaker property than the n-probability, i.e., every n-probable theory is n-consistent, but there is an n-consistent theory that is not n-probable (for arbitrary n). The semantical notion that turns out to be the probabilistic analogue of n-consistency is local variant of the n-consistency is given by

- T is locally n-probable if each subset of T with $n + 1$ elements is n-probable.

A theory T is locally n-probable if and only if it is n-consistent.

In [11] we also analyze inconsistent knowledge split into two parts: one part consists of consistent set of the "facts" $\{\phi_1, \ldots, \phi_k\}$, representing the knowledge of k experts, and the second part consisting of some statements believed to be probable by at least one expert, on the basis of his knowledge.

Conditional probabilities are used to introduce the notion of n-probability modulo "facts." The additional condition is that the expert's beliefs are highly compatible.

A theory T is said to be n-probable modulo $\{\phi_1, \ldots, \phi_k\}$, if there exists a probability measure μ with the following properties:

1. $\mu(\phi_1 \wedge \cdots \wedge \phi_k) > 0$.
2. $\mu\left(\phi_i | \phi_j\right) > 1 - \frac{1}{n}$ for all $i, j \in \{1, \ldots, k\}$.
3. For all $\psi \in T$, there exists $i \in \{1, \ldots, k\}$ such that $\mu\left(\psi | \phi_i\right) > 1 - \frac{1}{n}$.

Then the following result holds:

- If theory T is n-probable modulo $\{\phi_1, \ldots, \phi_k\}$, then T is $(n - k + 1)$-probable, hence $(n - k + 1)$-consistent.

A syntactic notion of n-consistency modulo a set of formulas is also introduced: T is n-consistent modulo S iff for any $\psi_1, \ldots, \psi_n \in T$, the theory $\{\psi_1, \ldots, \psi_n\} \cup S$ is consistent. For the finite S, the following result is proved:

- If T is n-probable modulo $\{\phi_1, \ldots, \phi_k\}$, then T is $(n - k + 1)$-consistent modulo $\{\phi_1, \ldots, \phi_k\}$.

2.4.3 *Measuring Inconsistency in Probabilistic Knowledge Bases*

In [79, 102], an approach to measuring degree of inconsistency of probabilistic knowledge bases is developed. Probabilistic knowledge base is any finite set of conditional statements, with probability values attached. The measures of inconsistency are based on distance from the knowledge base to the set of probability measures.

In [102], a probabilistic knowledge base is a set of probabilistic constraints of the form $(\phi|\psi)[d]$, where the pair of formulas $(\phi|\psi)$ is a conditional statement and $d \in [0,1]$. A knowledge base is said to be satisfiable, if there is a probability measure \mu on formulas, such that $\mu(\phi|\psi) = d$ for every element $(\phi|\psi)[d]$ of the base. In order to avoid case differentiation when $\mu(\psi) = 0$, it is assumed that $\mu(\phi|\psi) = d$ iff $\mu(\phi \wedge \psi) = d\mu(\psi)$.

Then, the distance from knowledge base $T = \{(\phi_i|\psi_i)[d_i] | i = 1, \ldots, n\}$ to a measure μ is defined as

$$\sum_{i=1}^{n} |c_i|,$$

where $\mu(\phi|\psi) = d_i$, for all $i = 1, \ldots, n$. Measure of inconsistency of T, denoted by $Inc^*(T)$, is obtained by minimization of the above sum, when μ ranges the set of probability measures. The following properties of Inc^* are shown in [102]:

- Consistency: If T is consistent, then $Inc^*(T) = 0$.
- Inconsistency: If T is inconsistent, then $Inc^*(T) > 0$.
- Monotonicity: $Inc^*(T) \leq Inc^*(T \cup \{(\phi|\psi)[d]\})$.
- Super-additivity: If $T \cap T' = \emptyset$, then $Inc^*(T \cup T') \geq Inc^*(T) + Inc^*(T')$.
- Weak Independence: If no propositional letter from ϕ, ψ appears in T, then $Inc^*(T) = Inc^*(T \cup \{(\phi|\psi)[d]\})$.
- Independence: If $(\phi|\psi)[d]$ does not appear in any minimal inconsistent subset of T, then $Inc^*(T) = Inc^*(T \cup \{(\phi|\psi)[d]\})$.
- Penalty: If $(\phi|\psi)[d]$ does appear in some minimal inconsistent subset of T, then $Inc^*(T) < Inc^*(T \cup \{(\phi|\psi)[d]\})$.
- Continuity: The function $Inc^*(\{(\phi_i|\psi_i)[x_i] | i = 1, \ldots, n\})$ is continuous in all variables x_i.

All of the above properties also hold for the normalization Inc_0^* of Inc^*, defined by $Inc_0^*(T) = \frac{Inc^*(T)}{|T|}$.

In [79], approach from [102] is extended in several ways:

- Interval-valued probabilistic constraints of the form $(\phi|\psi)[\ell, u]$ (where $0 \leq \ell \leq u \leq 1$) are allowed, with the standard probabilistic interpretation (if $\ell = u$, we write $(\phi|\psi)[\ell]$ as above).
- Inc^* is generalized to the family of measures DC^p (called "p-distances to consistency"), for $p \in [1, +\infty]$ (actually, DC^{p_Γ} is defined, where Γ represents correctly evaluated conditional statements, which should not be considered when assigning the p-distance of a theory), defined by the so-called p-norm. For the finite p the corresponding p-norm is defined by $\|x_1, \ldots, x_n\|_p = \left(\Sigma_1^n |x_i|^p\right)^{\frac{1}{p}}$, while for $p = +\infty$ the corresponding p-norm is defined by $\|x_1, \ldots, x_n\|_\infty = \max(|x_1|, \ldots, |x_n|)$.

- The extended definition of conditional probability from [102], which allowed null probability for the conditioning event, is replaced by standard definition.
- Inconsistency values range over the set $\{t, t+\varepsilon, 1/\varepsilon : t \in \mathbb{R}\}$, where $\varepsilon > 0$ is an infinitesimal.

Here, operations of nonstandard field of real numbers are not considered, and ε is only used to distinguish knowledge bases that are "closer to consistency" from the rest.

For $T = \{(\phi_i|\psi_i)[\ell_i, u_i] | i = 1, \ldots, n\}$ and $\vec{x} = (x_1, \ldots, x_{2n})$ such that $x_i \geq 0$ and $0 \leq \ell_i - x_i \leq u_i + x_i \leq 1$, let $T\vec{x} = \{(\phi_i|\psi_i)[\ell_i - x_i, u_i + x_i] : i = 1, \ldots, n\}$. If $F^p(T)$ denotes the set $\left\{ \left\| \vec{x} \right\|_p : T\vec{x} \text{ is satisfiable} \right\}$, then p distance to consistency DC^p is defined as follows:

$$DC^p(T) = \begin{cases} \min F^p(T), \min F^p(T) \text{ exists} \\ \varepsilon + \inf F^p(T), \text{otherwise, provided } F^p(T) \neq \varnothing \\ \dfrac{1}{\varepsilon}, F^p(T) = \varnothing \end{cases}.$$

Note that 1-norm is used in the definition of *Inc**. Let us try to clarify the definition of DC^p on the following CADIAG-2 example taken from [79]:

Example. Consider the unsatisfiable set of CADIAG-2 rules:

$$T = \{(D36|S157)[0.3], (D81|S157)[0.15], (D81|D36)[1]\}.$$

Let μ_n be the sequence of probability measures on the set of formulas generated by $D36$, $D81$ and $S157$, such that:

- $\mu_n(S157) = \frac{1}{n}$.
- $\mu_n(D36) = \frac{0.3}{n} + 1 - \frac{1}{n}$.
- $\mu_n(D81) = \frac{0.15}{n} + 1 - \frac{1}{n}$.
- $\mu_n(S157 \wedge D36) = \frac{0.3}{n}$.
- $\mu_n(S157 \wedge D81) = \frac{0.15}{n}$.
- $\mu_n(D36 \wedge D81) = \frac{0.15}{n} + 1 - \frac{1}{n}$.

Then the following equalities also hold:

- $\mu_n(D36|S157) = 0.3$.
- $\mu_n(D81|S157) = 0.15$.
- $\mu_n(D81|S157) = \frac{n-0.85}{n-0.7}$, and $\lim_{n \to \infty} \mu_n(D81|S157) = 1$.

Hence, $DC^p(T) = \varepsilon$ for all p.

The following properties of the measures DC^p are presented in [79]:

- $DC^p(T) \leq DC^p(T \cup T')$.
- $DC^p(T) = 0$ iff T is satisfiable.

- If $p < q$, then $DC^p(T) \geq DC^q(T)$.
- If $p \neq q$, then DC^p and DC^q induce distinct orderings, i.e., we can construct T and T' such that $DC^p(T) > DC^p(T')$ and $DC^q(T) < DC^q(T')$.

It is also shown that $DC^p(T) \leq |T|^{\frac{1}{p}}$, which enabled normalization $\overline{DC}^p$ of DC^p (see [79] for details), with the property.

- If $p < q$, then $\overline{DC}^p(T) \leq \overline{DC}^q(T)$, contrary to DC^p.

The same author presented similar approach for measuring inconsistency in fuzzy knowledge bases in [80]. Measures presented in this section are also used to develop strategies for repairing inconsistent knowledge bases.

2.5 Evidence and PST Logics

2.5.1 Reasoning About Evidence

In many cases, some probabilistic knowledge should be revised in the presence of new information. Evidence can be seen as that new information that proves or disproves certain beliefs – hypotheses. In the formalization of reasoning about evidence, we are particularly interested in the measuring of evidence.

Example. Suppose that the doctor suspects that a patient has hypertension. After five blood pressure readings, what can we say about the likelihood of the hypothesis that the patient has hypertension? If high blood pressure is detected in all cases, than we can say that this observation favors the hypothesis. In this particular case, the amount of evidence increases with the increment of the number of readings.

In many cases, knowing the prior probabilities is sufficient for the computation of the probability of hypotheses after observation. However, the prior probabilities are usually unknown, so we cannot compute posterior probabilities. Still, observations do provide some evidence in favor of one or several possible hypotheses. Intuitively, the probability of a hypothesis depends on:

- The prior probabilities of the hypothesis (previous belief of the doctor)
- To what extent the observations support the hypothesis

The second item is formalized by the weight of evidence – the function that assigns a number from the unit interval [0,1] to every observation and hypothesis. It should have value 0, if the observation fully disconfirms hypothesis, and value 1 if the observation fully confirms hypothesis.

In [37], Halpern and Pucella introduced a sound, complete and decidable logic for reasoning about evidence. This logic extends a logic from [21]. They defined an evidence space as a tuple $E = (H, O, \mu_1, \ldots, \mu\_m)$, where:

- $H = \{h_1, \ldots, h_m\}$ is the set of mutually exclusive and exhaustive hypotheses.
- $O = \{o_1, \ldots, o_n\}$ is the set of possible observations.
- For a hypothesis h_i, $\mu_i : O \rightarrow [0,1]$ is a likelihood function on O, i.e., $\mu_i(o_1) + \cdots + \mu_i(o_n) = 1$, such that for every $o \in O$ there is $i \in \{1, \ldots, m\}$ such that $\mu_i(o) > 0$.

For an evidence space E, the weight function $w_E : O \times H \rightarrow [0,1]$ is defined as

$$w_E(o_i, h_j) = \frac{\mu_j(o_i)}{\mu_1(o_i) + \cdots + \mu_m(o_i)}.$$

Intuitively, $w_E(o,h)$ is the likelihood that the hypothesis h holds, if the observation o is observed. If we also know the prior probability μ on hypotheses, we can calculate the probability of hypotheses after observation using Dempster's rule of combination (see [37]).

The formulas of the logic can express the relationship between prior probabilities, evidence, and posterior probabilities. Formally, operators P_0 and P_1 are introduced to reason about the prior and posterior probability of hypotheses, and w syntactically represents weight of evidence. Then, polynomial term is a linear combination of expressions of the form $P_0(\alpha)$, $P_1(\alpha)$ and $w(o,h)$, with rational coefficients (for technical reason, in [37] variables ranging over the set of real numbers are also included, and they are eliminated in [73]).

Formulas are built as Boolean combinations of observations, hypotheses, and formulas of the form $f \geq g$, where f and g are polynomial terms.

Example. Following the previous example, let has hypertension be a hypothesis h, and five readings of high blood pleasure an observation o. Then

$$(P_0(h) = 0.5 \wedge o \wedge w(o,h) \geq 0.8) \rightarrow P_1(h) \geq 0.7$$

has the meaning: "If doctor believed that patient has a hypertension with probability 0.5, and if five readings of high blood pleasure are observed, and if that observation supports diagnosis of hypertension with degree at least 0.8, then doctor should change his degree of belief to at least 0.7."

2.5.2 PST Logics

Nowadays, it is possible to locate a small object in the body (e.g., thrombus) and even to track their moving. Representing such information may involve not only space and time but also probability as there may be some uncertainty about the identity of an object, its exact location, or time value.

The PST (probabilistic spatiotemporal) framework is developed to provide a formalism for representing, querying, and updating such information using a simple syntax and an intuitive semantics, and the revision of a PST database is investigated in [32].

The basic PST framework of data concerning objects located in space and time is extended in [15], where the data is given with a probability interval to logics with propositional connectives as well as to the first-order logic. The axiomatization of a number of different logics, depending primarily on the allowed probability values as well as combinations of atomic formulas, is presented in [15]; soundness and completeness are proved and decidability is discussed. The logics depend on three sets:

- ID is a finite set of objects.
- S is a finite set of points in space.
- $T = \{1, \ldots, N\}$ is a finite set of time instances.

Semantics consists of finite sets of worlds and probability distributions over worlds, where each world describes one possible scenario for the location of each object for each time value.

Syntax consists of ST (spatiotemporal) formulas and PST formulas. An ST-atom is a formula of the form $loc(id,r,t)$ where $id \in ID$, $t \in T$, and $r \subseteq S$.

An ST formula is any Boolean combination of ST-atoms. The intended meaning of $loc(id,r,t)$ is that the object id is located in the region r in the time instant t.

Example. The ST formula

$$loc(thrombus, left - leg, 1) \wedge \neg loc(thrombus, lungs, 5)$$

can be read as "a thrombus is detected in patient's left leg on the first day, but it is not detected in his lungs on the fifth day."

Probability intervals are added to ST formulas, so a PST formula is any Boolean combination of formulas of the form $\alpha[\ell, u]$, where ℓ and u are rational probabilistic boundaries from the unit interval.

Example. The PST formula

$$(loc(thrombus, left - leg, 1) \wedge \neg loc(thrombus, lungs, 5))[0.8, 1]$$

formally expresses that the probability that thrombus will be located in the patient's left leg on the first day, but will be not located in his lungs on the fifth day is at least 0.8, while the formula

$$loc(thrombus, left - leg, 1)[0.9, 0.9] \wedge loc(thrombus, lungs, 5)[0, 0.1]$$

formally expresses that thrombus will be detected in patient's left leg on the first day with probability 0.9, and the probability that it will be detected in his lungs on the fifth day is at most 0.1."

2.6 MYCIN

MYCIN (see [96]) was the first medical expert system designed to handle inexact reasoning. Its development has been carried out throughout the 70s and early 80s as a part of the Stanford heuristic programming project. What distinct MYCIN from the other medical systems based on Bayesian analysis is its at the time novel (and rather revolutionary) inference engine. In order to handle imprecise and incomplete data, Edward Shortliffe has developed a rule-based inference system with the core containing around 200 (later that number has been increased to around 600) so-called if-then derivation rules. Though relatively narrow in the scope (MYCIN was designed to perform diagnostics and therapy recommendation for certain bacterial infections), the test has shown that MYCIN performed at the level of an average physician (around 69 % of correct diagnoses). This success has initiated significant interest in the AI community. MYCIN's architecture has become a theoretical foundation for the development of decision support systems for imprecise, incomplete, or even questionable data.

On Fig. 2.3 below is displayed the basic architecture of MYCIN. The static database contains if-then rules of the form

IF evidence list THEN conclusion.

We shall list two examples of MYCIN rules:

IF:

1. The stain of the organism is gram positive.
2. The morphology of the organism is coccus.
3. The growth conformation is chains.

THEN: There is suggestive evidence 0.7 that the identity of the organism is streptococcus.

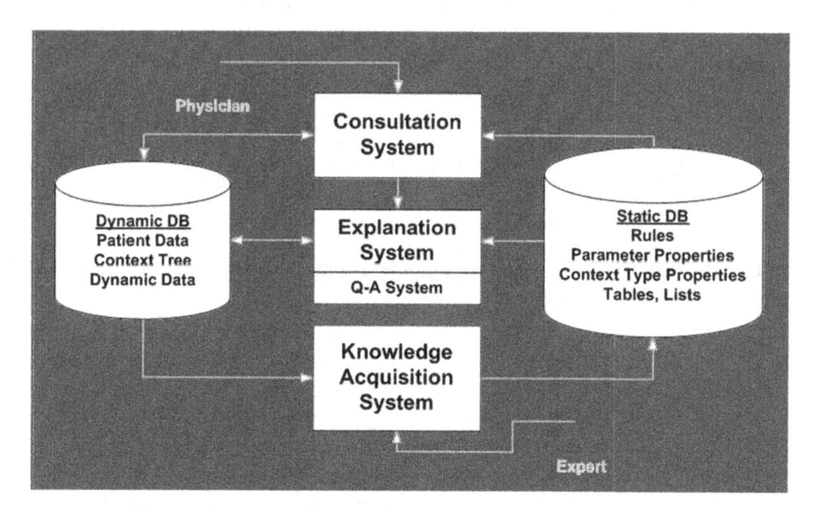

Fig. 2.3 Architecture of MYCIN

IF:

1. The site of the culture is blood.
2. The gram stain of the organism is gram negative.
3. The morphology of the organism is rod.
4. The portal of entry of the organism is urine.
5. The patient has not had a genitourinary manipulative procedure.
6. Cystitis is not a problem for which the patient has been treated.

THEN: There is suggestive evidence 0.6 that the identity of the organism is E. coli.

Each rule is completely modular; all relevant contexts are contained within the rule with explicitly stated premises. Conclusions are classified into nine categories called certainty factors. They are displayed in the following table:

−1 to −0.8	Definitely not
−0.8 to −0.6	Almost certainly not
−0.6 to −0.4	Probably not
−0.4 to −0.2	Possibly not
−0.2 to 0.2	Ignore
0.2 to 0.4	Possibly
0.4 to 0.6	Probably
0.6 to 0.8	Almost certainly
0.8 to 1	Definitely

Certainty factor is defined as a difference between the so-called measures of increased belief and disbelief in the following way:

$$CF(\alpha,\beta) = MB(\alpha,\beta) - MD(\alpha,\beta).$$

The measure $MB(\alpha,\beta)$ is a numerical representation of expert's belief in the hypothesis α given the evidence β. In terms of probability measure μ (μ reflect a priori medical knowledge), it is defined by

$$MB(\alpha,\beta) = \begin{cases} 1, \mu(\alpha) = 1 \\ \dfrac{\max(\mu(\alpha|\beta), \mu(\alpha)) - \mu(\alpha)}{1 - \mu(\alpha)}, \mu(\alpha) < 1 \end{cases} \cdot$$

Similarly, the measure $MB(\alpha,\beta)$ represents expert's disbelief in the hypothesis α given the evidence β; it is defined by

$$MD(\alpha,\beta) = \begin{cases} 0, \mu(\alpha) = 0 \\ \dfrac{\mu(\alpha) - \min(\mu(\alpha|\beta), \mu(\alpha))}{\mu(\alpha)}, \mu(\alpha) > 0 \end{cases} \cdot$$

MYCIN uses a modification of the method of reasoning called backward chaining to search its knowledge base. The diagnostic run starts with the interactive consultation, where the user is prompted to answer a series of approximately 60 questions in order to narrow down the search for the possible cause of the observed bacterial infection. A sample of the typical MYCIN consultation is given below:

1. Patient's name: (first – last) **FRED BRAUN**
2. Sex: **M**
3. Age: **55**
4. Are there any cultures for Fred Braun which may be related to the present illness, and from which organisms have been grown successfully in the microbiology laboratory? **Y**

CULTURE-1

5. From what site was the specimen for CULTURE-1 taken? **BLOOD**
6. Please give the date and time when CULTURE-1 was obtained. **JUN 20, 1977**

The first organism isolated from the blood culture of 20-JUN-77 (CULTURE-1) will be referred to as ORGANISM-1

7. Enter the laboratory-reported identity of ORGANISM-1: **UNKNOWN**
8. The strain (gram or Ziehl-Neelsen acid-fast) of ORGANISM-1: **NEG**
9. Is ORGANISM-1 rod or coccus (etc.)? **ROD**
10. What is the form of the individual organisms (e.g., lancet shaped for cocci, fusiform for rods)? **FUSIFORM**

. . .

23. What is the suspected portal of entry of the gram negative rod from the blood culture of 20-JUN-77 (ORGANISM-1) into this sterile site? **GI (6)**
24. Is Fred Braun a burn patient? **N**
25. Is Fred Braun a compromised host (e.g., alcoholic, sickle-cell disease, immunosuppressed)? **N**
26. Is Fred Braun febrile? **Y**

MYCIN selects antibiotics only on the basis of the identity of offending organisms. Thus, the program's first task is to decide, for each current organism deemed to be significant, which hypotheses regarding the organism's identity are sufficiently likely that they must be considered in choosing therapy. MYCIN uses the certainty factors of the various hypotheses in order to select the most likely identities. Each identity is then given as item number and the process is repeated for each significant current organism. The set of indications for therapy is then printed out, e.g.,

IF: the identity of the organism is pseudomonas,

THEN: I recommend therapy chosen from among the following drugs:

1. Colistin (0.98)
2. Polymyxin (0.96)

3. Gentamicin (0.96)
4. Carbenicillin (0.65)
5. Sulfisoxazole (0.64)

MYCIN was never used in clinical practice primarily due to objective technical difficulties. For example, in the 70s one MYCIN session lasted for 30 or more minutes, which is quite an unrealistic time commitment for any clinician. MYCIN's greatest influence was its demonstration of the power of its representation and reasoning approach. Rule-based systems in many nonmedical domains were developed in the years that followed MYCIN's introduction of the approach. In the 1980s, expert system "shells" were introduced, including one based on MYCIN, known as E-MYCIN, and supported the development of expert systems in a wide variety of application areas.

2.7 CADIAG-2

Another important example of the clinical decision support system that can handle imprecise, incomplete, and even false data is the expert system CADIAG-2, developed primarily by K.-P. Adlassing at the University of Vienna Medical School from the early 80s. Nowadays, the CADIAG project is the central research subject at the Institute for medical expert and knowledge-based systems and the Medical University of Vienna; see [2, 103]. The aim and the scope of CADIAG-2 is to provide support to clinicians in the interpretation of symptoms, signs, laboratory test results, clinical findings, and generation of complete clinical differential diagnosis.

The first version of the system has used three-valued logic, where the truth status of a particular medical entity (e.g., symptom, diagnosis) could be true, false, and undetermined. The corresponding inference engine resembled classical propositional logic.

It turns out that the usability of such a system is limited, mostly due to the fact that a symptom can be a very vague property. The typical example of this kind is "having a strong chest pain," where any numerical confirmation is questionable or even impossible.

Though this kind of situations may not interfere with the diagnostic performance of the expert physician, they pose a significant problem to the automatization of diagnostics. Namely, if not treated cautiously, borderline cases (cases where the presence or the absence of certain symptom is questionable) have a severe impact on the performance of the system. In mathematical terms, the problem lies in the discontinuity (small change of the input produces significant changes in the output) of the numerical representation of certain medical entity. For example, body temperature $t \geq 39^{\circ}C$ indicates high fever and (mandatory) application of antipyretics. However, physician will sometimes apply antipyretics for lower body temperatures, say $37.5^{\circ}C$. In the crisp (yes-no) case, the fever severity can be modeled as in Fig. 2.4.

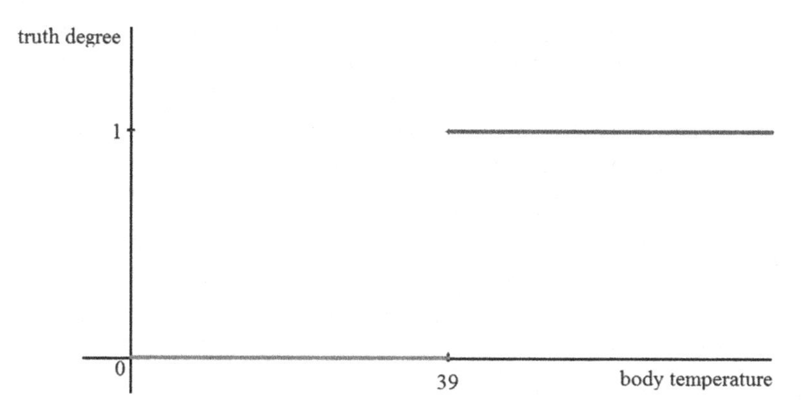

Fig. 2.4 A crisp (yes-no) representation of the fever severity. *Green* values ($t \geq 39°C$) indicate the absence of a high fever (the trigger indicator vale is 0), while the *red* values initiate severe fever

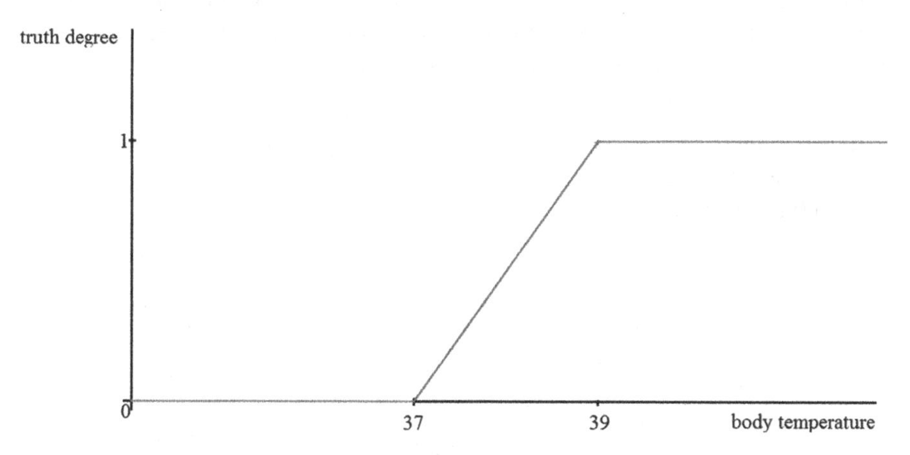

Fig. 2.5 A fuzzy set representing fever severity

In particular, any clinical diagnostic system that handles high fever as shown in Fig. 2.4 will treat the body temperature $39.5°C$ in the same way as $36.5°C$, which is obviously wrong.

A quite natural way to overcome this problem is to represent vague and imprecise notions by fuzzy sets. In the case of fever severity, a so-called "left shoulder" fuzzy set anchored in $37°C$ and $39°C$ would be the adequate representation of fever severity. It is shown in Fig. 2.5.

Fuzzy set *FS* (fever severity) displayed in Fig. 2.5 has the following mathematical representation:

$$FS(t) = \begin{cases} 0, t < 37 \\ \frac{1}{2}t - 18.5, 37 \leq t < 39 \\ 1, t \geq 39 \end{cases}.$$

For example, body temperature 38.5°C produces fever severity degree $FS(38.5) = 0.75$, which combined with other methods can lead to much better performance of the system in borderline cases.

In fuzzy logic terms, CADIAG-2 handles expressions of the form (α,t), where α is a medical entity (say symptom) and $t \in [0,1]$ represents its truth degree (emulates expert's confirmation of α). An adaptation of modus ponens (so-called generalized modus ponens) for graded formulas (pairs of the form (α,t)) is given by

$$\text{GMP}: \frac{(\alpha, t)(\alpha \to \beta, s)}{(\beta, t * s)}.$$

In principle, * can be any continuous t-norm, but it is best to think of * as either Lukasiewicz t-norm $x * y = \max(0, x + y - 1)$, Gödel t-norm $x * y = \min(x,y)$ or product t-norm $x * y = xy$.

Derivation rules of CADIAG-2 are based on fuzzy sets and they are divided in three categories. The first category consists of the so-called confirming degree rules. An example of this kind is the following rule:

IF suspicion of liver metastases by liver palpation
THEN pancreatic cancer
With the confirmation degree 0.55.

The next type of CADIAG-2 rules is the so-called mutually exclusion rules. An example of this kind is the following one:

IF positive rheumatoid factor
Then NOT seronegative rheumatoid arthritis

The last type of rules is so-called universal (always occurring) rules. An example of such a rule is given below:

IF NOT (rheumatoid arthritis AND splenomegaly AND leukopenia $\leq 4000/\mu l$)
THEN NOT Felty's syndrome.

In contrast to MYCIN, CADIAG-2 has a large number of inference rules and definitions in its core. The core contains 1761 symptom, 341 disease, 720 symptom-symptom rules (both premises and conclusion are symptoms), 218 disease-disease rules, and 17,573 symptom-disease rules (premises are symptoms, conclusions are diseases). This huge architecture has imposed question of global consistency of the inference engine. The number of papers has been published on this subject, suggesting various formalisms for representation of CADIAG-2 inference engine. All of them were based on fuzzy logics, mostly due to the fact that CADIAG-2 is actually based on principles very close to the basic principles known from t-norm-based logics. But there remain differences on a basic level that cannot be easily overcome.

Acknowledgments We would like to express our gratitude to professor Milan Stošović, MD, for his useful suggestions and comments. This work is partially supported by Serbian Ministry of Education and Science through grants III44006, III41103, ON174062, and TR36001.

References

1. M. Abadi and J. Y. Halpern. Decidability and expressiveness for first-order logics of probability. Information and Computation 112: 1–36, 1994.
2. K.-P. Adlassing, G. Kolarz. CADIAG-2: computer-assisted medical diagnosis using fuzzy subsets. In: M. Gupta, E. Sanchez (eds.), Approximate reasoning in Dexision analysis: 219–247, North-Holland 1982.
3. E. W. Adams. The logic of conditionals. Reidel, Dordrecht 1975.
4. N. Alechina. Logic with probabilistic operators. In Proc. of the ACCOLADE '94: 121–138, 1995.
5. L. Amgoud, S. Kaci. An argumentation framework for merging conflicting knowledge bases. International Journal of Approximate Reasoning 45(2007), pp. 321–340.
6. S. Benferhat, D. Dubois and H. Prade. Possibilistic and standard probabilistic semantics of conditional knowledge bases. Journal of Logic and Computation 9(6): 873–895, 1999.
7. S. Benferhat, A. Saffiotti, P. Smets. Belief functions and default reasoning. Artificial Intelligence (122): 1–69, 2000.
8. H. Bezzazi, R. Pino Pérez. Rational Transitivity and its models. Proceedings of the 26th International Symposium on Multiple-Valued Logic: 160–165, IEEE Computer Society Press, 1996.
9. H. Bezzazi, D. Makinson, R. Pino Pérez. Beyond rational monotony: some strong non-Horn rules for nonmonotonic inference relations. Journal of Logic and Computation 7(5): 605–631, 1997.
10. D. Doder, Z. Marković, Z. Ognjanović, A. Perović, M. Rašković: A Probabilistic Temporal Logic That Can Model Reasoning about Evidence. FoIKS 2010, LNCS 5956: 9–24, Springer 2010.
11. D. Doder, M. Rašković, Z. Marković, Z. Ognjanović. Measures of inconsistency and defaults. International Journal of Approximate Reasoning 51: 832–845, 2010.
12. D. Doder. A logic with big-stepped probabilities that can model nonmonotonic reasoning of system P. Publications de l'Institut Mathématique, ns. 90(104): 13–22, 2011.
13. D. Doder, A. Perović, Zoran Ognjanović: Probabilistic Approach to Nonmonotonic Consequence Relations. ECSQARU 2011, LNCS 6717: 459–471, Springer 2011.
14. D. Doder, Z. Ognjanović, Z. Marković. An Axiomatization of a First-order Branching Time Temporal Logic, Journal of Universal Computer Science 16(11): 1439–1451, 2010.
15. D. Doder, J. Grant, Z. Ognjanović. Probabilistic logics for objects located in space and time. To appear in Journal of logic and computation, doi: 10.1093/logcom/exs054.
16. D. Dubois, J. Lang, H. Prade. Possibilistic logic. In: Handbook of Logic in Artificial Intelligence and Logic Programming, Vol. 3, ed. D. M. Gabbay et al., Oxford University Press (1994), pp. 439–513.

17. D. Dubois, H. Prade. Possiblistic logic: a retrospective and prospective view. Fuzzy sets and systems 144: 3–23, 2004.
18. D. Dubois, L. Godo, H. Prade. Weighted logic for artificial intelligence: an introductory discussion. In ECAI-2012 workshop Weighted logic for artificial intelligence (eds L. Godo, H. Prade): 1–7, 2012.
19. P. M. Dung. On the acceptability of arguments and its fundamental role in nonmonotonic reasoning, logic programming and n-person games. Artificial Intelligence 77(1995), pp. 321–357.
20. R. Đorđević, M. Rašković, and Z. Ognjanović. Completeness theorem for propositional probabilistic models whose measures have only finite ranges. Archive for Mathematical Logic 43, 557–563, 2004.
21. R. Fagin, J. Halpern and N. Megiddo. A logic for reasoning about probabilities. Information and Computation 87(1–2): 78–128, 1990.
22. M. Freund. Injective models and disjunctive relations. Journal of Logic and Computation 3(3): 231–247, 1993.
23. M. Freund, D. Lehmann. On negation rationality. Journal of Logic and Computation 6(2): 263–269, 1996.
24. N. Friedman, J.Y. Halpern. Plausibility measures and default reasoning. Journal of the ACM 48: 648–685, 2011.
25. D. Gabbay. Theoretical foundations for non-monotonic reasoning in expert systems. Logics and models of concurrent systems: 439–485, Springer 1985.
26. D. M. Gabbay and A. Hunter. Making inconsistency respectable (Part 1). In: Fundamentals of Artificial Intelligence Research. Lecture Notes in Artificial Intelligence 535, Springer (1991), pp. 19–32.
27. P. Gardenfors and H. Rott. Belief revision. In: Handbook of Logic in Artificial Intelligence and Logic Programming Vol. 4, Oxford University Press (1995), pp. 35–132.
28. L. Godo, F. Esteva, P. Hájek. Reasoning about probability using fuzzy logic. Neural network world 10(5): 811–824, 2000.
29. L. Godo, E. Marchioni. Coherent conditional probability in a fuzzy logic setting. Logic journal of the IGPL 14(3): 457–481, 2006.
30. A. Gorry, O. Barnett. Sequential Diagnosis by Computer. The Journal of the American Medical Association 205(12): 849–854, 1968.
31. J. Grant, A. Hunter. Measuring inconsistency in knowledge bases. Journal of Intelligent Information Systems 27: 159–184, 2006.
32. J. Grant, F. Parisi, A. Parker, and V.S. Subrahmanian. An agm-style belief revision mechanism for probabilistic spatio-temporal logics. Artificial Intelligence 174: 72–104, 2010.
33. J. Grant, A. Hunter. Measuring the good and the bad in inconsistent information. Proc. of IJCAI 2011: 2632–2637, 2011.
34. T. Hailperin. Sentential Probability Logic, Origins, Development, Current Status, and Technical Applications. Lehigh University Press, 1996.
35. P. Hajék. Metamathematics of Fuzzy Logic. Kluwer 1998.
36. J. Halpern. Reasoning about uncertainty. The MIT Press, 2003.
37. J. Halpern, R. Pucella. A logic for reasoning about evidence. Journal of Artificial Intelligence Research 26 (2006), pp. 1–34.
38. A. Hunter and S. Konieczny. Approaches to measuring inconsistent information. In: Inconsistency Tolerance. Lecture Notes in Computer Science 3300, Springer (2005), pp. 191–236.
39. A. Hunter and S. Konieczny. Shapely inconsistency values. In: Proceedings of the 10th International Conference on Knowledge Representation (2006), pp. 249–259.
40. A. Hunter and S. Konieczny. Measuring inconsistency through minimal inconsistent sets. In: Proceedings of the 11th International Conference on Knowledge Representation), AAAI Press (2008), pp. 358–366.
41. A. Hunter. Paraconsistent logics. In: Handbook of Defeasible Reasoning and Uncertainty Management Systems, Vol. 2, ed. D. M. Gabbay and Ph Smets, Kluwer (1998), pp. 11–36.

42. A. Ilić-Stepić, Z. Ognjanović, N. Ikodinović, A. Perović: A p-adic probability logic. Math. Log. Q. 58(4–5): 263–280 (2012)
43. N. Ikodinović. Craig interpolation theorem for classical propositional logic with some probability operators. Publications de L'Institut Mathematique Ns. 69(83): 27–33, 2001.
44. N. Ikodinović and Z. Ognjanović. A logic with coherent conditional probabilities. ECSQARU 2005, LNAI 3571: 726–736, Springer 2005.
45. D. Jovanović, N. Mladenović and Z. Ognjanović. Variable Neighborhood Search for the Probabilistic Satisfiability Problem. In Metaheuristics: Progress in Complex Systems Optimization (edts. K.F. Doerner, M. Gendreau, P. Greistorfer, W. Gutjahr, R.F. Hartl, M. Reimann), Springer Series: Operations Research/Computer Science Interfaces Series vol. 39:173–188. Springer 2007.
46. B. Jaumard, P. Hansen, and M. P. de Aragao. Column generation methods for probabilistic logic. ORSA Journal on Computing 3: 135–147, 1991.
47. H.J. Keisler. Model theory for infinitary logic. North-Holland 1971.
48. H. J. Keisler. Hyperfinite model theory. In: Logic Colloquium 76. (eds. R. O. Gandy, J. M. E. Hyland): 5–110. North-Holland 1977.
49. H. J. Keisler. Probability quantifiers. In: Model Theoretic Logics, (eds. J. Barwise and S. Feferman): 509–556. Springer 1985.
50. K. Knight. Measuring inconsistency. Journal of Philosophical Logic 31 (2001), pp. 77–91.
51. S. Konieczny, J. Lang and P. Marquis. Quantifying information and contradiction in propositional logic through epistemic actions. In: Proceedings of the 18th International Joint Conference on Artificial Intelligence (2003), pp. 106–111.
52. S. Konieczny, J. Lang and P. Marquis. Reasoning under inconsistency: the forgotten connective. In: Proceedings of the 19th International Joint Conference on Artificial Intelligence (2005), pp. 484–489.
53. S. Kraus, D. Lehmann and M. Magidor. Nonmonotonic reasoning, preferential models and cumulative logics. Artificial Intelligence 44 (1990), pp. 167–207.
54. D. Lehmann and M. Magidor. What does a conditional knowledge base entail? Artificial Intelligence 55 (1992), pp. 1–60.
55. T. Lukasiewicz. Probabilistic Default Reasoning with Conditional Constraints. Annals of Mathematics and Artificial Intelligence 34: 35–88, 2002.
56. D. Makinson. General patterns in nonmonotonic reasoning. In: Handbook of Logic in Artificial Intelligence and Logic Programming, Vol. 3, Clarendon Press, Oxford (1994), pp. 35–110.
57. Z. Marković, Z. Ognjanović, and M. Rašković. A probabilistic extension of intuitionistic logic. Mathematical Logic Quarterly 49: 415–424, 2003.
58. Miloš Milošević, Zoran Ognjanović. A first-order conditional probability logic, Logic Journal of IGPL 20(1): 235–253, 2012.
59. N. Nilsson. Probabilistic logic. Artificial Intelligence, 28: 71–87, 1986.
60. N. Nilsson. Probabilistic logic revisited. Artificial Intelligence, 59: 39–42, 1993.
61. Z. Ognjanović, M. Rašković. A logic with higher order probabilities. Publications de L'Institute Matematique, n.s. 60(74): 1–4, 1996.
62. Z. Ognjanović, M. Rašković. Some probability logics with new types of probability operators. Journal of Logic and Computation 9(2): 181–195, 1999.
63. Z. Ognjanović, M. Rašković. Some first-order probability logics. Theoretical Computer Science 247(1–2): 191–212, 2000.
64. Z. Ognjanović, J. Kratica and M. Milovanović. A genetic algorithm for satisfiability problem in a probabilistic logic: A first report. LNCS/LNAI 2143: 805–816, Springer 2001.
65. Z. Ognjanović, U. Midić and J. Kratica. A genetic algorithm for probabilistic SAT problem. In: Artificial Intelligence and Soft Computing ICAISC 2004 LNCS/LNAI 3070: 462–467, Springer 2004.

66. Z. Ognjanović, Z. Marković and M. Rašković. Completeness theorem for a Logic with imprecise and conditional probabilities. Publications de L'Institut Mathematique Ns. 78(92): 35–49, 2005.
67. Z. Ognjanović, U. Midić and N. Mladenović. A Hybrid Genetic and Variable Neighborhood Descent for Probabilistic SAT Problem. LNCS 3636: 42–53, Springer 2005.
68. Z. Ognjanović, N. Ikodinović and Z. Marković. A logic with Kolmogorov style conditional probabilities. In Proceedings of the 5th Panhellenic logic symposium, Athens, Greece, July 25–28, 2005: 111–116.
69. Z. Ognjanović. Discrete Linear-time Probabilistic Logics: Completeness, Decidability and Complexity. Journal of Logic Computation 16(2): 257–285, 2006.
70. Z. Ognjanović, A. Perović, M. Rašković. Logics with the Qualitative Probability Operator. Logic Journal of IGPL 16(2):105–120, 2008.
71. Z. Ognjanović, M. Rašković, Z. Marković. Probability logics. In Logic in computer science (ed Z. Ognjanović): 35–111, Matematički Institut SANu, 2009, ISBN 978-86-80593-40-1.
72. Z. Ognjanović, D. Doder, Z. Marković. A Branching Time Logic with Two Types of Probability Operators SUM-2011, LNCS 6929: 219–232, Springer 2011.
73. Z. Ognjanović, Z. Marković, M. Rašković, D. Doder, A. Perović: A propositional probabilistic logic with discrete linear time for reasoning about evidence. Ann. Math. Artif. Intell. 65(2–3): 217–243, 2012.
74. J. Pearl. System Z: a natural ordering of defaults with tractable applications to default reasoning. Proceedings of the 3rd Conference on the Theoretical Aspects of Reasoning About Knowledge: 121–135, 1990.
75. A. Perović, Z. Ognjanović, M. Rašković, Z. Marković: Qualitative Possibilities and Necessities. ECSQARU 2009, LNCS 5590: 651–662, Springer 2009.
76. A. Perović, Z. Ognjanović, Miodrag Rašković, D. G. Radojević: Finitely additive probability measures on classical propositional formulas definable by Gödel's t-norm and product t-norm. Fuzzy Sets and Systems 169(1): 65–90, 2011.
77. A. Perović, D. Doder, Z. Ognjanović: On Real-Valued Evaluation of Propositional Formulas. FoIKS 2012, LNCS 7153: 264–277, Springer 2012.
78. A. Perović, A. Takači, Srđan Škrbić: Formalising PFSQL queries using $L\prod\frac{1}{2}$ fuzzy logic. Mathematical Structures in Computer Science 22(3): 533–547, 2012.
79. D. Picado-Muino. Measuring and repairing inconsistency in probabilistic knowledge bases. International Journal of Approximate Reasoning 52 (2011), pp. 828–840.
80. D. Picado-Muino. Measuring and repairing inconsistency in knowledge bases with graded truth. Fuzzy Sets and Systems 197 (2012), pp. 108–122.
81. G. Priest. Paraconsistent logics. In: Handbook of Philosophical Logic, Vol. 6, Kluwer (2002).
82. M. Rašković. Completeness theorem for biprobability models. Journal of Symbolic Logic 51(3): 586–590, 1986.
83. M. Rašković. Completeness theorem for singular biprobability models. Proc. of Amer. Math. Soc. 102: 389–392, 1988.
84. M. Rašković. Completeness theorem for a monadic logic with both ordinary first-order and probability quantifiers. Publications de L'Institut Mathematique Ns. 47(61): 1–4, 1990.
85. M. Rašković. A completeness theorem for an infinitary intuitionistic logic with both ordinary and probability quantifiers. Publications de L'Institut Mathematique Ns. 50(64): 7–13, 1991.
86. M. Rašković. Classical logic with some probability operators. Publications de l'Institut Mathematique, n.s. 53(67): 1–3, 1993.
87. M. Rašković, R. Đorđević. Continuous time probability logic. Publications de L'Institut Mathematique Ns. 57(71): 143–146, 1995.
88. M. Rašković, R. Đorđević. Probability Quantifiers and Operators. VESTA, Beograd, 1996.
89. M. Rašković, Z. Ognjanović. A first order probability logic LP_Q. Publications de L'Institute Matematique, ns. 65(79): 1–7, 1999.
90. M. Rašković, R. Đorđević, Z. Marković. A logic of approximate reasoning. Publications de L'Institut Mathematique Ns. 69(83): 8–12, 2001.

91. M. Rašković, Z. Ognjanović, and Z. Marković. A Logic with Conditional Probabilities. JELIA 2004, LNCS 3229: 226–238, Springer 2004.
92. M. Rašković, Z. Marković and Z. Ognjanović. A logic with approximate conditional probabilities that can model default reasoning. International Journal of Approximate Reasoning 49(1):52–66, 2008.
93. K. P. S. Bhaskara Rao, M. Bhaskara Rao. Theory of charges. Academic press 1983.
94. R. Reiter. A theory of diagnosis from first principles. Artificial Intelligence 32: 57–95, 1987.
95. N. Rescher and R. Manor. On inference from inconsistent premises. Theory and Decision 1 (1970), pp. 179–219.
96. E. Shortliffe, B. Buchanan. A model of inexact reasoning in medicine. Mathematical Biosciences 23: 351–379, 1975.
97. A. Sistla and E. Clarke. The complexity of propositional linear temporal logic. Journal of the ACM, 32(3): 733–749, 1985.
98. R. Sorensen. Blindspots. Claredon Press, Oxford, 1988.
99. R. C. Stalnaker. A theory of conditionals. In: Studies in Logical Theory (ed. N. Rescher), American Philosophical Quarterly Monograph Series, Vol. 2, Blackwell, Oxford, 1968.
100. R. C. Stalnaker. Nonmonotonic consequence relations. Fundamenta Informaticae (21): 7–21, 1994.
101. K. D. Stroyan, W. A. J. Luxemburg. Introduction to the theory of infinitesimals. Academic Press 1976.
102. M. Thimm. Measuring inconsistency in probabilistic knowledge bases. Twenty-fifth Conference on Uncertainty in Artificial Intelligence, AUAI Press (2009), pp. 530–537.
103. T. Vetterlein, K.-P. Adlassing. The medical expert system CADIAG-2, and the limits of reformulation by means of formal logics. In: Tagungsband der eHealth2009 und eHealth Benchmarking 2009: 123–128, 2009.
104. W. van der Hoeck. Some consideration on the logics $P_F D$. Journal of Applied Non-Classical Logics 7(3): 287–307, 1997.
105. Z. Zhu, D. Zhang, S. Chen, W. Zhu. Some contributions to nonmonotonic consequence. Journal of Computer Science and Technology, 16(4): 297–314, 2001.

Chapter 3
Transforming Electronic Medical Books to Diagnostic Decision Support Systems Using Relational Database Management Systems

Milan Stosovic, Miodrag Raskovic, Zoran Ognjanovic, and Zoran Markovic

At present clinical decision support systems (CDSS) are not widely used (unlike Internet or information systems) in spite of their great capability to support decision making [1, 2]. There are many problems in realisation of these systems including difficulties with domain selection, knowledge base construction and maintenance, and problems with diagnostic algorithms and user interface [2–5]. Basic doctors fear of missing some diagnosis, and therapy data is not always in the focus of CDSS. It is necessary to involve best medical researchers from a narrow field of science to support constant changes of knowledge. Programmers are needed to maintain and update CDSS. Finally, potential users are not eager to use them because they are more familiar with some well-known sources of knowledge such as electronic books, medical citation databases and medical articles. Since the doctors are responsible for their patient's lives, they are reluctant to rely on some 'mechanism' they do not completely understand and which, to many of them, seems like a 'black box'. Aiming to resolve some of these problems, we consider possible linking of a part of CDSS, diagnostic decision support systems (DDSS), to well-known medical books. The books became electronic in the past but could they evolve further to DDSS?

Different methods of reasoning were considered in selecting adequate model for DDSS [6] including associations, probabilities, causality, functional relationships, temporal constrains, locality, similarity and clinical practice. These relationships provide organising principles which influence knowledge base and the inference mechanism. The association between diseases and findings was one of the first principles used in DDSS. For each disease there is a list of findings, so when we know findings, we can find the disease. This procedure is better if we know strengths of the associations which are usually frequency of appearance of some

M. Stosovic (✉)
Clinic of Nephrology, Clinical Center of Serbia, Belgrade, Serbia
e-mail: milanst@eunet.rs

M. Raskovic • Z. Ognjanovic • Z. Markovic
Mathematical Institute of the Serbian Academy of Sciences and Arts, Belgrade, Serbia

G. Rakocevic et al. (eds.), *Computational Medicine in Data Mining and Modeling*,
DOI 10.1007/978-1-4614-8785-2_3,

finding in particular disease (they are used as prior probability). For example, QMR [7] uses three properties for diagnostic reasoning: frequency of finding with disease, frequency of disease with finding and the importance of the finding. Association reasoning provided large knowledge base that is simply and easily understood even by doctors who are not experienced in medical informatics. Another similar reasoning uses criteria tables to associate diseases with cluster of findings (e.g. CASNET [7]) and can better distinguish between similar findings than simple associations. Although association as organising principle is well known and defined many years ago, it seems suitable for this DDSS but has to be adapted. We also considered production rules (e.g. Mycin [7]) as another organising principle that can be used in this form of DDSS. Other organising principles such as Bayesian networks are not suitable because there are no prior probabilities defined in the books and they have a lot of information that can be a problem for this type of network. Diagnostic reasoning based on causality has very complex knowledge representation system (e.g. CASNET). Functional relationships are very effective for reasoning but only when they are available. Because of that neither causality nor functional relationships are considered since they are not suitable for planning DDSS. The same holds for location and similarity and case-based reasoning. Nowadays, temporal relationships are intensively explored for use in CDSS and include dynamic decision networks and multiagent systems [8]. However, temporal relationships are not so important in most books and the amount of information can be a problem. So we decided to use classic association reasoning but with many adaptations and modifications.

Nowadays, data-intensive systems are more dominant than knowledge-intensive systems which were used in the past [9], and data mining, statistical analysis and self-learning systems tend to be used instead of expert knowledge. Graphical visualisation is used rather than tables and lists. Automation is another tendency including acquiring of the knowledge [10, 11]. There is a requirement that CDSS has to be a part of a normal doctor's workflow [12]. But could this new approach provide 'evidence-based medicine' that is now obligatory and satisfy many best practice guidelines that modern medicine is required?

What do doctors really need when using DDSS? Do they need complex knowledge base and inference engine or it could be only information-retrieving system? Some researchers believe that it could be a weighted information-retrieval approach to provide a context-dependent ranking of likely diseases through matching the patient's symptom information to typical disease symptomatology [13].

For transforming books to DDSS, the most applicable is well-known association reasoning that uses classic knowledge base, and instead of special inference mechanism, it may use only information-retrieving system. That mechanism can be easily understood by doctors so they can trust the results. Why should we use some complicated concept that was unsuccessful in this domain when simpler is real and applicable? But new elements should be included. So sequence of actions that we have done begins with an analysis of the main problems in functioning of modern DDSS [1–13]; some recommendations and new solution are proposed. Logical aspect is also analysed trying to mimic natural doctor's thinking by DDSS [1]. Further, analysing text in medical books, basic structural forms were determined. Finally, knowledge-based DDSS was proposed trying to meet the following criteria: (a) knowledge base should correspond to the structure of the data in textbooks; (b) inference mechanism should follow the logic of the books and evidential human thinking; (c) every DDSS screen and output results should fulfil expectations of the reader when they consult the books; (d) knowledge acquisition tools should enable easy creation of knowledge base and (e) consulting DDSS should be achieved through a doctor-friendly user interface.

Probabilities are presented using sensitivity, specificity and predictive value [14].

$$\text{Sensitivity}\,(\%) = \frac{\text{True Positives (TP)}}{\text{TP} + \text{False Negatives (FN)}} \times 100$$

$$\text{Specificity}\,(\%) = \frac{\text{True Negatives (TN)}}{\text{TN} + \text{False Positives (FP)}} \times 100$$

$$\text{Predictive value}\,(\%) = \frac{\text{TP} + \text{TN}}{\text{Number of predictions}} \times 100$$

3.1 Conceptual Accents

Because medical books are a set of knowledge data, they could be transformed to knowledge-based DDSS. It is necessary to convert unstructured text into structured database. That is the way to put data in more active and useful form.

However, new approach is needed. Authors of the books have to define diseases, findings and relations among findings and diseases when writing books. At present, experts are engaged to create a knowledge base. They are expensive and duplicate the process when creating a knowledge base.

Another problem is permanently changing knowledge, even from day to day. Using experts to maintain and update knowledge base is quite expensive and requires frequent updates. So, why duplicate the same process? Better idea is to use authors to update knowledge base when they write books. In addition, best books have their editions, so they do not need to be updated monthly or on daily basis. It would be necessary to have DDSS editor to unify different styles and approaches of the authors.

Particular books have a large number of editions (for example, Harrison's principles of internal medicine had 18 editions, [15]). Students and physicians believe them. They have a practice to consult these books. So, why somebody should not use DDSS based on these books for diagnosis support? Why not consult the book in form of DDSS for a particular case?

DDSS application has to be made by programmers in a form of empty knowledge database that could be used with different medical books. Empty knowledge databases were introduced long ago (eMycin) [7].

Should those databases be object oriented or relational? There is an enormous number of facts in the books so relational databases may be more suitable.

Publishers of the books could become more interested in financing DDSS as a part of their books edition like they finance electronic books together with paper edition.

3.2 Basic Logic Outline

First we will consider doctor's 'natural' decision making and reasoning. This process is complex and is not fully understood. However, it is clear that some forms of shortcuts are commonly in use [1]. Also, we are talking here about reasoning under uncertainty [2]. However, doctors learn to reduce uncertainty to a great extent. They know how to make evidence from uncertainty. First of all, they look for the facts such as subjective feelings, objective examination and various tests and scans that all could be designated as 'disease manifestations' or 'findings'. Findings are clues for diagnosis of disease. Doctors record every finding in memory together with their properties and their importance in form of sensitivity and specificity for some disease [14]. For example, when a man with oedema (finding) comes to a nephrologist, doctor will determine pretibial pitting oedema (properties). He knows from experience that this type of oedema is almost always present in nephrotic syndrome (high sensitivity) but it can also be found in heart, liver or other diseases (low specificity). This is a very simple process and requires only 'recording' of facts to memory and 'retrieving' them back. At this point, a doctor has clues for diagnosis (nephrotic syndrome) and differential diagnosis (heart and liver diseases). Now he will ask himself which findings are related with nephrotic syndrome. That is protein in urine (finding) more than 3.5 g/24 h (properties). It is found in almost all nephrotic syndromes (high sensitivity) but can be found in some other disease (low specificity). Another finding is serum albumin less than 35 g/l (properties) and is found in almost all nephrotic syndromes (high sensitivity) but can be found in some other disease (low specificity). However, he also knows from experience that taken together, oedema, protein in urine more than 3.5 g/24 h and serum albumin less than 35 g/24 h are found almost only in nephrotic syndrome (high sensitivity and high specificity). So he makes a diagnosis of nephrotic syndrome without much reasoning but just using 'retrieving from his database' in his mind. He transformed uncertainty to certainty (evidence) using simple 'shortcut reasoning processes', sensitivity and specificity, that could be easily mimicked by DDSS using only database retrieval. At least two tables are needed for recording these processes: 'Findings' (see Table 3.1) and Relations (see Table 3.2). So we can retrieve both tables for some finding (oedema), and from 'Relation' table, we can 'offer' user to enter other findings (amount of protein in urine and serum albumin level), and because of high sensitivity and specificity, we can make a diagnosis (nephrotic syndrome). This is 'evidence level' of making diagnosis where

Table 3.1 Simplified example of database table 'Findings'

IDfind	Finding	Disease	Type of data	Property	Value	Low value	High value	Sensit	Specif	Predict	Text link
100001	Pretibial oedema	Nephrotic syndrome	Examination	Pitting	None	None	None	High	Medium	Medium	
100002	U-Protein	Nephrotic syndrome	Urinalysis		>4.5 g/24 h	None	None	High	Medium	High	
100003	S-Albumin	Nephrotic syndrome	Laboratory		<35 g/l	None	None	High	Medium	Medium	
100004	S-Cholesterol	Nephrotic syndrome	Laboratory		>6.5 g/l	None	None	Low	Low	Low	
100005	S-Triglyceride	Nephrotic syndrome	Laboratory		>1.9 g/l	None	None	Low	Low	Low	
100006	Pretibial oedema	Congestive heart failure	Examination	pitting	None	None	None	High	Medium	Medium	
100007	Pretibial oedema	Hepatic cirrhosis	Examination	pitting	None	None	None	Medium	Medium	Medium	
100008	Ischemic heart pain	Ischemic heart disease	Symptom	[a]	None	None	None	Medium	High	Medium	

IDfind identification number of Findings table; *Sensit* sensitivity; *Specif* specificity; *Predict* predictive value

[a]Chest pain spreading in left arm, itching or burning... (see text)

Table 3.2 Simplified example of database table 'Relations'

IDrel	Disease	Relation	TypeRelation	NumbFind	Sensit	Specif	Predict	Text Link
10001	Nephrotic syndrome	100002 and 100003	Causal	2	High	High	Medium	
10002	Nephrotic syndrome	100001 and 100002 and 100003	Causal	3	High	High	High	
10003	Nephrotic syndrome	100001 and 100002 and 100003 and 100004 and 100005	Causal	5	Medium	Very high	High	

IDrel identification number of Relation table; *TypeRelation* type of relation; *NumberFind* number of findings; *Sensit* sensitivity; *Specif* specificity; *Predict* predictive value uncertainty is diminished. It also represents answer to the question that doctor asks himself – which disease is it?

But if doctor could not find evidence for some disease because sensitivity and/or specificity of the findings remain low, he will ask himself – which disease could it be? He will add sensitivity of all findings for one disease and for another, etc. However, he knows that there is no logical process to find out cumulative evidence of a number of findings for some disease because it is not predictable and can change over time. For example, disease can change clinical manifestation from time to time or from one population to another. The only way to find cumulative evidence is to measure sensitivity and specificity of some combination of findings in a given place and given time interval. Without experience or such measuring of the sensitivity and specificity of the relation of a set of findings for some disease, diagnosis remains hypothetical and uncertain. In such situations, course of disease or therapy gives evidence for diagnosis after disease is cured and a doctor does not expect from CDSS to make a diagnosis. He expects from CDSS only to offer possibilities.

What facts will doctor analyse and how in such situations? He will use advanced reasoning. Which type of relation exists among findings? If it is etiologic, pathogenic or causal relation, the sensitivity or specificity could be high. On the other hand, he may look for some experiments with animals to find out the type of relations. The facts used in this case frequently are not findings because it cannot be determined in patients. However, they represent clues for diagnosis. The basic doctor's fear is to miss some facts and some possible diagnosis. That leads to a wrong direction in diagnostic process and because of that he needs a book or a DDSS to find all possibilities. Every technique and logical approach in analysis will be appreciated by the doctor, but DDSS should list not only possibilities and facts but also methods of analysis.

3.3 Structure of the Data

Further analysis will be directed to the text forms and contents, i.e. structure of the data in the books. For example, Harrison's textbook of internal medicine has a number of chapters dedicated to systems of organs, and they are divided to groups of diseases. Some disease can be standalone but can also be a part of other disease. Some diseases never appear alone but are a part of a few diseases and are designated as syndromes. So, we need another table in knowledge database for disease classification, Table 3.3. It seems best to classify diseases in form of a tree. For example, if 'kidney disease' is a point in the tree, one branch leads from it to 'glomerular diseases' and from there one branch leads to 'glomerulonephritis'. Disease classification is always presented in contents of the book.

Analysis of the text that depicts diseases reveals many parts such as symptoms and signs, clinical course, epidemiology, aetiology, pathogenesis of disease and a number of tests (laboratory, films, histology, functional, etc.) where findings can be

Table 3.3 Simplified example of database table 'Disease'

IDdis	Disease	Classification	Frequency	Syndrome	Text link
1001	Nephrotic syndrome	Kidney diseases/ glomerular diseases	Frequent	Yes	

IDdis identification number of Disease table

found. All these 'findings' have some properties. For example, oedema could be pitting or elastic. So the simplest form to describe findings is to relate them to their 'properties'. All facts could be extracted from the book in form of diseases, findings and properties related to findings. But, for example, there are many types of pain and different localisations and duration. Ischemic heart pain typically is chest pain spreading in left arm, itching or burning and is localised behind the sternum and could have short or long duration. Intensity of the pain may be different. Laboratory analyses always have numeric value and normal ranges. Should we separate these facts in special properties? Should we separate part of the body, time of emerging, duration, etc.? Some properties are so important that they can be defined together with findings, for example, chest pain, left arm pain, abdominal pain, pretibial oedema and sacral oedema. So localisation of the finding can be put together with findings. On the other hand, some properties always appear together so there is no need to separate them as we can see for properties of ischemic heart pain. Because of that it should be made possible that the editor of DDSS may define separate new 'special properties'.

Textbooks are rich in tables that offer lists of many possibilities. Doctors need them to be sure that they are not missing some diagnosis. They have to be also sure that DDSS lists all possibilities for that question and situation. Using DDSS should be easier and faster than using tables in books.

Textbooks are also rich in algorithms. They depict exact roadmap for diagnosis and frequently are based on guidelines. It is hard to believe that all medical knowledge could be one day transformed into a set of algorithms, but they would be very useful for entering data into knowledge base.

The basic feature of the doctor's thinking is 'importance' of some findings for the diagnosis of some disease. When analysing text in the books, we can see that they rarely have sensitivity and specificity, so doctors have to acquire them through experience. During acquisition of data to knowledge base, authors of the text should express their experience and define sensitivity and specificity in a few categories such as very high ($p < 0.01$), high ($p < 0.05$), medium ($p < 0.15$) and low ($p > 0.15$). Development of DDSS would lead to regular measuring of sensitivity and specificity for every combination of findings and that could be easy if it is extracted from an information system.

Textbooks are rich in synonyms. Vocabulary thesaurus is a basic feature of DDSS and could be an enormous problem [4]. But there are also many prefixes and suffixes such as *hyper*-tension, *hypo*-tension, leukocyt-*osis* and leuko-*penia*. Because of that, another table of synonyms is required (Table 3.4). For example, leukocytosis could be defined as leukocytes count greater than 10×10^9/l etc.

Table 3.4 Simplified example of database table 'Thesaurus'

IDthes	Medical term	Synonymous
1000001	Leukocyt-*osis*	Leukocytes $> 10 \times 10^9$/l
1000002	*Hyper*-tension	Systolic tension > 140 mmHg and diastolic tension >90 mmHg

IDthes identification number of Thesaurus table

3.4 Structure of Relations Among Sets of Findings and Diseases

Now we have diseases together with findings, their properties and their importance in form of sensitivity and specificity, but findings are rarely independent and texts in books are full of 'relations among sets of findings' that determine diseases. For example, damage of the glomerular basement membrane in the kidney causes loss of albumin in the urine. That is the cause of increased protein in the urine (finding) and decreased protein in the serum (finding). Decreased albumin in serum may cause oedema (finding). So, findings of some disease are not only a set of finding but a set of related findings. As discussed earlier, low level of serum albumin and protein in urine more than 3.5 g/24 h are different findings with high sensitivity but low specificity. However, taken together they have high sensitivity and specificity for the diagnosis of nephrotic syndrome.

Instead of two, there could be, for example, eleven important findings for some disease [16] but any four findings among them are sufficient for diagnosis, so not only sensitivity and specificity for the diagnosis should be noted but also a number of findings sufficient for diagnosis.

However, previous examples are simple relations, and there are examples for complex relation – 'relation of the relations'; major and minor findings for diagnosis of rheumatic fever could be an example [17]. In this case, first two simple relations (two subsets of related findings) should be made for major and for minor findings, and then a complex relation between these two simple relations with new sensitivity and specificity for the disease should be added.

As it was discussed earlier, it is necessary to record a type of relations among the findings and facts (causal, pathophysiologic, guidelines recommended, etc.).

3.5 Multiple References Relations to Other Publications

Some books, such as 'Harrison's principles of the internal medicine', have no references within the text and most often only the dominant opinion is presented. However, books such as 'Oxford textbook of nephrology' [18], for example, frequently cite a number of references for the same finding with different values. How to handle this problem? The solution could be to input all of these references for the same finding and form relation among them that may be designated as 'multiple references'. During processing, median opinion could be used, and all of the opinions may be presented in the output explanation.

3.6 Structure of the Knowledge Base

This is a simple description of relational knowledge base but the actual solution may be much more complicated. At least four tables are necessary (Tables 3.1, 3.2, 3.3 and 3.4). The first table is 'Disease' with fields IDdis, disease, classification three, frequency, text link and syndrome (Table 3.3). Field 'syndrome' is useful to know could some disease be a syndrome in some other disease. Field 'text link' contains links to the text in the electronic book. The second table is 'Findings' with fields IDfind, disease, data type, finding, property, value, low value, high value, sensitivity, specificity, predictive value and text link (Table 3.1). Intensity or frequency could be added but it is better to allow the editor to add some fields with special property. This table could have a field 'relation type' that describes type of relation of findings and disease (etiologic, pathophysiologic, causal, etc.). The third table is 'Relations' with fields IDrel, disease, relation, relation type, number of findings, sensitivity, specificity, predictive value and text link (Table 3.2). The field relation contains names or IDs of the related findings. Relation type could be (1) simple, (2) reference difference, (3) text table, (4) algorithm, (5) guidelines, (6) causal, (7) pathophysiologic, etc. Finding number contains number of findings from the maximal number of findings that are necessary for diagnosis. Table 'Thesaurus' is added to this simple knowledge base and fields could be IDthes, medical term and synonymous (Table 3.4).

3.7 Input

There are two modes of operation. The first of them enables creation of knowledge base (knowledge acquisition) (Figs. 3.1, 3.2 and 3.3). This input means conversion of text from the book to data base. This process theoretically could not be fully automated because some facts simply do not exist and have to be added by authors (e.g. sensitivity and specificity). Authors create only 10–20 pages in a textbook; frequently they analyse only one disease and the part dedicated to diagnosis is about 1–2 pages, so creation of the input to knowledge base resembles creation of abstract or key words from article. However, it is possible to make a semi automat input, i.e. search titles for nouns and the corresponding adjectives and present to authors as a disease possibilities. Similarly, searching of the text could present nouns as candidates for findings. In any way, authors of the book should look into their electronic texts from their title page. First input will be disease and classifications (under constant monitoring of spelling and synonyms). Reading the text, authors will locate facts and transform them into 'findings' and 'relations' until the end of their text. For example, when defining heart pain five findings will be input: chest

New diseases and findings
X
Search text
Paste
Check defined diseases / findings
Check Spelling
Check Synonyms
Disease Classification
Text Link
Rules
Nephrotic syndrome
Disease
Finding
Cancel

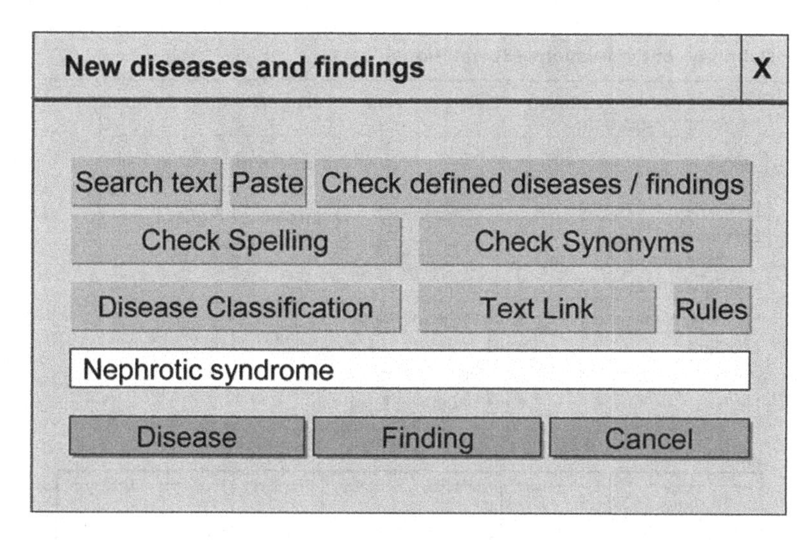

Fig. 3.1 Simplified example of data acquisition screen – introducing new diseases and findings

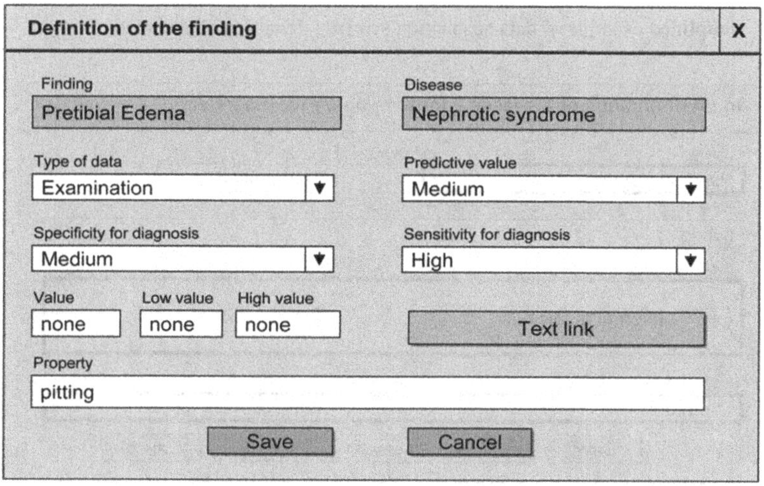

Definition of the finding
X
Finding
Pretibial Edema
Disease
Nephrotic syndrome
Type of data
Examination
Predictive value
Medium
Specificity for diagnosis
Medium
Sensitivity for diagnosis
High
Value
Low value
High value
none
none
none
Text link
Property
pitting
Save
Cancel

Fig. 3.2 Simplified example of data acquisition screen – definition of the finding

pain with properties of ischemic heart pain short duration, chest pain with properties of ischemic heart pain long duration, chest pain with properties non-specific, left arm pain with properties non-specific and abdominal pain with properties epigastric non-specific. Should heart pain be defined as syndrome – stenocardia? So the editor of DDSS has to prepare a plan for definingsymptoms and signs before data acquisition begin, keeping in mind future retrieving and needs of users. A pilot phase can be done on a part of text before the beginning of acquisition of data. Tables in the text could also be transforming into findings and relations.

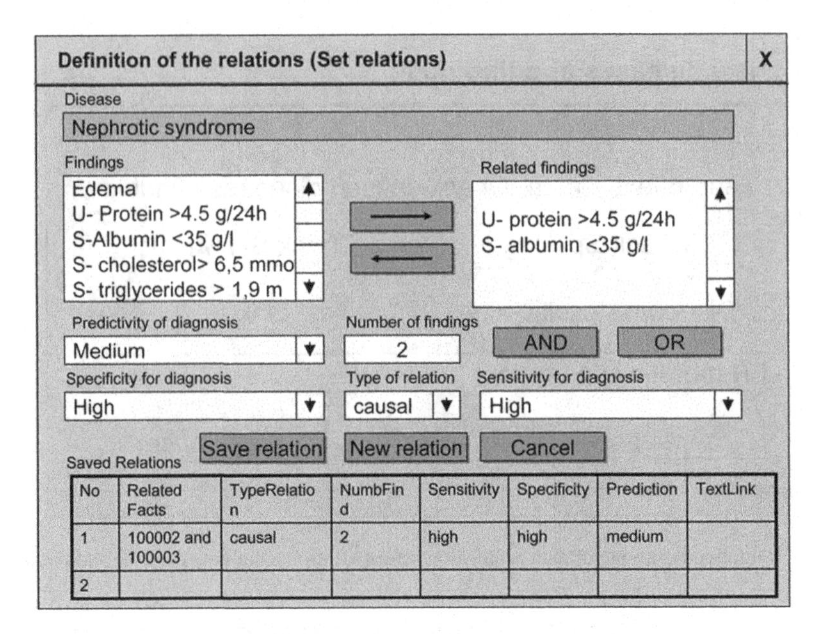

Fig. 3.3 Simplified example of data acquisition screen – definition of the relation

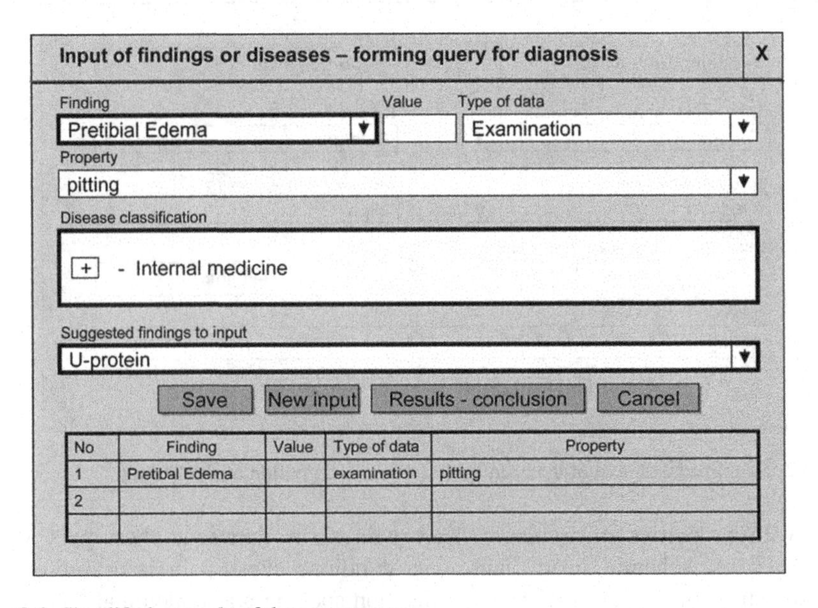

Fig. 3.4 Simplified example of data query screen

The second mode of operation is 'consultation' of DDSS by doctors or students (Fig. 3.4). It can be done using two styles. The first of them is input of disease, when they are interested in what findings are needed for some disease. They could enter the name of disease (under constant monitoring of spelling and synonyms) or through disease classification. The second style is to input findings, when they are interested in what diseases could result from their findings. They could enter a finding (under constant monitoring of spelling and synonyms) and during typing a letter popup would offer names of whole findings in a style of Google search. For example, when user types word 'pain', the system should offer chest pain, left arm pain, abdominal pain, etc. from the list of findings. When this finding is entered, other findings will be suggested through relations with diseases for previous finding in a form of popup. However, one could continue to input new findings without selecting any of the offered options. At the moment when the user does not have another finding, he would demand output.

3.8 Output

Output should be classic – which means a list of diseases ranked by sensitivity and specificity selected by user (Fig. 3.5). Part of the output should be the records from the knowledge base, which were the basis for selection. However, graphs can be also generated to list diseases according to sensitivity and specificity parameters. It is possible to generate graphs in the form of trees. For example, when somebody selects finding oedema, all diseases with oedema formation will be selected. When he selects significant proteinuria in another step, then previous diseases can be displayed as a tree graph according to whether they have or not significant proteinuria, etc. Links to text in the book should also be available for the selected diseases.

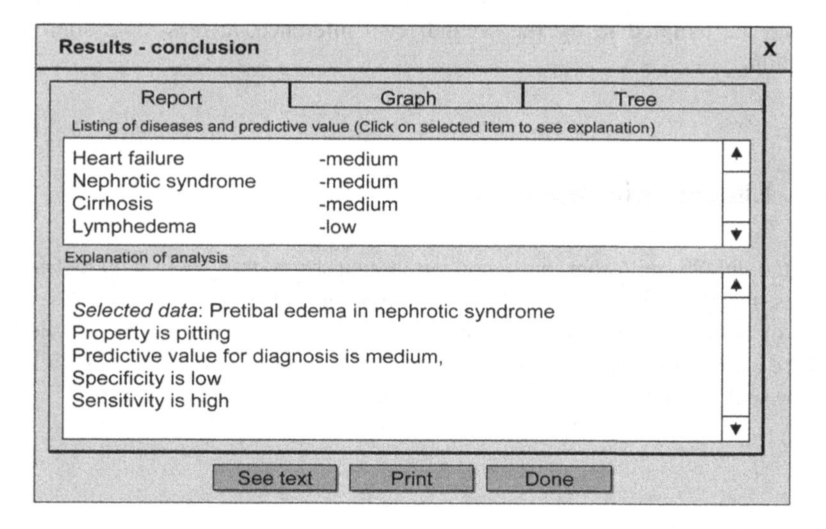

Fig. 3.5 Simplified example of data output screen

3.9 User-Friendly Interface

Developing a user-friendly interface is crucial. This interface should enable authors of the books and articles to define diseases and findings and relations among them in a simple way. So, the main task of this project is developing a simple interface that can mark and copy some 'words' in the electronic text of the book and after that transfer them to the input screen of DDSS. Examples of simple interface are presented in Figs. 3.1 3.2, 3.3, 3.4 and 3.5.

3.10 Inference Mechanisms

Two levels of inference are proposed in this model of DDSS: evidence and nonevidence levels. The evidence level uses SQL, so there is no special inference mechanism. As it was discussed earlier using example of oedema, safe evidence level could be achieved using only retrieval from database that simulates human reasoning. Confidence of this process depends only on the quality of the knowledge base. Because this model uses only existing or 'positive' facts, if there are no findings, relations and diseases in the selected domain, then the diagnosis will be uncertain. That is the same as the natural thinking: if a doctor does not know some fact, he will establish wrong diagnosis.

The second, advanced, level could use any method which derives possible diagnosis and could be adapted to work with this knowledge base [19–24]. For a survey of such methods, see Chap. 2. As discussed earlier, doctors are reluctant to use such methods, whose functioning is not transparent to them, as the primary system for deriving diagnosis. However, when the evidence level inference fails, they may be tempted to try the second level inference, at least as a source of inspiration.

3.11 Linkage with the Text

Presented DDSS represent only converted text from the book with extracted diseases, facts and their relationships. This type of DDSS could be used independently of book and instead of authors a team of experts could make knowledge database in a classic way. However, if some DDSS is created from the book, linkage with the text in the book would be very useful for additional explanations.

3.12 Linkage with Information System

Although this DDSS in basic form might not contain patients' information, it is crucial to connect it with local database. Such an information system then becomes a source of experience for exact calculation of sensitivity, specificity and overall predictive value. In that way, this model becomes 'active decision support system' and gives new piece of evidence for every combination of findings. However, extensions of information system may be necessary to make this connection functional.

3.13 Discussion

Presented model of diagnostic decision support system could enable the conversion of a book to a knowledge base, i.e. conversion of text to database. The electronic books could be transformed to decision support systems that would be published together with paper and electronic editions and even could be linked to local information system. Authors of the book should be the creators of this decision support system, guided by the editor. In the centre of this diagnostic decision support system is knowledge database which would enable manipulating data using SQL language and probability assessment through sensitivity, specificity and predictive value as basic processes.

To our knowledge diagnostic decision support system with those characteristics does not exist now. There are no 'active' books that could be 'asked' to give some advice on professional level although some relationship between DDSS and text-book has been suggested [25, 26]. Elements of presented DDSS are not new but they are used in a new concept which fulfills the aim of this study to suggest the most safe and reliable model, which is easy to understand by the authors who are not information technology professionals. Fagan and his team evaluated automated acquisition, i.e. conversion of text to knowledge base, but they did not evaluate the whole concept of creating CDSS system including structure of knowledge base, relative probability information and output of the results [10, 11]. In addition, automated acquisition would not persuade doctors to use CDSS and specially DDSS that is the most rarely used among CDSS. Knowledge base and design of different CDSS can be specific, and because of that, we evaluate here only DDSS. Textbooks are a platform to precisely define domain and scope of facts that doctors need and to persuade them to believe and, most importantly, to accept DDSS.

Using of sensitivity and specificity for probability measure of the findings and group of findings is crucial for acceptance of DDSS by doctors because it is well proven in medical practice and represents a way of doctors thinking in establishing diagnosis. If sensitivity and specificity is high for some combination of findings (that can be extracted from local database), then the diagnosis is made on evidence base level and decision is certain. If sensitivity and specificity is not high, then clear evidence for certain disease does not exist and other techniques should be used. To our knowledge, there are no DDSS that use these parameters for relative probability information. The cornerstone of the confidence and wide usage of DDSS is reliability in helping to make diagnosis and the best way is to use the same tools that doctors use in their practice – that is, sensitivity and specificity and the principle of evidence-based medicine. In addition, the proposed structure of the knowledge base is flexible enough to allow different forms of knowledge to be entered.

The possibility that authors could create knowledge base was mentioned earlier, but the concept was not clear [7, 27]. It is necessary to motivate authors of the textbook to create structured knowledge applicable for knowledge bases when they create their texts. Automated acquisition could help them but that is not crucial. If they accept to create key words and abstracts, why should they not also accept to create excerpts in a form of structured text? So, the problem, as we see it is to produce structured knowledge together with text creation. Then we can easily convert structured knowledge to text. As we learn to read and write and to use computers and Internet, we have to learn to think database way.

At present it is recommended that clinical decision systems should be integrated within underlying electronic health records, computerised physician order entry system or e-prescribing system and provide decision support automatically as a part of clinical workflow [28–30]. However, books still exist and represent very important source of consultation, so better structured knowledge from the books could be very useful for the users, clinicians and students.

Described model of diagnostic decision support system is intended to be supplemental to electronic editions of major books. Authors of the books could use this model to transform their texts into a knowledge base. The main motive to do that would be the fact that decision support systems could enable much faster and better searching of the books and can help in making decisions.

Acknowledgments The work presented here was supported by the Serbian Ministry of Education and Science (project III44006).

Appendix

Example 1

Text from the book

Establishing the Diagnosis of Chronic Kidney Disease (CKD)

The most important initial diagnostic step in the evaluation of a patient is elevated serum creatinine that can help to distinguish newly diagnosed CKD from acute or subacute renal failure. "Previous measurements of plasma creatinine concentration are particularly helpful..." Normal values

(continued)

(continued)

from recent months suggest that the current renal dysfunction could be more acute, and hence reversible. "In contrast, elevated plasma creatinine concentration in the past suggests that the renal disease represents the progression of a chronic process. Even if there is evidence of chronicity, there is the possibility of a superimposed acute process..." If the history suggests multiple systemic manifestations of recent onset (e.g., *fever, polyarthritis, and rash) renal insufficiency can be acute process.*

Some of the laboratory tests and imaging studies can be helpful. "Evidence of metabolic bone disease with hyperphosphatemia, hypocalcemia, and elevated PTH and bone alkaline phosphatase levels suggests chronicity. Normochromic, normocytic anemia suggests that the process has been ongoing for some time. The finding of bilaterally reduced kidney size (<8.5 cm in all but the smallest adults) favors CKD." However, once the CKD is advanced the kidneys are small and scarred.

Table 3.5 Table 'Disease' for the first example

IDdis	Disease	Classification	Frequency	Syndrome	Text link
1002	Chronic kidney disease	Disease of kidney and urinary tract	Frequent	No	
1003	Metabolic bone disease	Disorders of bone and mineral metabolism	Frequent	Chronic kidney disease	
1004	Normochromic, normocytic anaemia	Haematology	Frequent	Chronic kidney disease	

IDdis, identification number of Disease table

The above text is extracted from Chap. 280 Chronic Kidney Disease (15, page 2318) that deals with establishing diagnosis. However, some data from previous parts of this chapter may be important for diagnosis (e.g. imaging studies). As we can see there are no probabilities in this text. Because of that some expert has to add it, and the best solution is that the author transforms this text to knowledge base. There is only one disease in this text: *chronic kidney disease* that is a part of *disorders of the kidney and urinary tract*. However, *metabolic bone disease* (part of *disorders of bone and mineral metabolism*) and *normochromic and normocytic anaemia* (part of *haematology*) can be defined as syndromes in this text. There are many findings such as *serum creatinine, imaging studies, small and scarred kidneys, hyperphosphatemia, hypocalcaemia, PTH, bone alkaline phosphatase and kidney size.*

Now we can acquire previous data in the knowledge base. Diseases and syndromes are acquired in table 'Disease' (Table 3.5). *Metabolic bone diseases*

Table 3.6 Table 'Findings' for the first example

IDfind	Finding	Disease	Type of data	Property	Value	Low value	High value	Sensit	Specif	Predict	Text link
100009	S-creatinine	Chronic kidney disease	Laboratory	Elevated for 6 months	>150	None	None	High	High	High	
100010	Renal ultrasound	Chronic kidney disease	Imaging	Small and scarred kidneys	None	None	None	High	High	High	
100011	S-creatinine	Chronic kidney disease	Laboratory	Elevated less than 6 months	>150	None	None	High	Medium	Medium	
100012	Kidney size	Chronic kidney disease	Imaging	Small	None	None	None	High	Medium	Medium	
100013	Kidney size	Chronic kidney disease	Imaging	Normal or enlarge	None	None	None	Low	Low	Low	
100014	S-phosphate	Chronic kidney disease	Laboratory	Elevated	>1.7 mmol/l	None	None	Low	Medium	Low	
100015	S-calcium	Chronic kidney disease	Laboratory	Decreased	<2.1 mmol/l	None	None	Low	Low	Low	
100016	S-PTH	Chronic kidney disease	Laboratory	Elevated	None	None	None	Low	Low	Low	
100017	Bone alkaline phosphatase	Chronic kidney disease	Laboratory	Elevated	None	None	None	Low	Low	Low	

IDfind identification number of Findings table; *Sensit* sensitivity; *Specif* specificity; *Predict* predictive value

Table 3.7 Table 'Thesaurus' for the first example

IDthes	MedicalTerm	Synonymous
1000003	Hyperphosphatemia	Elevated S-phosphate; S-phosphate > 1.7 mmol/l
1000004	Hypocalcaemia	Decrease S-calcium; S-calcium < 2.1 mmol/l

IDthes identification number of Thesaurus table

Table 3.8 Table 'Relations' for the first example

IDrel	Disease	Relation	Type-Relation	NumbFind	Sensit	Specif	Predict	Text Link
10003	Chronic kidney disease	100011 and 100012	Causal	2	High	High	High	
10004	Chronic kidney disease	100011 and 1003 and 1004	Causal	3	High	Medium	Medium	

IDrel identification number of Relation table; *TypeRelation* type of relation; *NumberFind* number of findings; *Sensit* sensitivity; *Specif* specificity; *Predict* predictive value; Column 'Relations' shows IDfind from Table 3.5 – Findings or IDdis from Table 3.6 – Diseases

and *normochromic and normocytic anaemia* are diseases, but here in *chronic kidney disease* they appear as syndromes. Findings are showed in Table 3.6. Findings are entered without prefix (hyper or hypo) but properties contains these meaning (elevated or decreased). Kidney size are used two times with properties *small* and *normal and enlarged*. It is also possible to use *small kidney size* and *normal or enlarged kidney size* as different findings.

This differentiation is necessary because kidney size has different meaning for the diagnosis of chronic kidney failure and as we can see sensitivity, specificity and predictive values are different. Solving this dilemma in style is on editor. Different meanings have to be explained in Thesaurus table (Table 3.7) where hyperphosphatemia is explained as elevated S-phosphate.

As we can see from Findings table, diagnosis of chronic kidney disease could be established certainly in two cases: if the kidneys are small and scarred or if S-creatinine is elevated more than 6 months. However, is there some combination of findings that can establish diagnosis certainly? Yes, for example, if we found small kidneys and elevated creatinine for less than 6 months in a person who is not small, then the diagnosis of chronic kidney disease is certain (Table 3.8). Another combination that can raise importance is elevated serum creatinine less than 6 months together with *metabolic bone disease* and *normochromic and normocytic anaemia*. If DDSS is connected to information system, sensitivity, specificity and predictive value can be measured.

Consider now retrieval of knowledge base. If somebody wants to know about findings for the chronic kidney disease, then that is very simple. He will find *chronic kidney disease* in *disease of kidney and urinary tract*. Popup with findings will list findings in chronic kidney disease according to sensitivity, specificity and predictive value so it will be similar to table Findings and will include fields finding + property together with syndromes in chronic kidney disease.

If query is raised typing the word creatinine, it will list all creatinine findings in all diseases together with properties. If somebody selects elevated creatinine for more than 6 months, diagnosis of chronic kidney disease will appear. If somebody selects elevated creatinine less than 6 months, then findings and properties for chronic kidney disease will appear. However, if somebody selects elevated phosphorus, then some endocrine disease findings will appear in popup. Finding for chronic kidney disease will be at the end of popup. Typing another finding that is specific for kidney diseases will now show another list of findings where kidney disease findings appear first.

Example 2

Text from the book

Confirming the Diagnosis of Acute Rheumatic Fever (ARF)

"Because there is no definitive test, the diagnosis of ARF relies on the presence of a combination of typical clinical features together with evidence of the precipitating group A streptococcal infection..." Experts of the World Health Organization clarified the use of the Jones criteria in ARF. These criteria include a preceding streptococcal type A infection as well as some combination of major and minor manifestations.

Table 215 *Criteria for diagnosis ARF (simplified)*

Diagnostic Categories	*Criteria*
Primary episode of Rheumatic fever	*Two major or one major and two minor manifestations plus evidence of preceding group A streptococcal infection*
Rheumatic chorea	*Other evidence not required*
Insidious onset of rheumatic carditis	*Other evidence not required*
Chronic valve lesions of rheumatic heart disease	*Other evidence not required*
Major manifestations	*Carditis* *Polyarthritis* *Chorea* *Erythema marginatum* *Subcutaneous nodules*
Minor manifestations	*Clinical: fever, polyarthralgia* *Laboratory: elevated erythrocyte sedimentation rate or leukocyte count* *Electrocardiogram: prolonged P-R interval*
Evidence of a preceding streptococcal infection within the last 45 days	*Elevated or rising anti-streptolysin O or other streptococcal antibody, or A positive throat culture, or rapid antigen test for group A streptococcus, or recent scarlet fever*

Table 3.9 Table 'Diseases' for the second example

IDdis	Disease	Classification	Frequency	Syndrome	Text link
1005	Acute rheumatic fever	Disorders of connective tissue and joints	No	No	
1006	*Rheumatic chorea*	Neurological disorders	No	Disorders of connective tissue and joints	
1007	*Rheumatic carditis*	Disorders of cardio-vascular disease	No	Disorders of connective tissue and joints	
1008	*Rheumatic heart disease*	Disorders of cardio-vascular disease	No	Disorders of connective tissue and joints	
1009	*Carditis*	Disorders of cardio-vascular disease	No	Disorders of connective tissue and joints	
1010	*Polyarthritis*	Disorders of connective tissue and joints	No	Yes	
1011	*Chorea*	Neurological disorders	No	Disorders of connective tissue and joints	
1012	*Erythema marginatum*	Dermatologic disorders	No	Disorders of connective tissue and joints	
1013	*Subcutaneous nodules*	Dermatologic disorders	No	Disorders of connective tissue and joints	

IDdis identification number of Disease table

The above text is extracted from Chap. 322 Acute Rheumatic Fever (15, page 2755) that deals with establishing diagnosis. However, some data from previous parts of this chapter may be important for diagnosis. As we can see this is special case were diagnostic criteria are clear and could be easily transformed without expert.

There is only one disease – *acute rheumatic fever* – but several other diseases are used as syndromes: *rheumatic chorea, rheumatic carditis, rheumatic heart disease, carditis, polyarthritis, chorea, erythema marginatum, subcutaneous nodules and scarlet fever*. There are next findings: *fever, polyarthralgia, erythrocyte sedimentation rate (ESR), leukocyte, P-R interval, anti-streptolysin O (ASTO) and throat culture*.

Now we can acquire previous data in the knowledge base. Diseases and syndromes are acquired in table Disease (Table 3.9) and findings in table Findings (Table 3.10).

Synonymous are acquired in Thesaurus table (Table 3.11)

Complex relationship among the findings is showed in table Relations (Table 3.12). Here syndromes are acquired and then relations among findings.

Table 3.10 Table 'Findings' for the second example

IDfind	Finding	Disease	Type of data	Property	Value	Low value	High value	Sensit	Specif	Predict	Text link
100018	*Fever*	Acute rheumatic fever	Examination		>37	None	None	High	Low	Low	
100019	*Polyarthralgia*	Acute rheumatic fever	Symptom		None	None	None	High	Low	Low	
100020	*ESR*	Chronic kidney disease	Laboratory	Elevated	>20	None	None	High	Low	Low	
100021	*leukocyte*	Acute rheumatic fever	Laboratory	Elevated	>10,000	None	None	Medium	Low	Low	
100022	*P-R interval*	Acute rheumatic fever	ECG	*Prolonged*	None	None	None	Low	Low	Low	
100023	*ASTO*	Acute rheumatic fever	Laboratory	Elevated	>200	None	None	High	Medium	Medium	
100024	*Throat culture*	Acute rheumatic fever	Laboratory	Streptococcus	>100,000/ml	None	None	High	Medium	Medium	

IDfind identification number of Findings table; *Sensit* sensitivity; *Specif* specificity; *Predict* predictive value

Table 3.11 Table 'Thesaurus' for the second example

IDthes	Medical term	Synonymous
1000005	*ESR*	*Erythrocyte sedimentation rate*
1000006	ASTO	*Anti-streptolysin O*

IDthes identification number of Thesaurus table

Table 3.12 Table 'Relations' for the second example

IDrel	Disease	Relation	TypeRelation	NumbFind	Sensit	Specif	Predict	Text link
10005	Acute rheumatic fever	1006	Causal	0	High	High	High	
10006	Acute rheumatic fever	1007	Causal	0	High	High	High	
10007	Acute rheumatic fever	1008	Causal	0	High	High	High	
10008	Acute rheumatic fever	1009 or 1010 or 1011 or 1012 or 1013	Causal	1	High	Medium	Medium	
10009	Acute rheumatic fever	100018 or 100019 or 100020 or 100021 or 100022		2	High	Medium	Medium	
10010	Acute rheumatic fever	100023 or 100024		1	High	Medium	Medium	
10011	Acute rheumatic fever	10008 and 10009 and 10010			High	High	High	
10012	Acute rheumatic fever	1009 or 1010 or 1011 or 1012 or 1013	Causal	2	High	Medium	Medium	
10013	Acute rheumatic fever	10010 and 10012			High	High	High	

IDrel identification number of Relation table; *TypeRelation* type of relation; *NumberFind* number of findings; *Sensit* sensitivity; *Specif* specificity; *Predict* predictive value; Column 'Relations' shows IDdis from Table 3.9 – 'Disease', IDfind form Table 3.10 – 'Findings' or IDrel from this table

There are two levels of relations: primary and secondary that is formed from primary. Because of that, these secondary relations are 'relations of the relations' and they have the same format of IDrel in field relations. For example, 10011 is complex and is derived from relations 10008 (necessary one findings-syndrome), 10009 (necessary two findings) and 10010 (necessary one findings). Complex relation 10013 is derived from 10012 which is the same as 10008 but necessary are two findings-syndromes and 10010 (necessary one findings) as can be seen from the text that was transformed.

Retrieval from this knowledge base is the same as it was discussed in previous example. However, in this case there are more syndromes than in the first example and Relation table is complex. In addition, there are no separate knowledge base tables for some disease. They are unique for all disease as can be seen from ID numbers in the tables but they are showed separately in order to simplify presentation.

References

1. Decision-Making in Clinical Medicine. In Fauci, Branuwald, Kasper, Hauser, Longo, Jameson and Lorenco eds. Harrison's Principles of Internal Medicine. Seventeenth edition. (The McGraw-Hill Companies, Inc. New York: 2008: 16–23).
2. Kong Guilan, Xu Dong-Ling, Yang Jian-Bo. Clinical decision support systems: A review on knowledge representation and inference under uncertainties. International Journal of Computational Intelligence Systems 1 (2008) 159–167.
3. Berner E and La Lande T. Overview of Clinical Decision Support Systems in Berner ES. Clinical Decision Support Systems: Theory and Practice (Health Informatics) (Springer, New York, 2007: 3–22).
4. Carter J. Design and Implementation Issues in Berner ES. Clinical Decision Support Systems: Theory and Practice (Health Informatics) (Springer, New York, 2007: 64–98)
5. Miller R and Geissbuhler A. Diagnostic Decision Support Systems Systems in Berner ES. Clinical Decision Support Systems: Theory and Practice (Health Informatics) (Springer, New York, 2007: 99–125).
6. Long WJ. Medical informatics: reasoning methods. Artif Intell Med. 2001 23:71–87.
7. Perry CA. Knowledge bases in medicine: a review. Bull Med Libr Assoc. 1990 78:271–82.
8. Bennett CC, Hauser K. Artificial intelligence framework for simulating clinical decision-making: A Markov decision process approach. Artif Intell Med. 2012 Dec 31. doi:pii: S0933-3657(12)00151-0. 10.1016/j.artmed.2012.12.003.
9. Horn W. AI in medicine on its way from knowledge-intensive to data-intensive systems. Artif Intell Med. 2001 Aug;23(1):5–12.
10. Berrios DC, Kehler A, Fagan LM. Knowledge requirements for automated inference of medical textbook markup JAMIA (1999) S676–S80
11. Shankar RD, Tu SW, Martins SB et al. Integration of textual guideline documents with formal guideline knowledge bases JAMIA (2001) S617–21
12. Osheroff JA, Teich JM, Middleton B et al. A roadmap for national action on clinical decision support . JAMIA. 14 (2007) 141–145
13. Oberkampf H, Zillner S, Bauer B and Hammon M. Towards a ranking of likely diseases in terms of precision and recall. Proceedings of the 1st International Workshop on Artificial Intelligence and NetMedicine, Montpellier, France, 2012. 11–20.

14. Vecchio TJ. Predictive value of a single diagnostic test in unselected populations. N Engl J Med 274 (1966) 1171–73
15. Longo, Fauci, Kasper, Hauser, Jameson and Lorenco eds. Harrison's Principles of Internal Medicine. Eighteenth edition. (The McGraw-Hill Companies, Inc. New York, 2012).
16. Systemic lupus erythematosus. In Fauci, Branuwald, Kasper, Hauser, Longo, Jameson and Lorenco eds. Harrison's Principles of Internal Medicine. Seventeenth edition. (The McGraw-Hill Companies, Inc. New York, 2008, 2075–2082).
17. Acute rheumatic fever. In Fauci, Branuwald, Kasper, Hauser, Longo, Jameson and Lorenco eds. Harrison's Principles of Internal Medicine. Seventeenth edition. (The McGraw-Hill Companies, Inc. New York, 2008, 2092–2095).
18. Davison A, Cameron S, Rits E, Grunfelt J, Wenearls C, Ponticelli C and Ypersele C eds. Oxford Textbook of Clinical Nephrology. Third edition. (Oxford University Press, Oxford, 2005).
19. Marković Z, Rašković M, Ognjanović Z. A Logic with Approximate Conditional Probabilities that can Model Default Reasoning. International Journal of Approximate Reasoning 49 (2008) 52–66
20. Perović A, Ognjanović Z, Rašković m, Radojević D. Finitely additive probability measures on classical propositional formulas definable by Godel's t-norm and product t-norm, Fuzzy Sets and Systems 169 (2011) 65–90
21. D. Doder. A logic with big-stepped probabilities that can model nonmonotonic reasoning of system P. Publications de l'Institut Mathématique, ns. 90(104): 13–22, 2011.
22. D. Doder, J. Grant, Z. Ognjanović. Probabilistic logics for objects located in space and time. To appear in Journal of logic and computation, doi: 10.1093/logcom/exs054.
23. A. Ilić-Stepić, Z. Ognjanović, N. Ikodinović, A. Perović, A p-adic probability logic, Mathematical Logic Quarterly, vol. 58 (4–5), 263–280, 2012.
24. Z. Ognjanović, Z. Marković, M. Rašković, D. Doder, A. Perović: A propositional probabilistic logic with discrete linear time for reasoning about evidence. Ann. Math. Artif. Intell. 65(2–3): 217–243, 2012.
25. Kawamata F, Kondoh M, Mori C, Endoh J, Takahashi T. Computer-aided clinical laboratory diagnosis in conjunction with the electronic medical textbook. Medinfo. 8 (1995) 955
26. Miller RA, McNeil MA, Challinor SM et al. The internist-1/quick medical reference project–status report. West J Med. 145 (1986) 816–22
27. Berrios DC, Kehler A, Kim DK et al. Automated text markup for information retrieval from an electronic textbook of infectious disease JAMIA (1998) S975–5
28. Garg AX, Adhikari NK, McDonald H et al. Effects of computerized clinical decision support systems on practitioner performance and patient outcomes: a systematic review. JAMA. 293 (2005) 1223–38.
29. Kawamoto K, Houlihan CA, Balas EA et al. Improving clinical practice using clinical decision support systems: a systematic review of trials to identify features critical to success. BMJ. 330 (2005) 765–73
30. Patwardhan MB, Kawamoto K, Lobach D et al. Recommendations for a clinical decision support for the management of individuals with chronic kidney disease. Clin J Am Soc Nephrol. 4 (2009) 273–83

Chapter 4
Text Mining in Medicine

Slavko Žitnik and Marko Bajec

4.1 Introduction

Text (data) mining defines various techniques to derive high-quality structured information from unstructured textual sources. Research estimates that 90 % of all data is in unstructured format, such as emails, voice or video records, data streams, and Word documents. In the last decade, the estimated growth of unstructured data is about 62 %, whereas the amount of structured data has grown only by 22 %.

Researchers in the field of medicine, biologists, etc., spend a lot of time searching for the available information about a specific research area or already conducted experiments. For example, if one was trying to develop a new drug, he might be interested in finding all gene products that are involved in a specific bacterial process and have specific sequences or structures. The information researchers in the field of medicine often need is mainly written in unstructured format in some scientific articles, which means it cannot be directly processed by computers. Therefore, natural language processing techniques are becoming significantly important to uncover structured data from biomedical literature.

Natural language processing is a research field that combines techniques to automatically analyze, understand, or generate human-readable texts. Due to increasing number of genomes, sequences, proteins, individual gene studies, etc., it is difficult and time consuming to manually collect and interpret large-scale experimental results. Therefore, methods should automatically enable efficient data retrieval, processing, and integration. Typical text mining tasks therefore include text categorization, text clustering, information extraction, taxonomy/ontology generation, sentiment analysis, and document summarization.

In this chapter, we cover basic introduction and review of accompanying methods for the following main text mining fields:

S. Žitnik (✉) • M. Bajec
Faculty of Computer and Information Science, University of Ljubljana, Tržaška cesta 25, SI-1000 Ljubljana, Slovenia
e-mail: slavko.zitnik@fri.uni-lj.si; marko.bajec@fri.uni-lj.si

G. Rakocevic et al. (eds.), *Computational Medicine in Data Mining and Modeling*, DOI 10.1007/978-1-4614-8785-2_4, © Springer Science+Business Media New York 2013

Information retrieval deals with indexing, searching, and retrieval of relevant documents given an input query.

Information extraction tries to automatically extract structured data from unstructured sources with the main tasks of named entity recognition, relationship extraction, and coreference resolution.

Data integration solves the problem of merging and redundancy elimination in data.

There are a lot of examples where text mining techniques directly helped in medicine-related research. Some recent examples are improving prediction of protein functional sites [58], leveraging computational predictions to improve literature-based Gene Ontology [59], or linking ontologies and protein-protein interactions to literature [60].

The rest of the chapter is organized as follows. In Sect. 4.2, we introduce examples of medicine linguistic resources. Some of them are direct result of text mining algorithms and others serve as collections of unstructured sources yet to be processed. All the resources we introduce are further categorized as a raw database or an ontology. Section 4.3 presents publicly available platforms, tools, and libraries that can be used for text mining over the medical data. First, we introduce general frameworks that provide additional plug-ins to work with medical domain and then we present some domain-specific tools that were initially built to work with medical domain. In the following section, we give an overview of information retrieval task and present a model, which is most commonly used to search over biomedical data. Next, we review information extraction techniques, present data transformations, and explain standard evaluation metrics. We also give an overview of commonly used machine-learning methods and more thoroughly explain Conditional Random Fields algorithm. In Sect. 4.6, we focus on data integration, where we present data transformation and explain general network-based merging and redundancy elimination techniques. Lastly, in Sect. 4.7, we summarize the chapter.

4.2 Medicine Linguistic Resources

Linguistic resources in the field of text mining for biomedicine are very important. They contain useful information regarding conducted experiments and results. By integration of these results, new findings can be uncovered, and more importantly, the research is not repeated and therefore a lot of time is saved. For example, if you were trying to develop a new drug, you might want to find all gene products that are responsible for bacterial protein synthesis in selected genome (e.g., human). These findings are most commonly written in scientific papers and therefore we present some linguistic sources that contain such extracted data.

We further categorize language resources into *databases* and *ontologies* (see definition in Sect. 4.2.2). Sometimes it is hard to differ between the two, so therefore we classify resource as an ontology if it is encoded into a standard ontology description format (e.g., RDF/XML, OWL), follows a structure of taxonomy, or contains relationships or associations between concepts. We do not explicitly review datasets, created specifically for shared tasks or challenges, but they play an important role in developing and evaluating new text mining algorithms. Some examples of such shared tasks are KDD Cup 2002, TREC Genomics Tracks 2003–2007, and still active BioNLP and BioCreative shared tasks.

4.2.1 Scientific Literature Databases

MEDLINE [17] is one of the largest medical databases that indexes academic journal papers from the broad fields of medicine, health, life sciences, behavioral sciences, chemical sciences, and bioengineering. It contains more than 20 million documents from more than 5,000 worldwide journals, and each day, a few thousand of new references are added. In order to access this large database, PubMed information retrieval system was developed. The system is still actively supported and provides Boolean-style search (Sect. 4.4) with combination of field names from MeSH ontology (see Sect. 4.2.2).

Online Mendelian Inheritance in Man (OMIM) [18] is also a public bibliography database that contains a comprehensive extraction of human genes, genetic phenotypes, and all known Mendelian disorders from bibliographic data. The database was constituted in 1960s and is still daily updated. In contrast to other databases, which are extensively generated using automatic extraction algorithms, it is curated manually. Researchers use OMIM data to research the causalities between phenotypes and genotypes.

4.2.2 Ontologies

Ontology was primarily defined as a specification of conceptualization [19]. Ontologies are used in various fields, but maybe they are mostly known to support the idea of Semantic Web [20]. An ontology represents knowledge about a domain which is generally represented as a set of concepts (e.g., protein, gene, genome) and relationships (e.g., is expressed, inhibits, regulates) among them. An instance (e.g., BRCA1) in a semantic database (i.e., a database that uses an ontology as a domain) therefore represents a data item of a specific concept type. Furthermore, ontology specification also provides a definition of rules and axioms, which enables the use of inference mechanisms to uncover unseen data. Lastly, one of the most important things that ontologies provide is a seamless interconnection of multiple data sources. Therefore, a direct connection (e.g., *owl:sameAs*, *rdfs:seeAlso*, *owl: equvivalentClass* relationship types) to concepts in other semantic databases can be added to enrich a single ontology-based data without explicit copying. A special case of an ontology is taxonomy, which only defines hierarchical structure of the concepts (i.e., categorization).

GENIA [21] is a corpus of 2,000 MEDLINE abstracts that are linguistically annotated with part-of-speech tags, syntactic dependencies, and most importantly semantic annotations. The ontology contains concepts of biological entities like proteins or genes that were manually tagged during the phase of term annotation. GENIA is therefore a taxonomy of 47 biological nominal categories (e.g., biological source, substance, or organism type).

BioInfer [22] is, similarly as GENIA, an information extraction corpus that additionally contains relationships between entities. The data was extracted from 1,100 biomedical research abstracts. Additionally, two separate taxonomies were built, one containing entity concepts and the second relationships. Lastly, both were integrated into a single ontology.

Medical Subject Headings (MeSH) [23] is a manually curated hierarchical thesaurus. It consists of naming descriptors, from general (e.g., anatomy, healthcare) to specific (e.g., colloid cysts). Each descriptor contains a unique identifier, short description, and one or more tree numbers that identify locations within the tree structure. MeSH is therefore mainly used for indexing articles within MEDLINE database (see Sect. 4.2.1).

Systematic Nomenclature Of Medicine Clinical Terms (SNOMED CT) [24] is one of the most comprehensive multilingual clinical healthcare terminology sources. It consists of more than 300,000 taxonomic objects which are linked with more than 1,300,000 relationships (e.g., is-a, is-synonym). The whole ontology also enables uncovering specific patterns using simple descriptive logic (e.g., transitivity). Similarly to MeSH, it offers structured collection of medical terms and therefore focuses to be used to index, retrieve or store clinical data, record clinical details of individuals, and support decision making using descriptive logic inference techniques.

Gene Ontology (GO) [25] is an ontology that describes how gene products interact in a cellular context and covers the domains of cellular component, molecular function (i.e., activity of a gene product), and biological process (i.e., operations or events with defined beginning and end). Furthermore, it also includes association links to other ontologies (e.g., FlyBase or geneDB) and offers querying across them (e.g., a user can query for all the gene products in the human genome that are involved in a specific molecular process).

MicroRNA-cancer association database (miRCancer) [26] is regularly updated database using text mining algorithms against PubMed. Human cancers are correlated to microRNAs, so the database could be used for cancer indicators identification and treatment purposes. The text mining algorithms are based on manual constructed rules, based on typical microRNA expressions in literature. Lastly, all miRNA-cancer associations are confirmed manually, so therefore miRCancer is one of few databases that stores exact empirical results.

Lastly, we need to mention the type of ontologies that focus just on integrating different databases or ontologies by providing mutual associations (sometimes also called metathesaurus). They are very useful in natural language processing as systems can use just one interface to query multiple contexts. Some examples of such ontologies are Unified Medical Language System (UMLS), Open Biological and Biomedical Ontologies (OBO), and Disease Ontology. They combine hundreds of public datasets and also the ones we have mentioned earlier.

4.3 Text Mining Platforms and Tools

During the text mining research on biomedical data, a number of techniques have been developed. Some research work has further evolved into publicly available services, libraries, or larger projects. In this section, we present some general text mining tools that have been successfully used on biomedicine domain or were already initially developed to solve a biomedicine problem.

It is common that the researchers from the field of general text mining often relearn their approaches on different domain and evaluate them without having deeper knowledge about data [1]. Some of the general tools, successfully applied to a biomedicine domain, are listed in Sect. 4.3.1. Presented frameworks therefore also include special biomedical plug-ins to handle specific data formats and visualizations.

In Sect. 4.3.2, we present text mining tools that were initially developed to work with biomedical data. In contrast to general approaches, most of them include large corpuses of processed data and data manipulation tools.

4.3.1 General Tools

General Architecture for Text Engineering (GATE) [2] is an open-source platform containing a number of text analysis tools. It has been actively developed for more than 20 years by academic and industry contributors. The community is still constantly including new state-of-the-art techniques and technologies. Recently, they also provided a cloud-based service for large-scale text processing.

The family of products offers a variety of tools to work with. GATE Developer (Fig. 4.1) is a special integrated development environment for end users. It provides a graphical user interface, which enables the user to read new data, construct a pipeline of text processing modules, and visualize and analyze the input data. Each component also offers customizations of special parameters or an ability to define custom extraction rules. Next, the GATE Teamware provides a web collaboration framework to annotate and curate new text corpora. Having a large number of tagged corpora is essential to improve or build new text mining techniques, and therefore it is very useful to use the same semiautomatic tagging technique for a group of manual annotators. Lastly, GATE Embedded provides an open-source library that can be directly used by software developers when building new text

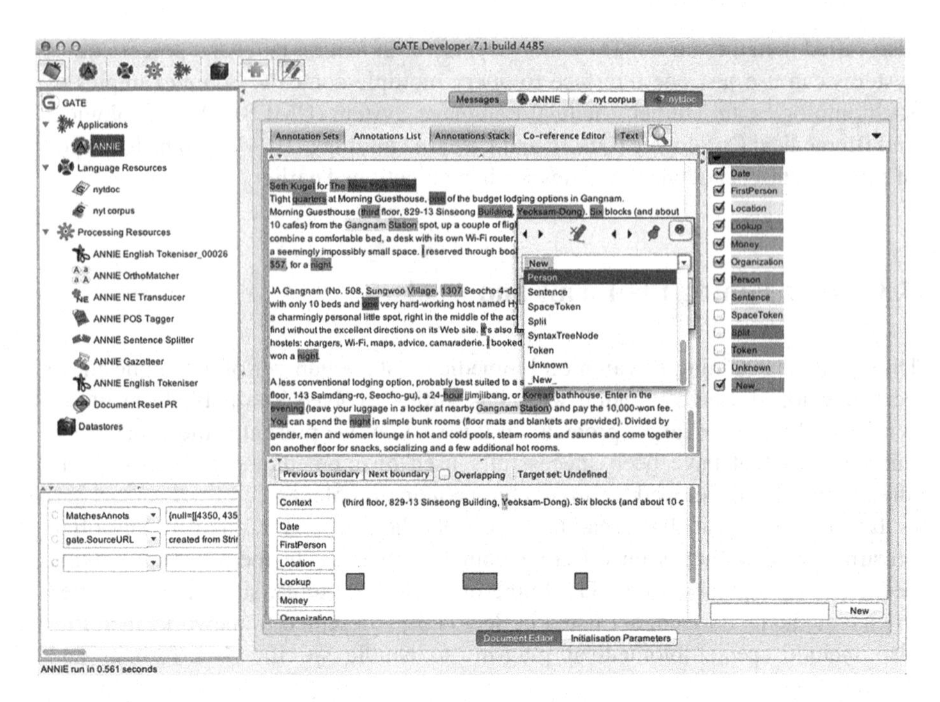

Fig. 4.1 GATE Developer. On the *left side*, selected applications are shown with a number of processing resources that can process language resources. On the *right side*, a number of known annotation sets are listed. In the *center*, a processed document is visualized and user is given a possibility to refine the results

analysis supported systems. Moreover, using the library, new custom plug-ins for GATE Developer tool can be implemented.

In the field of biomedicine, GATE tool was used to detect a head and neck cancer mutation association, medical records analysis, richer drug-related searching, protein-protein interactions, etc. [3]. There are also a lot of biomedical-specific plug-ins already developed. Furthermore, it allows to structure extracted data using an ontology language, or it allows using an ontology to guide the extraction process.

Unstructured Information Management Architecture (UIMA) [4] is a similar framework to GATE, but not so focused into providing a nice graphical user interface (Fig. 4.2). By itself it is an empty framework for the analysis of unstructured data like video, audio, and text data. It provides a general framework for data acquisition, representation, processing, and storing. The UIMA main goal was to develop a lot of reusable components, for example, annotators or external resources that can be easily plugged into the system. Next to general components, also specialized biological annotators or medical knowledge extraction have been developed. The framework became better known after IBM's system Watson, which was built on top of UIMA, and had won the 2011 Jeopardy challenge.

In contrast to GATE and UIMA frameworks, a number of natural language processing libraries exist. Their main focus is to solve a specific task and to

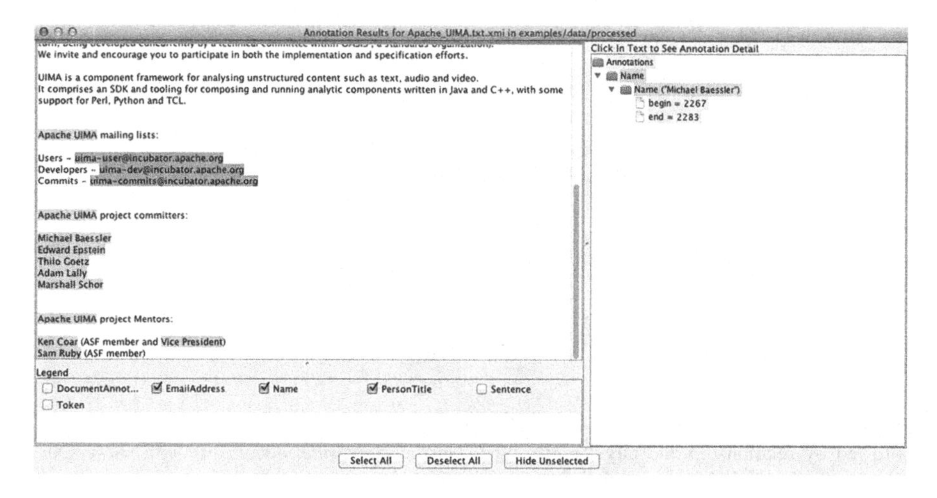

Fig. 4.2 UIMA example annotation results. On the *left side*, the source document is annotated with selected tags, and on the *right side*, annotation details are shown

be used in third-party applications. *Stanford Core NLP* [5] is one of the most comprehensive set of tools that next to the text preprocessing techniques provides named entity recognizer, simple relationship extractor, and coreference resolution system. By functionality, similar library is *Breeze* [6], and other more general are also *OpenNLP* [7], *NLTK* [8], *DepPattern* [9], or *FreeLing* [10]. One of the key differences among them is also the programming language a specific library is available in.

4.3.2 Medicine-Specialized Tools

Turku Event Extraction System (TEES) is one of the best performing systems at biological shared tasks [11]. It uses a classification-based machine-learning approach to detect events from which it further identifies relationships. The whole TEES text mining process was designed to uncover biomedical interactions (i.e., relationships between proteins/genes and corresponding processes) in research articles. Additionally, the system includes standard text preprocessing tools and is adaptable to various tasks, like speculation and negation detection (i.e., similar to general sentiment analysis), protein/gene coreference resolution, or synonym detection.

Extraction of Classified Entities and Relations from Biomedical Texts (EXCERBT) [12] system is a result of a Ph.D. thesis in the field of bioinformatics. Its main goal is to use a simple machine-learning algorithm with shallow semantic role labeling in contrast to slow full-text parsing. Next to EXCERBT, the authors also perform large-scale network analysis for extracted relationship ranking using neighboring documents similarities. Lastly, they use big data database

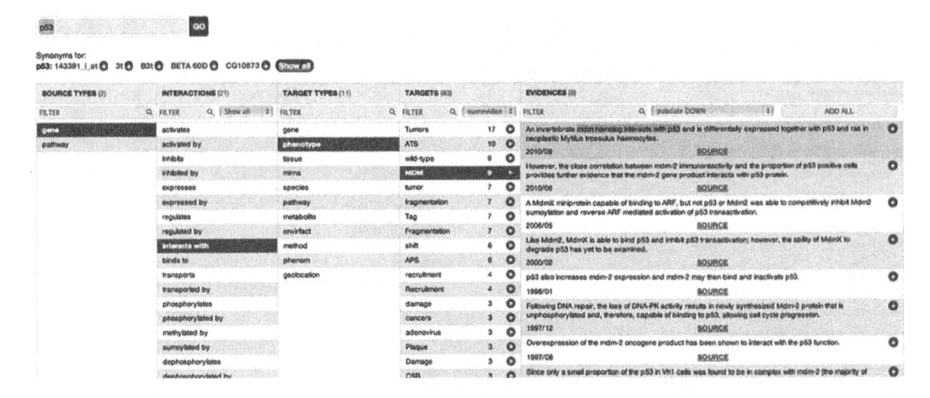

Fig. 4.3 EXCERBT search engine user interface. On the *left side*, we select the source entity type, followed by relationship and target entity type. On the *right side*, documents with short text snippets, from which the whole relationship was extracted, are presented

technologies to efficiently store, index, and retrieve data from few ten million PubMed documents. In Fig. 4.3, we show a graphical user interface built on top of EXCERBT with a simple information retrieval example.

Search Tool for the Retrieval of Interacting Genes/Proteins (STRING) is an integration and information retrieval tool along with a database of known and predicted protein interactions [13]. The tool offers a successful confidence scoring function and merely focuses on an interactive user interface. The STRING database covers more than thousand types of organisms, ranging from bacteria to humans, which are represented by fully sequenced genomes. In Fig. 4.4, we show a sample STRING result when searching for *RAD51* protein in context of a human genome. The integration methods underneath the graphical user interface use standard approaches with some minor domain modifications.

Multiple Association Network Integration Algorithm (GeneMANIA) [14] is a real-time network integration algorithm for gene function prediction. The first part of the algorithm creates a single-gene functional association network using other genomic data sources. In the second part, a technique of label propagation (i.e., an approach that uses node values and their connections to achieve converged network) is used to identify gene functions from previous step. On the top of proposed algorithm, a network-based web interface has been developed that enables the user to find association data for a set of input genes. The association data includes protein/gene interactions, protein domain similarity, pathways, co-expressions, and co-localizations.

Genie [15] is a tool that evaluates biomedical literature and identifies genes within the texts from a number of databases. Its goal is to provide a ranked set of genes given target genome and a biomedical topic. Moreover, it supports natural language query input and provides high precision results. The researchers can further use that information to more thoroughly focus on higher ranked genes.

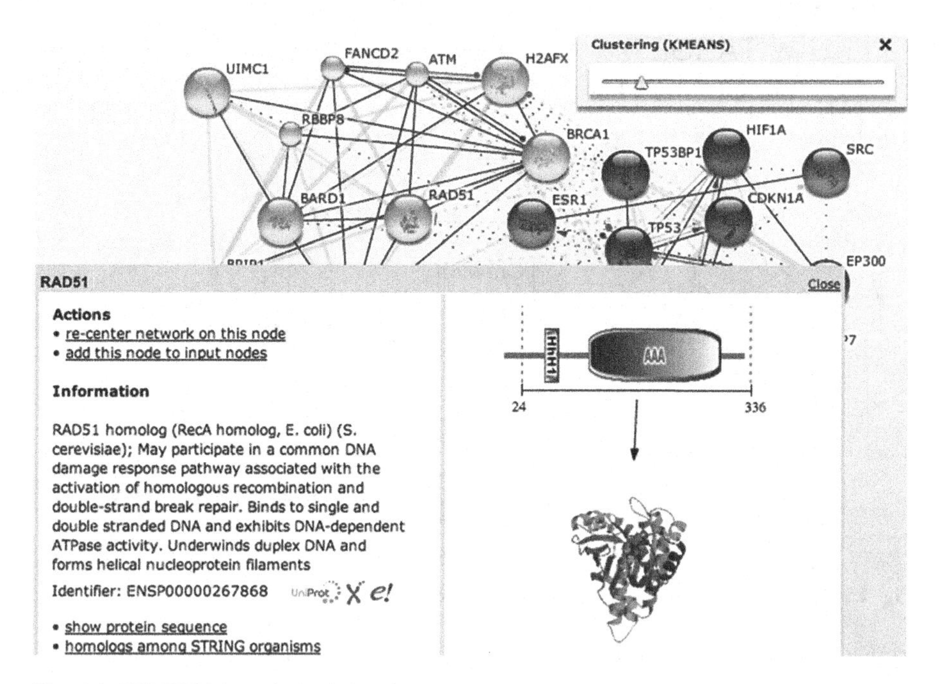

Fig. 4.4 STRING interactive web interface. The tool provides search interface and a selection of target organism. Results are shown as a network of interactions, which a user can deeply analyze and investigate

GENIA Tagger [16] is an example of information extraction tool that provides text preprocessing taggers (i.e., part-of-speech tagger, parser) with named entity recognizer. From general tools, it differs by being trained on biomedical data with accordingly tuned features.

4.4 Information Retrieval

Information retrieval [27] is a task of obtaining relevant information resources to a query from a document database. Documents can be represented in different formats, for example, web pages, videos, and images. A query on the other hand can also be of arbitrary type, entered by a user or prepared by a program.

In 1945, Vannevar Bush designed a concept of a system called *memex* that would be able to store and retrieve data [29]. The proposed system was naive comparing to modern information retrieval systems, but his idea was revolutionary in those days. He also designed a Wikipedia-like retrieval system and focused especially on book retrieval. The earliest information retrieval systems were then really implemented in libraries.

In 1960s, Professor Gerard Salton with a research group at Cornell University developed SMART (System for the Mechanical Analysis and Retrieval of Text).

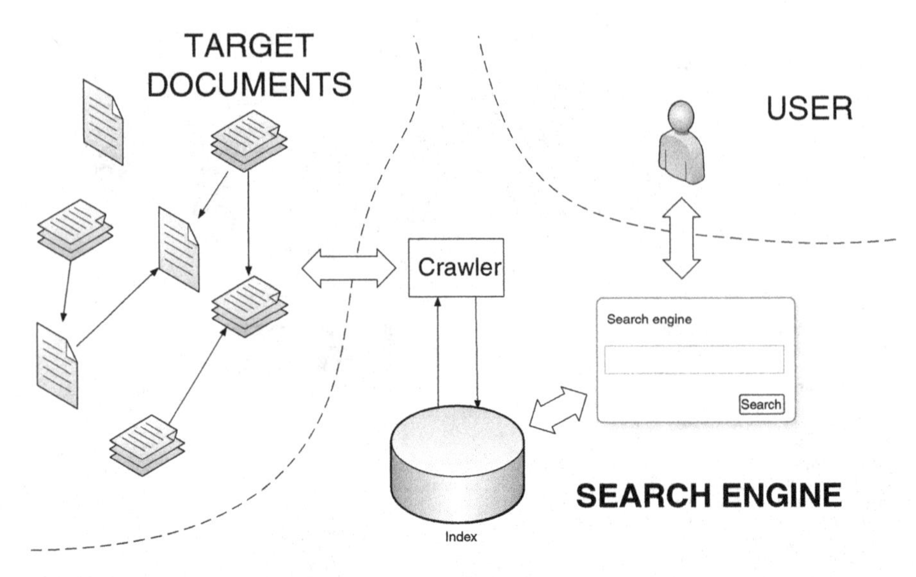

Fig. 4.5 General information retrieval system schema

This is identified as one of the first information retrieval systems. Nevertheless, it used TF-IDF vector-based index and a simple algorithm for query transformation. As Professor Salton contributed a lot to the research field, every 3 years, the *Gerard Salton Award* is given to an individual who makes significant contribution to information retrieval.

The last revolution to information extraction happened in 1990s with the development of the Internet. Before the Internet, most people were trying to get information from other people than from information retrieval system. However, in the last decade, most of the traditional information retrieval systems transformed into search engines over the World Wide Web. Newer studies therefore found that more than 90 % of Internet users use search engines to get everyday information [30].

The efficiency of an information retrieval system depends not only on the system but also on a user. A user needs to correctly define a relevant search query for his information need. According to the need, we identify the following information retrieval system types [32]: (1) *navigational* (user needs to directly access the target document or web page), (2) *informational* (user wants to get information that is contained in an unstructured document), and (3) *transactional* (user needs to access a specific resource in order to make a transaction) (e.g., shopping, document upload, watching a video).

In Fig. 4.5, we show a general information retrieval system (i.e., search engine) architecture. It's main parts consist of *user interface*, *crawler,* and *index*. User interface must provide an intuitive way for the user to express his information need and appropriate design to show relevant results. At this step, modern search engines additionally preprocess the query using query reformulation or expansion techniques to provide best results (e.g., expansion of a query using synonyms).

Table 4.1 A collection of five raw documents

Doc ID	Content
1	BRCA1 encodes a nuclear phosphoprotein and maintains genomic stability
2	BRCA1 acts as a tumor suppressor
3	The encoded protein combines with other tumor suppressors
4	Mutations in BRCA1 are responsible for approximately 40 % of breast cancers
5	A related pseudogene, which is also located on chromosome 17, has been identified

Table 4.2 Example of preprocessed documents

Doc ID	Content
1	brca1 encode nuclear phosphoprotein maintain genome stability
2	brca1 act tumor suppressor
3	encode protein combine tumor suppressor
4	mutations brca1 responsible approximate 40 % breast cancer
5	related pseudogene locate chromosome 17 identify

Crawler is the essential part of a search engine that retrieves new data from the web or other information pool that a search engine focuses on. Generally, it is built as a set of independent agents that automatically look for new documents or identify their changes. Lastly, the most important part is the index. It contains crawled documents that are stored in a specialized format for quick retrieval. In contrast, some search engines – metasearch engines – do not have their own infrastructure, but just use a number of other information retrieval systems and then show the user combined and re-ranked results.

4.4.1 Data Representation

Typical procedure for adding a new document into a search engine is (1) document retrieval, then (2) document processing, and lastly (3) adding a document into a local index structure. The part of document processing is especially important as it cleans the document and selects the most relevant parts that represent the document, which are further used at indexing. The preprocessing methods include stemming (i.e., removing common endings: *ants/ant*), lemmatization (i.e., semantic transformation of inflected words into a basic form: *better/good*), stop words removal (i.e., removing of highly frequent and therefore unimportant words: *in*, *an*, *the*, *a*), or case folding (i.e., representing words in lower case: *Slovenia/slovenia*).

For example, in Table 4.1, we show a collection of five raw documents. Before storing the documents into information retrieval system index, we perform preprocessing on the documents. After preprocessing, we get cleaned documents with words in basic forms (Table 4.2). These documents representations are now appropriate to be indexed by specific indexing model.

In the following sections, we review some of the models that are mostly used for indexing biomedical data resources (see Sect. 4.3).

Table 4.3 Example of index using posting lists

Word	Posting list
brca1	1, 2, 4
tumor	2, 3
suppressor	2, 3
cancer	4
...	...

4.4.2 Models

Information retrieval model defines how documents D are represented (i.e., indexed) within the information retrieval system.

Most of the models index documents by the containing words. For example, *bag-of-words* approach represents each document as an unordered collection of words. Let $V = \{t_1, t_2, \ldots, t_{|V|}\}$ be a collection of all known words. For every word t_i in a document $d_j \in D$, we define $w_{ij} \geq 0$ as a weight of word within the document. Therefore, we can model each document as a vector:

$$d_j = \left(w_{1j}, w_{2j}, \ldots, w_{|V|j}\right).$$

When the user enters a query, it is transformed using the same principles as documents and then compared to all document vectors within the index. Further, ranking is performed to select best matching documents and then a limited set is returned as a result.

Boolean model is one of the oldest and simplest document retrieval techniques. The weights in documents have only binary values and therefore the results ranking is not possible for this model type:

$$w_{ij} = \begin{cases} 1, & t_i \in d_j \\ 0, & |otherwise \end{cases}$$

The query is formulated by words that can be combined using logical operators *AND*, *OR*, or *NOT*. The system then returns all the documents that comply to the query. For example, a result for query "BRCA1 AND tumor" (logical operators are usually implicitly added by the system) would return the document 2.

If we had a large number of documents to index, it would be very time consuming to compare the query to every internal document. Therefore, we should build inverted indexes or posting lists. An example of posting list is presented in Table 4.3. For every word in vocabulary V, it contains a list of all documents where a specific word occurs. Using this technique, we quickly retrieve the lists for every word of the query and calculate the intersection. All documents in the intersection are then returned as a result. For previous query, we would take the intersection of lists for "brca1" and "tumor" and therefore return document 2 as the result.

Nevertheless, the Boolean indexing model is one of the most naive ones, but still a lot of biomedicine resources (e.g., PubMed) depend on it. That is why generally researchers are interested into all documents containing specific words (e.g., protein names) with some constraints (e.g., does not contain BRCA1 gene), so thus there is no need to use sophisticated models in this case.

Vector space model is a more sophisticated but still simple indexing model. It indexes each document as a vector of weights w_{ij} for each word. In comparison to Boolean model, it can return a ranked list of documents ranked by similarity to a user query.

The weight values are most commonly calculated using TF-IDF (Term Frequency–Inverse Document Frequency) scheme. The intuition of TF-IDF is that a word is more important if it occurs in few documents, rather to a number of documents. Let f_{ij} be a number of occurrences of the word t_i in d_j and *term frequency* tf_{ij} it's accompanying normalized value. Let n be a number of documents within the system. The df_i is a number of documents, which contain the word t_i. By computing *inverse document frequency* idf_i, we identify how common or rare is a term across all documents. The final TF-IDF weight w_{ij} is therefore a product of both frequency values:

$$tf_{ij} = \frac{f_{ij}}{max\left\{f_{1j}, f_{2j}, \ldots, f_{|V|j}\right\}},$$

$$idf_i = \log\frac{n}{df_i},$$

$$w_{ij} = \text{TF-IDF}_{ij} = tf_{ij} \times idf_i.$$

The final document similarities are then calculated using cosine similarity, which is a normalized scalar product of two documents (e.g., query q and document d_j) vectors:

$$cosine\_sim\left(\vec{q}, \vec{d_j}\right) = \frac{\vec{q} \cdot \vec{d_j}}{\left|\vec{q}\right|\left|\vec{d_j}\right|}$$

For example, let us take the previous query "BRCA1 tumor" and preprocessed documents from Table 4.2. Document 2, which matches the most, should have the highest similarity value. From the theory or intuition, we know that words which occur many times are not so important. So therefore, the second-ranking document should be document 3 because "tumor" is less frequent than "BRCA1." Full ranking for the query is shown in Table 4.4.

Table 4.4 Vector space model ranking according to TF-IDF cosine similarity

Doc ID	TF-IDF cosine similarity
2	0.49
3	0.29
1	0.07
4	0.06
5	0.0

4.5 Information Extraction

Machine understanding of textual documents has been challenging since early computer era. Information extraction [28] therefore attempts to analyze text and extract its structured semantic contents. The extracted results therefore enable new ways to query, organize, analyze, or visualize data. Firstly, information extraction techniques were used to work only on general domain data, like person and company names. There were also a lot of systems developed that ease web searching by the use of structured data, automatically extract opinions, structurally compare products from unstructured reviews, etc. Recently, the same techniques were adopted in bioinformatics field to extract biological objects (e.g., proteins, genes), their interactions, and experiment results from the vast biomedical databases. Since biological names are very different from classical ones, this task is also responsible for conducting new research in the information extraction field.

The most important information extraction tasks consist of named entity recognition (e.g., extraction of person names, locations, organizations), relationship extraction (i.e., identification of relationships among entities), and coreference resolution (i.e., clustering of mentions to an entity). Thus, information extraction techniques have roots in the natural language processing community, as text was one of the first and still is highly important unstructured information source. Nevertheless, the term information extraction is used also to extract structured data from arbitrary source types like videos, images, and sound.

The early information extraction research was strongly driven by Message Understanding Conference (MUC) competitions from 1987 (MUC-1) to 1997 (MUC-7) [44]. Initial competitions focused merely into named entity recognition. Later, newer important competitions, supporting more tasks, and containing larger data corpuses emerged, like Automatic Content Extraction (ACE) [45], Semantic Evaluation (SemEval) [46], and Conference on Natural Language Learning (CoNLL) Shared Tasks [47].

With the advent of Semantic Web, ontologies gained importance, and researchers try to represent all knowledge using an ontology (see Sect. 4.3.2). As ontology is a highly adaptable schema that defines entities, their attributes, and connections among them, it is an ideal model to represent extracted data. Therefore, a special subfield, ontology-based information extraction emerged, where ontologies are used for

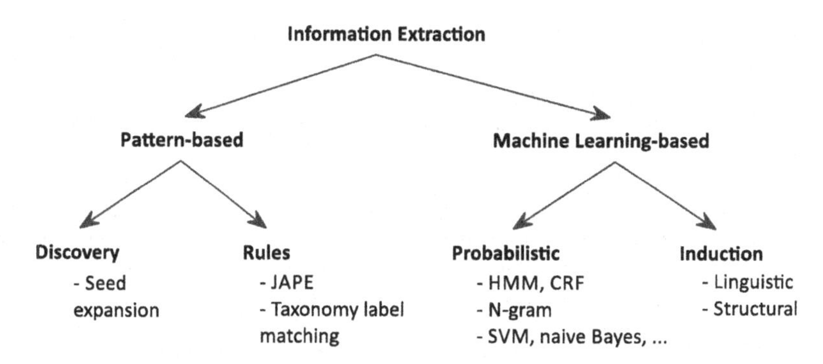

Fig. 4.6 Information extraction methods classification

guiding the whole information extraction process, represent data schema, or are themselves a result of the process.

We categorize the methods for information extraction into two main dimensions, as presented in Fig. 4.6. These are pattern-based and machine-learning-based approaches. Pattern-based methods extract data using some template that can be automatically extracted from seed examples (e.g., a set of document containing identified phone numbers is given as input. Then, the algorithm learns some simple rules on how phone numbers occur in text – e.g., text before and after the number) or rules that are created by the user (e.g., a user creates a rule that if a set of numbers appear directly after the word "tel:", it is a phone number). On the other hand, machine-learning-based approaches are typically given a tagged data corpus, a set of features, and then a training algorithm learns the specific information extraction model. These two approaches further divide into induction and probabilistic methods. The induction methods generally build models on linguistic or structural data. Recently, the probabilistic methods have become successful and therefore researchers use all types of data mining methods to learn the information extraction models.

4.5.1 Data Representation

It is essential for information extraction tasks to transform the unstructured data into representation that our model can understand and therefore infer appropriate results.

Table 4.5 shows an example of unstructured source and accompanying tagged data for named entity recognition, relationship extraction, and coreference resolution. We can observe that named entities are typically noun phrases of one or more tokens (i.e., protein names within *PRO* tags). Relationships are similarly associated

Table 4.5 Example of information extraction dataset from BioNLP 2013 gene regulation ontology shared task

Activin binds directly ActR-IIB, and this complex associates with ActR-IB. In the resulting complex, ActR-IB becomes hyperphosphorylated, and this requires the kinase activity of ActR-IIB
The unstructured source
<PRO> **Activin**</PRO> binds directly <PRO> **ActR-IIB**</PRO>, and this complex associates with <PRO> **ActR-IB**</PRO>. In the resulting complex, <PRO> **ActR-IB**</PRO> becomes hyperphosphorylated, and this requires the kinase activity of <PRO> **ActR-IIB**</PRO>
Annotated entities
Activin <REL> **binds**</REL> directly ActR-IIB, and this complex <REL> **associates**</REL> with ActR-IB. In the resulting complex, ActR-IB <REL> **becomes hyperphosphorylated** </REL>, and this <REL> **requires the kinase activity**</REL> of ActR-IIB
Annotated relationships
<1> **Activin**</1> binds directly <2> **ActR-IIB**</2>, and <3> **this complex**</3> associates with <4> **ActR-IB**</4>. In <3> **the resulting complex**</3>, <4> **ActR-IB**</4> becomes hyperphosphorylated, and this requires the kinase activity of <2> **ActR-IIB**</2>
Annotated coreferences

with verb phrases (i.e., words within *REL* tags). Usually after identifying relationship verb phrase, it needs to be classified into a particular relationship type because in the text it can be stated using multiple ways. Lastly, coreferences tag each entity mention with a number. Mentions that have the same number therefore represent the same underlying entity. Mentions are all references to an entity and can be of named (e.g., *John Doe*), nominal (e.g., *the guy with the glasses*), or pronominal (e.g., *he*) type.

Most often, the text document is represented as a set of sequences, where each sequence represents one sentence. Furthermore, each sequence consists of a list of observable tokens, which are words or symbols that form the sentence. Therefore, additional modifiers or target classes (e.g., named entities) can be stored as attributes for each token. The most widespread sequence training data representation is BIO (i.e., begin/inside/outside) notation. There, each token is labeled with one of the following: (1) *B*, when it is the first word in a phrase (e.g., *B-PRO*); (2) *I*, when it is second or later word within a phrase (e.g., *I-PRO*); and (3) *O*, when it is unclassified word or token. For example, a part of previous example "*Activin binds directly ActR-IIB*" would be tagged as "*B-PRO B-REL O B-PRO O*."

Additional attributes are usually added to the dataset in the phase of text preprocessing. In this phase, researchers would like to enrich the dataset in order to use richer feature functions during the information extraction algorithm execution. Some examples of such attributes are part-of-speech tags, lemmas, parse trees, and synonyms. Part-of-speech tags mark a role of a word, based on its definition or context [57]. They are only an extension of a simplified form of identification of words as nouns, verbs, adjectives, adverbs, etc., that are taught in elementary school. Parse trees are another commonly used type of attributes, which represent the syntactic structure and dependencies between parts of the sentence. In Fig. 4.7, we show an example parse tree of the first example sentence

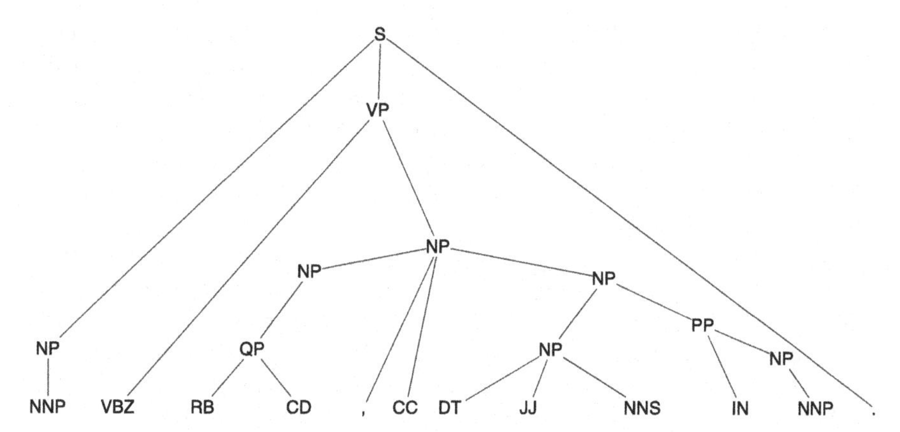

Fig. 4.7 Dependency parse tree for the first sentence in Table 4.5

from Table 4.5. From the figure, we can see the sentence structure with some dependencies. Each node within a tree defines a constituent, which is a word or a group of words that function as a single unit within a hierarchical structure. For example, the node *QP* defines a constituent "directly ActR-IIB."

Formally, we represent the sequence of observable tokens as $\overline{x}^{k_i} = \{\overline{x}_1^{k_i}, \overline{x}_2^{k_i}, \ldots, \overline{x}_n^{k_i}\}$. Index k_i stands for token type that can be a word or additional attribute such as part-of-speech tag, lemma, parse tree, and constituent. Each observable sequence is associated with corresponding target labeling sequence $\overline{y}^{l_i}$, where $l_i \in \{EE, REL, COR\}$ identifies entity extraction, relationship extraction, and coreference resolution label types. An example dataset containing two sentences is shown in Table 4.5. Sequences $\overline{x}_1^{k_1}$ and $\overline{x}_2^{k_1}$ represent input words from both sentences. Respectively, sequences $\overline{y}_1^{EE,REL,COR}$ and $\overline{y}_2^{EE,REL,COR}$ can therefore represent target BIO notation labels for all three tasks. When we additionally introduce lemmatized text "*Activin bind direct ActR-IIB, and this complex associate with ActR-IB. In the result complex, ActR-IB become hyperphosphorylate, and this require the kinase activity of ActR-IIB*," we store it into accompanying sequences $\overline{x}_1^{k_2}$ and $\overline{x}_2^{k_2}$ which are used to enrich data to improve prediction capabilities.

4.5.2 Methods for Extraction

Rule-based methods are one of the oldest and easiest to implement, but mostly do not achieve best performances (some tools and frameworks support their own ways to design user-defined patterns – Sect. 4.4). Therefore, in this section, we focus on machine-learning methods for information extraction, especially on sequence classifier methods.

Traditionally, standard classifiers, like multinomial naive Bayes, SVM, decision trees, random forest, and neural networks [31], were used to extract information. Therefore, the data is needed to be transformed for direct classification. For example, a named entity classifier decided for each word or phrase if it represents an entity or not. Similarly, for relationships, a classifier was trained to decide if there is a relationship between the two mentions within a sentence. Lastly, coreference resolution problem was transformed into a pairwise decision problem and classifier decided for every pair of mentions when the two are coreferent or not [40–42].

However, typically in information extraction, tokens are rarely independent of each other. This observation therefore led to a number of models that enable capturing dependencies between the labels. One of the first popular algorithms were hidden Markov models [54], and later maximum entropy Markov models [55], conditional Markov models [56], and the state-of-the-art methods for sequence classification Conditional Random Fields [39] were developed. In contrast to previous direct methods, these try to predict the most probable labeling sequence for the whole sequence and not only for one token. Next, we elaborate Conditional Random Fields.

Conditional Random Fields (CRF) [39] is a discriminative model that estimates the joint distribution $p(\overline{y}|\overline{x}, w)$ over the target sequence $\overline{y}$ conditioned on the observed sequence $\overline{x}$. For example, i th sentence x_i is represented as a sequence of words $\overline{x}_i$ with additional corresponding sequences that represent attribute values such as part-of-speech tags $\overline{x}_i^{k_2}$, lemmas $\overline{x}_i^{k_3}$, relationships $\overline{x}_i^{k_4}$, or other observable features $\overline{x}_i^{k_j}$. These observable values and attributes are used by feature functions f_l that are weighted during CRF training in order to model target sequence $\overline{y}_i$. Target sequence contains labels that we would like to automatically label. For example, at information extraction tasks, these are named entity tags (e.g., *PRO*, *GEN*), relationships tags (e.g., *INTERACTS-WITH*, *INHIBITS*), or coreference entity numbers.

Training a CRF is therefore finding a weight vector w that predicts most probable sequence $\hat{y}$ given $\overline{x}$. Hence,

$$\hat{y} = \text{argmax}_{\overline{y}} p(\overline{y}|\overline{x}, w),$$

where the conditional distribution equals

$$p(\overline{y}|\overline{x}, w) = \frac{exp\left[\sum_{l=1}^{m} w_l \sum_{i=1}^{len(\overline{x})} f_l(\overline{y}, \overline{x}, i)\right]}{C(\overline{x}, w)}.$$

Here, m is the number of feature functions and $C(\overline{x}, w)$ is a normalization constant computed over all possible sequences $\overline{y}$.

An example of feature function is as follows:

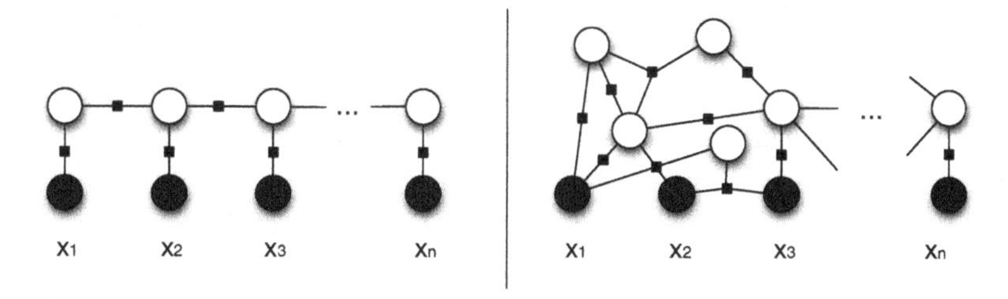

Fig. 4.8 Linear-chain CRF model on the *left* and an example of arbitrary structured CRF model on the *right*

$$f_l(\overline{y}, \overline{x}, i) = \mathit{if}\ (\overline{y}_i == PRO \vee \overline{x}_{i-1} == \text{protein})\ \mathit{then}$$

$$\mathit{return\ True}$$

$$\mathit{else}$$

$$\mathit{return\ False}$$

The structure of a CRF defines how dependencies with target labels are modeled. General *graphical model* can depend on many labels, and therefore it is not tractable to train it or inference using it without using complex approximation algorithms. On the other hand, linear-chain CRF is a simple bigram CRF model and therefore easier and faster to train and label. In Fig. 4.8, a simple comparison between the two is presented.

The selection of informative feature functions is the main source of increase of precision and recall (see Sect. 4.5.3) when training machine-learning classifiers. CRF feature functions are usually implemented as templates and then final features are generated by scanning the entire training data. In natural language processing tasks, it is common to have few thousands or more features. Furthermore, with the development of well-designed feature functions also a rather simple pairwise algorithm can achieve state-of-the-art results. Some researchers like to categorize feature functions, so therefore we briefly introduce some main categories and give few feature function examples per category:

Preprocessing. These feature functions use standard preprocessing labelings. Generally, before employing any further text analysis, we first detect sentences, tokens, enrich data with lemmas, part-of-speech (POS) tags, chunks, parse trees, etc. These additional labelings are then used within feature functions, for example, *do POS tags match, distribution of POS tags, length between mentions in a parse tree, depth of mention within a parse tree, parse tree parent value match, and is pronoun of demonstrative/definitive type*.

Location. Sometimes it is important to know where does a specific word or phrase reside. Location feature functions deal with a phrase location compared to the whole document, sentence, or other phrases. Some examples of feature functions are *sentence/mention/token distance between the two, is first/last mention*, and *are within the same sentence*.

String Shape. Input data is represented as word phrases and thus we are interested if they share some property. These feature functions depend on raw input text, like, for example, *starts with upper case*, *do both start with upper case*, *prefix/postfix/whole of left/right on distance x match*, *do exactly match*, *is appositive of another*, *is prefix/suffix*, *edit distance similarity*, *do comply to Hearst rules*, *is in quoted string*, *contains substring*, *contains head/extend words*, and *length difference*.

Semantic. This class of feature functions captures semantic relationships. Mostly, additional semantic sources like WordNet [43], specialized lexicons, semantic gazetteer lists, or ontologies are used. For example, some of the feature functions are *do named entity types match*, *are relationships tags between mentions connected*, *do both mentions represent an animate object*, *do mentions agree on gender/number*, *do both mentions speak* (taking context words into account), *heuristic if a mention is an alias of another*, *mention word sense*, *do mentions share same WordNet synset*, *is one hypernym/hyponym/synonym of another*, *pronoun types within a mention*, and *mention type*.

There is no silver bullet model that would outperform all others on a specific task. Therefore, current state-of-the-art approaches use a hybrid approach combining rule-based techniques with an ensemble of machine-learning models. The biggest advantage of using rules seems to capture first the most confident extractions before others.

4.5.3 Evaluation Metrics

Most commonly, the performance of named entity recognition and relationship extraction is evaluated with the standard *precision*, *recall*, and *F-score* values. Precision measures the ratio of correctly identified objects and those that were extracted. Recall on the other side is the ratio of correctly identified objects and all correct objects in a dataset.

$$Precision = \frac{number\ of\ correctly\ extracted\ objects}{number\ of\ extracted\ object}$$

$$Recall = \frac{number\ of\ correctly\ extracted\ objects}{number\ of\ correct\ objects\ in\ a\ dataset}$$

We can always achieve 100 % recall by extracting every possible object from the dataset, but in contrast, precision value will be low. The function of recall is therefore monotonically increasing with the number of extracted objects. On the other hand, the more objects we extract, the more possible it is to make an error and thus precision value may be lower. To take both measures into account, the measure F-score was proposed:

$$F_\beta = \left(1 + \beta^2\right) \frac{Precision \times Recall}{\beta^2 \times Precision + Recall}.$$

In general, F_1 score is used to evaluate the systems, which is a harmonic of precision and recall. The final F_1 score is therefore closer to the lower value between precision and recall.

Coreference resolution is a slightly different approach to object extraction. The goal is to achieve best mention clustering and therefore appropriate measures must be proposed. In the early 1990s, a graph-based scoring algorithm was used that produced very unintuitive results (these old measures were similar to entity resolution evaluation techniques – see Sect. 4.6) [48, 49]. Later some more metrics were proposed, but there is still no general agreement which one to use for full evaluation. Therefore, at newer shared tasks, an arithmetic mean of some standard measures is calculated to compare the systems. The most commonly used are the following:

MUC [50]. The key idea to developing a new MUC measure was to give an intuitive explanation of results for coreference resolution systems. The measure was developed for evaluating competing systems at shared task in the sixth Message Understanding Conference (MUC-6). It is a *link-based* metric (focuses on pairs of mentions) and most widely used. MUC precision counts precision errors by computing the minimum number of links that need to be added in order to connect all the mentions referring to an entity. Recall, on the other hand, measures how many of the links must be removed that no two mentions referring to different entities are connected in the graph. Thus, MUC metric prefers systems having more mentions per entity (e.g., a system that creates a single entity over all mentions will receive 100 % recall along with high precision). Yet another MUC downside is that it ignores entities with only one mention (singleton entities).

BCubed [51]. The BCubed metric tries to solve MUC shortcomings by focusing on mentions and measures the overlap between predicted and true clusters. It computes recall and precision values for each mention. If k is the key entity and r response entity containing mention m, the recall for mention m is calculated as $|k \cap r|/|k|$ and precision for the same mention as $|k \cap r|/|r|$. This score has the advantage to measure the impact of singleton entities and gives more weight to the splitting or merging of larger entities. It also gives equal weight to all types of entities and mentions.

CEAF [52]. CEAF metric tries to solve both MUC and BCubed shortcomings. Furthermore, another goal was to achieve better interpretability. The result therefore reflects the percentage of correctly recognized entities. It is an *entity-based* (a mention-based version also exists) metric that tries to match response entity with at most one key entity. This is maximum bipartite matching problem that can be solved by Kuhn-Munkres algorithm. Here, recall is the $(total\ similarity)/|k|$, and precision is the $(total\ similarity)/|r|$.

BLANC [53]. The most recently proposed BiLateral Assessment of Noun Phrase Coreference (BLANC) uses a method of Rand index to evaluate coreference problem. Bilateral comes from the idea that measure takes into account coreferent and also non-coreferent links. Previous metrics reward links to entities equally no matter what is its size, but in principle, assigning it to a large one is making a larger number of equally important pairwise decisions. BLANC should model coreference resolution better since it assigns equal importance to every decision of coreferentiality. To calculate the final score, we first compute precision and recall separately for coreferent and non-coreferent links (some boundary cases that may contain only singletons or a single set are defined separately). Then, BLANC score is an average of both F-scores.

4.6 Data Integration

Heterogeneous data matching and merging is due to increasing amount of linked and open (online) data sources rapidly becoming a common need in various fields. Different scenarios demand for analyzing heterogeneous datasets collectively, enriching data with some online data source or reducing redundancy among datasets by merging them into one. Literature provides several state-of-the-art approaches for matching and merging, although there is a lack of general solutions combining different dimensions arising during the matching and merging execution.

Data sources commonly include not only network data (e.g., entities and connections between them) but also data with semantics (e.g., entities are classified as genes, proteins). Thus, a state-of-the-art solution should employ semantically elevated algorithms (i.e., algorithms that can process data with semantics according to an ontology) to fully exploit the data at hand. In particular, the architecture should support all types and formats of data and provide appropriate data for each algorithm. As algorithms favor different representations and levels of semantics behind the data, architecture should be structured appropriately.

Due to different origin of (heterogeneous) data sources, the trustworthiness (or accuracy) of their data can often be questionable. Especially, when many such datasets are merged, the results are likely to be inexact. A common approach for dealing with data sources that provide untrustworthy or conflicting statements is the use of trust management systems and techniques. Thus, matching and merging should be advanced to a trust-aware level to jointly optimize trustworthiness of data and accuracy of matching or merging. Such collective optimization can significantly improve over other approaches.

4.6.1 Data Representation

Most natural representation of any related data are *networks* [33]. They are based upon mathematical objects called *graphs*. Informally speaking, graph consists of a

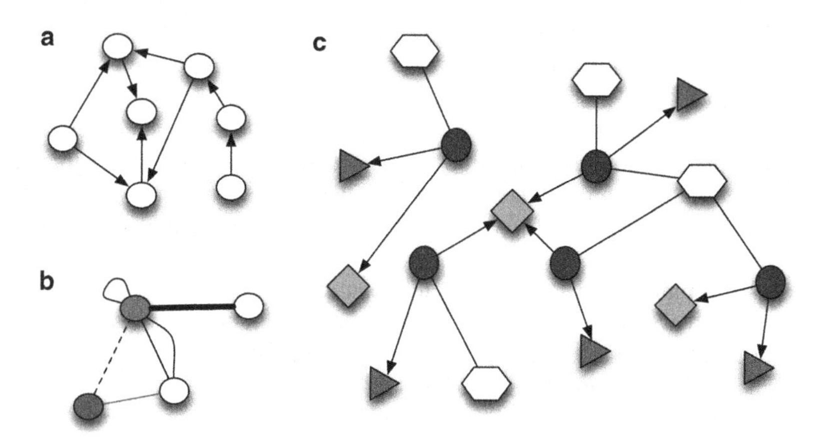

Fig. 4.9 Integration data representation as a network: (**a**) directed graph, (**b**) labeled undirected multigraph, and (**c**) network representing a group of related proteins and genes

collection of points, called *vertices*, and links between these points, called *edges*. An example of network representation is shown in Fig. 4.9. Let V_N, E_N be a set of vertices, edges for some graph N, respectively. We define graph N as $N = (V_N, E_N)$ where

$$V_N = \{v_1, v_2, \ldots, v_n\},$$

$$E_N \subseteq \{\{v_i, v_j\} |\ v_i, v_j \in V_N \wedge i < j\}.$$

Edges are sets of vertices; hence, they are not directed (*undirected graph*). In the case of directed graphs, the edge equation reformulates into

$$E_N \subseteq \{(v_i, v_j) |\ v_i, v_j \in V_N \wedge i \neq j\},$$

where (v_i,v_j) is an edge from v_i to v_j. The definition can be further generalized by allowing multiple edges between two vertices and *loops* (edges that connect vertices with themselves). Such graphs are called *multigraphs* (e.g., graph (b) in Fig. 4.9).

In practical biomedicine applications, we commonly strive to store some additional information along with the vertices and edges, similarly to additional attributes at information extraction (see Sect. 4.5). Therefore, we formally define *labels* or *weights* for each node and edge in the graph – they represent a set of properties that can also be described using two attribute functions:

$$A_{V_N} : V_N \rightarrow \Sigma_1^{V_N} \times \Sigma_2^{V_N} \times \ldots,$$

$$A_{E_N} : E_N \rightarrow \Sigma_1^{E_N} \times \Sigma_2^{E_N} \times \ldots,$$

$A_N = (A_{V_N}, A_{E_N})$, where $\Sigma_i^{V_N}$, $\Sigma_i^{E_N}$ are sets of all possible vertex, edge attribute values, respectively.

Generally, most useful attributes are semantic data and trust values. Semantic data defines connection to ontology-based features. Nodes that represent entities are connected to concepts and edges can be of object or data property types. Using ontology attributes, we can form an overlay network and therefore use more sophisticated similarity measures for merging execution. Trust values are especially important at redundancy elimination step because the system needs to select the most appropriate representative value from all the values within merged data.

Trust is a complex psychological-sociological phenomenon. Despite of, people use term trust in everyday life widely and with very different meanings [34]. In the context of computer networks, trust is modeled as a related data between entities. Formally, we define a trust between related data as

$$\omega_E : E \times E \rightarrow \Sigma^E$$

where E is a set of entities and Σ^E a set of all possible numerical or descriptive trust values. ω_E thus represents one entity's attitude towards another and is used to model trust(worthiness) of all entities in E. To this end, different trust modeling methodologies and systems can be employed, from qualitative to quantitative [34–36].

4.6.2 General Data Integration Framework

In Fig. 4.10, we show a general end-to-end integration framework [37]. At first, the data needs to be preprocessed and transformed into network-based format as described in Sect. 4.6.1. Then, attribute resolution is performed, followed by entity resolution (i.e., merging). Entity resolution module takes a network with duplicated nodes as input and returns merged network, where new nodes (i.e., clusters) consist of a set of old duplicated nodes. The attribute resolution technique uses the same approach as entity resolution but works on attributes from different data sources and identifies which attributes represent the same data. Lastly, redundancy elimination step selects one representative value from each cluster and returns cleaned network. Post-processing step transforms the result network into selected format (e.g., attribute-value pairs, ontology-based) and returns it as a final result of integration execution.

Entity Resolution. A naive approach for entity resolution is simple pairwise comparison of attribute values among different entities. Although such approach could be already sufficient for flat data, this is not the case for network data, as the approach completely discards related data between the entities. For instance, when two entities are related to similar entities, they are more likely to represent the same entity. However, only the attributes of the related entities resolve to the same entities when their related entities resolve to not only similar but the same

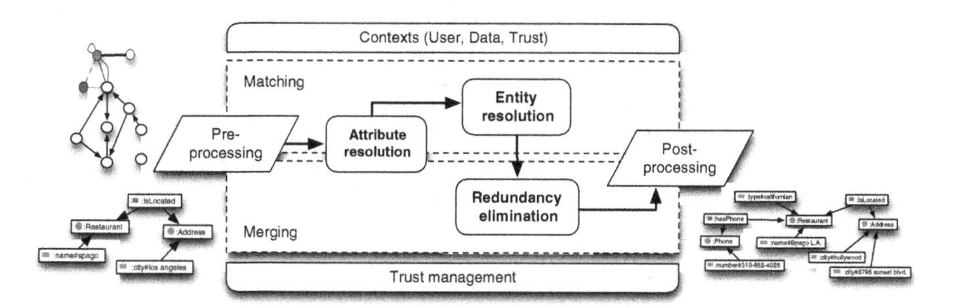

Fig. 4.10 General end-to-end integration framework

Table 4.6 Collective entity resolution algorithm

	Collective entity resolution algorithm
1	Initialize clusters as $C = \{\{k\} \mid k \in K\}$
2	Initialize priority queue $Q = \varnothing$
3	**for** $c_i, c_j \in C$ **and** $sim(c_i, c_j) \geq \theta_S$ **do**
4	$Q.\ insert(sim(c_i, c_j), c_i, c_j)$
5	**end for**
6	**while** $Q \neq \varnothing$ **do**
7	$(sim(c_i, c_j), c_i, c_j) \leftarrow Q.\ pop()$
8	**if** $sim(c_i, c_j) < \theta_S$ **then**
9	**return** C
10	**end if**
11	$C \leftarrow C - \{c_i, c_j\} \cup \{c_i \cup c_j\}$
12	**for** $(sim(c_x, c_k), c_x, c_k) \in Q$ **and** $x \in \{i,j\}$ **do**
13	Q.remove $(sim(c_x, c_k), c_x, c_k)$
14	**end for**
15	**for** $c_k \in C$ **and** $sim(c_i \cup c_j, c_k) \geq \theta_S$ **do**
16	Q.insert $(sim(c_i \cup c_j, c_k), c_i \cup c_j, c_k)$
17	**end for**
18	**for** $c_n \in nbr(c_i \cup c_j)$ **do**
19	**for** $c_k \in C$ **and** $sim(c_n, c_k) \geq \theta_S$ **do**
20	Q.insert$(sim(c_n, c_k), c_n, c_k)$
21	**end for**
22	**end for**
23	**end while**
24	**return** C

entities. An approach that uses information, and thus resolves entities altogether, is denoted *collective* entity resolution algorithm.

As an example, we show a state-of-the-art collective data clustering algorithm, proposed by Bhattacharya and Getoor [38]. The algorithm (Table 4.6) is actually a greedy agglomerative clustering. Entities are represented as a group of clusters C, where each cluster represents a set of entities that resolve to the same entity.

Table 4.7 Redundancy elimination algorithm

	Redundancy elimination algorithm
1	Initialize merged cluster nodes K^C
2	**for** $c \in C$ **and** $a \in A$ **do**
3	$k^c.a = \underset{v}{\text{argmax}} \prod_{k \in c \wedge k.a = v} T(k.a) \prod_{k \in c \wedge k.a \neq v} (1 - T(k.a))$
4	**end for**
5	**return** K^C

At the beginning, each entity resides in a separate cluster. Then at each step, the algorithm merges two clusters in C that are most likely to represent the same entity. During the algorithm, similarity of clusters is computed using a joint similarity measure, combining attribute, and related data similarity. First is a basic pairwise comparison of attribute values, while second introduces related information into the computation of similarity (i.e., data accessible using cluster neighbors in a network).

The algorithm (Table 4.6) first initializes clusters C and priority queue of similarities Q, considering the current set of clusters (lines 1–5). Each cluster represents at most one entity as it is composed out of a single knowledge chunk. Algorithm then, at each iteration, retrieves currently the most similar clusters and merges them (i.e., matching of resolved entities), when their similarity is greater than threshold θ_S (lines 7–11), which represents minimum similarity for two clusters that are considered to represent the same entities. In line 11, clusters are simply concatenated. Next, lines 12–17 update similarities in the priority queue Q, and lines 18–22 insert (or update) also neighbors' similarities (required due to related similarity measure). When the algorithm terminates, clusters C represent a sets of data resolved to the same entity. These clusters are then used to merge data at the redundancy elimination step.

After the entities have been resolved by entity resolution, the next step is to eliminate the redundancy and merge the data. Let $c \in C$ be a cluster representing some entity, $k_1, k_2, \ldots, k_n \in c$ be its merged references, and $k^c \in K^C$ be the merged data within cluster. Furthermore, for some attribute $a \in A$, we have precalculated values per data source. The algorithm (Table 4.7) first initializes merged network K^C. Then for each attribute $k^c. a$, it finds the most probable value among all given references k_i within cluster c (line 3). When the algorithm unfolds, K^C represents a merged dataset with resolved entities and eliminated redundancy.

In Fig. 4.11, we show example of data integration execution. First part represents input data in form of networks from three different data sources. Secondly, the result of entity resolution contains merged network in which some nodes contain more values (from each data source). Lastly, after redundancy elimination step, the final result contains a cleaned network and the most appropriate value for each node.

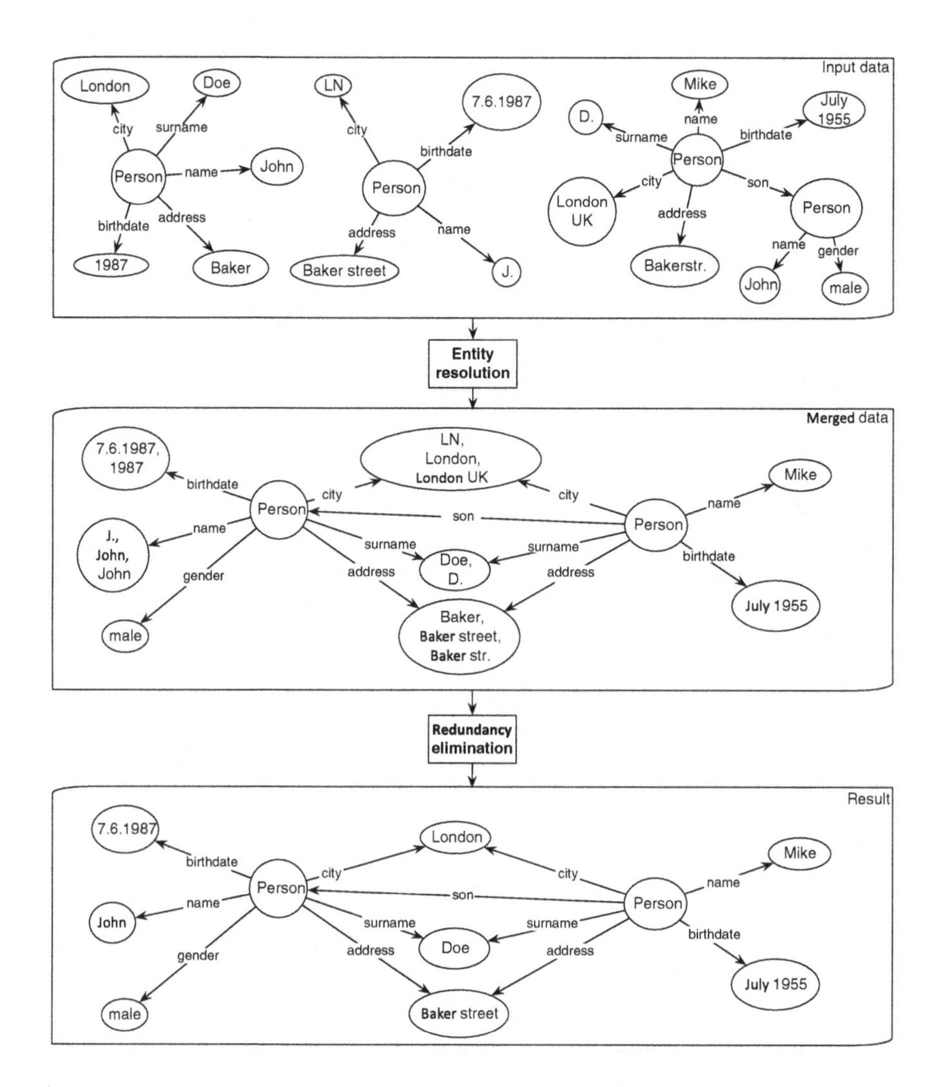

Fig. 4.11 Example of data integration execution on person domain

4.7 Summary

Many medical applications and current ongoing medical research depend on text mining techniques. A lot of research work has already been done, and therefore in this chapter, we have overviewed some methods that enable researchers to automatically retrieve, extract, and integrate unstructured medical data. Due to increasing number of unstructured documents, the automatic text mining methods ease access to relevant data, already conducted research along with its results, and save money by trying to eliminate repeated research experiments.

In the last decade, the text mining field has been generally fast evolving, and still, there is a lot of research to be done. In information retrieval, biomedical language resources typically use simple query models, which seem sufficient when enough of relevant data is extracted. Information extraction is currently receiving a lot of attention because researchers are trying to adapt techniques from other domains to work on biomedical data. Further, these techniques are essentials for automatic research texts processing and extraction of findings from research literature. Lastly, also very important topic of data integration still needs to improve models to merge data and select representative values. The latter is especially important as a reference to the same entity can be represented using many different forms.

References

1. Nguyen NLT, Kim JD, Miwa M et al (2012) Improving protein coreference resolution by simple semantic classification. BMC bioinformatics 13:304–325
2. Cunningham H, Maynard D, Bontcheva K et al (2011) Text Processing with GATE (Version 6). University of Sheffield Department of Computer Science, Sheffield
3. Cunningham H, Tablan V, Roberts A, Bontcheva K (2013) Getting More Out of Biomedical Documents with GATE's Full Lifecycle Open Source Text Analytics. PLoS Computational Biology 9:1–16
4. Ferrucci D, Lally A (2004) UIMA: an architectural approach to unstructured information processing in the corporate research environment. Natural Language Engineering 10:327–348
5. Toutanova K, Klein D, Manning C et al (2011) Stanford Core NLP. The Stanford Natural Language Processing Group. http://nlp.stanford.edu/software/corenlp.shtml. Accessed 20 March 2013
6. Hall D, Ramage D (2013) Breeze. Berkeley NLP Group. http://www.scalanlp.org. Accessed 20 March 2013
7. Kottmann J, Margulies B, Ingersoll G et al (2010) Apache OpenNLP. The Apache Software Foundation. http://opennlp.apache.org. Accessed 20 March 2013
8. Bird S, Loper E, Klein E (2009) Natural Language Processing with Python. O'Reilly Media, Sebastopol
9. Gamalo P (2009) DepPattern. Grupo de Gramatica do Espanol. http://gramatica.usc.es/pln/tools/deppattern.html. Accessed 20 March 2013
10. Padró L, Stanilovsky E (2012) FreeLing 3.0: Towards Wider Multilinguality. Proceedings of the Language Resources and Evaluation Conference. Turkey, Istanbul 2473–2479
11. Björne J, Ginter F, Salakoski T (2012) University of Turku in the BioNLP'11 Shared Task. BMC Bioinformatics, 13:1–13
12. Barnickel T, Weston J, Collobert R et al (2009) Large Scale Application of Neural Network Based Semantic Role Labeling for Automated Relation Extraction from Biomedical Texts. PLoS ONE 4:1–6
13. Szklarczyk D, Franceschini A, Kuhn M et al (2010) The STRING database in 2011: functional interaction networks of proteins, globally integrated and scored. Nucleic Acids Researc 39:561–568
14. Mostafavi S, Ray D, Warde-Farley D et al (2008) GeneMANIA: a real-time multiple association network integration algorithm for predicting gene function. Genome Biology 9:1–15
15. Fontaine JF, Priller F, Barbosa-Silva A, Andrade-Navarro MA (2011) Genie: literature-based gene prioritization at multi genomic scale. Nucleic Acids Research 39:455–461

16. Tsuruoka Y, Tsujii J (2005) Bidirectional Inference with the Easiest-First Strategy for Tagging Sequence Data. Proceedings of Human Language Technology Conference/EMNLP 2005. Vancouver, Canada 467–474
17. Allison JJ, Kiefe CI, Carter J, Centor RM (1999) The art and science of searching MEDLINE to answer clinical questions. Finding the right number of articles. International Journal of Technology Assess in Health Care 15:281–296
18. Hamosh A, Dcott AF, Amberger JS et al (2005) Online Mendelian Inheritance in Man (OMIM), a knowledgebase of human genes and genetic disorders. Nucleic Acids Research 33:514–517
19. Gruber TR (1993) A translation approach to portable ontologies. Knowledge Acquisition 5: 199–220
20. Berners-Lee T, Hendler J, Lassila O (2001) The Semantic Web. Scientific American 284: 28–37
21. Jin-Dong K, Ohta T, Teteisi Y, Tsujii J (2003) GENIA corpus - a semantically annotated corpus for bio-text mining. Bioinformatics 19:180–182
22. Pyysalo S, Ginter F, Heimonen J et al (2007) BioInfer: a corpus for information extraction in the biomedical domain. BMC bioinformatics, 8:50–74
23. Rogers FB (1963) Medical Subject Headings. Bulletin of the Medical Library Association 51:114–116
24. Spackman KA, Campbell KE (1998) Compositional concept representation using SNOMED: towards further convergence of clinical terminologies. Proceedings of the AMIA Symposium. Orlando, Florida 740–744
25. Ashburner M, Ball CA, Blake JA et al (2000) Gene Ontology: tool for the unification of biology. Nature genetics 25:1–25
26. Xie B, Ding Q, Han H, Wu D (2013) miRCancer: a microRNA–cancer association database constructed by text mining on literature. Bioinformatics 29:638–644
27. Manning CD, Raghavan P, Schütze H (2008) Introduction to Information Retrieval. Cambridge University Press, Cambridge
28. Sarawagi S (2008) Information Extraction. Foundations and Trends in Databases 1:261–377
29. Bush V (1945) As We May Think. The Atlantic Monthly 176:101–108
30. Fallows D (2004) The internet and daily life. Pew/Internet and American Life Project. http://www.pewinternet.org/Reports/2004/The-Internet-and-Daily-Life.aspx. Accessed 21 March 2013
31. Witten IH, Frank E (2005) Data Mining: Practical Machine Learning Tools and Techniques (Second Edition). Morgan Kaufmann Publishers, San Francisco
32. Broder A (2002) A taxonomy of web search. ACM SIGIR Forum 36:3–10
33. Newman MEJ (2010) Networks: an introduction. Oxford University Press, Oxford
34. Trček D, Trobec R, Pavešić N, Tasić J (2007) Information systems security and human behaviour. Behaviour and Information Technology 26:113–118
35. Nagy M, Vargas-Vera M, Motta E (2008) Managing conflicting beliefs with fuzzy trust on the semantic web. Proceedings of the Mexican International Conference on Advances in Artificial Intelligence 827–837
36. Richardson M, Agrawal R, Domingos P (2003) Trust management for the semantic web. Proceedings of the International Semantic Web Conference 351–368.
37. Žitnik S, Šubelj L, Lavbič D et al (2013) General Context-Aware Data Matching and Merging Framework. Informatica 24:1–34
38. Bhattacharya I, Getoor L (2007) Collective entity resolution in relational data. ACM Transactions on Knowledge Discovery from Data 1:5–40.
39. Lafferty JD, McCallum A and Pereira FCN. Conditional random fields: Probabilistic models for segmenting and labeling sequence data, Proceedings of the Eighteenth International Conference on Machine Learning, San Francisco: Morgan Kaufmann, 2001, pp. 282–289.
40. Soon WM, Ng HT and Lim DCY. A machine learning approach to coreference resolution of noun phrases, Computational linguistics, 2001, 27: 521–544.

41. Ng V, Cardie C (2002) Improving machine learning approaches to coreference resolution. Proceedings of the 40th Annual Meeting on Association for Computational Linguistics 104–111
42. Bengtson E, Roth D (2008) Understanding the value of features for coreference resolution. Proceedings of the Conference on Empirical Methods in Natural Language Processing 294–303
43. Miller GA (1995) WordNet: A Lexical Database for English. Communications of the ACM 38:39–41
44. Grishman R, Sundheim B (1996) Message understanding conference-6: A brief history. Proceedings of the 16th Conference on Computational Linguistics. Morristown, USA 466–471
45. NIST (1998-present) Automatic Content Extraction (ACE) Program
46. Recasens M, Marquez L, Sapena E et al (2010) Semeval-2010 task 1: Coreference resolution in multiple languages. Proceedings of the 5th International Workshop on Semantic Evaluation. Uppsala, Sweden 1–8
47. Pradhan S, Moschitti A, Xue N, Uryupina O, Zhang Y (2012) CoNLL-2012 Shared Task: Modeling Multilingual Unrestricted Coreference in OntoNotes. Proceedings CoNLL '12 Joint Conference on EMNLP and CoNLL - Shared Task. Pennsylvania, USA 129–135
48. Chincor N (1991) MUC-3 Evaluation metrics. Proceedings of the 3rd conference on Message understanding. Pennsylvania, USA 17–24
49. Chincor N, Sundeheim B (1993) MUC-5 Evaluation metrics. Proceedings of the 5th conference on Message understanding. Pennsylvania, USA 69–78
50. Vilain M, Burger J, Aberdeen J, Connolly D, Hirschman L (1995) A model-theoretic coreference scoring scheme. Proceedings of the sixth conference on Message understanding. Pennsylvania, USA 45–52
51. Bagga A, Baldwin B (1998) Algorithms for scoring coreference chains. The first international conference on language resources and evaluation workshop on linguistics coreference. Pennsylvania, USA 563–566
52. Luo X (2005) On coreference resolution performance metrics. Proceedings of the conference on Human Language Technology and Empirical Methods in Natural Language Processing. Vancouver, Canada 25–32
53. Recasens M, Hovy E (2011) BLANC: Implementing the Rand index for coreference evaluation. Natural Language Engineering 17:485–510
54. Rabiner L (1989) A tutorial on Hidden Markov Models and selected applications in speech recognition. Proceedings of the IEEE 77:257–286
55. McCallum A, Freitag D, Pereira F (2000) Maximum entropy markov models for information extraction and segmentation. Proceedings of the International Conference on Machine Learning. Palo Alto, USA 591–598
56. Klein D, Manning CD (2002) Conditional structure versus conditional estimation in NLP models. Workshop on Empirical Methods in Natural Language Processing. Philadelphia, USA 1–8
57. DeRose SJ (1988) Grammatical category disambiguation by statistical optimization. Computational Linguistics 14:31–39
58. Verspoor KM, Cohn JD, Ravikumar KE, Wall ME (2012) Text Mining Improves Prediction of Protein Functional Sites. PLoS ONE 7:e32171.
59. Park J, Costanzo MC, Balakrishnan R et al (2012) CvManGO, a method for leveraging computational predictions to improve literature-based Gene Ontology annotations. Database, doi:10.1093/database/bas001
60. Krallinger M, Leitner F, Vazquez M et al (2012) How to link ontologies and protein–protein interactions to literature: text-mining approaches and the BioCreative experience. Database, doi:10.1093/database/bas017

Chapter 5
A Primer on Information Theory with Applications to Neuroscience

Felix Effenberger

5.1 Introduction

Neural systems process information. This processing is of fundamental biological importance for all animals and humans alike as its main (if not sole) biological purpose is to ensure the survival of an individual (in the short run) and its species (in the long run) in a given environment by means of perception, cognition, action, and adaption.

Information enters a neural system in form of sensory input representing some aspect of the outside world, perceivable by the sensory modalities present in the system. After processing this information or parts of it, the system may then adjust its state and act according to a perceived change in the environment.

This general model is applicable to very basic acts of cognition as well as to ones requiring higher degrees of cognitive processing. Yet, the underlying principle is the same. Thus measuring, modeling, and (in the long run) understanding information processing in neural systems is of prime importance for the goal of gaining insight to the functioning of neural systems on a theoretical level.

Note that this question is of theoretical and abstract nature so that we take an abstract view on information in what follows. We use Shannon's theory of information [97] as a tool that provides us with a rigid mathematical theory and quantitative measures of information. Using information theory, we will have a conceptual look at information in neural systems. In this context, information theory can provide both explorative and normative views on the processing of

F. Effenberger (✉)
Max-Planck-Institute for Mathematics in the Sciences,
Inselstr. 22, 04103 Leipzig, Germany
e-mail: felix.effenberger@mis.mpg.de

G. Rakocevic et al. (eds.), *Computational Medicine in Data Mining and Modeling*,
DOI 10.1007/978-1-4614-8785-2_5,

information in a neural system as we will see in Sect. 5.6. In some cases, it is even possible to gain insights on the nature of the "neural code," i.e., the way neurons transmit information via their spiking activity.

Information theory was originally used to analyze and optimize man-made communication systems, for which the functioning principles are known. Nonetheless, it was soon realized that the theory could also be used in a broader setting, namely, to gain insight into the functioning of systems for which the underlying principles are far from fully understood, such as neural systems. This was the beginning of the success story of information-theoretic methods in many fields of science such as economics, psychology, biology, chemistry, and physics.

The idea of using information theory to quantitatively assess information processing in neural systems has been around since the 1950s; see the works of Attneave [6], Barlow [9], and Eckhorn and Pöpel [32, 33]. Yet, as information-theoretic analyses are data intensive, these methods were rather heavily restricted by (a) the limited resources of computer memory and computational power available and (b) the limited accuracy and amount of measured data that could be obtained from neural systems (on the single cell as well as at the systems level) at that time. However, given the constant rise in available computing power and the evolution and invention of data acquisition techniques that can be used to obtain data from neural systems (such as magnetoencephalography (MEG), functional magnetic resonance imaging (fMRI), or calcium imaging), information-theoretic analyses of all kinds of biological and neural systems became more and more feasible and could be carried out with greater accuracy and for larger and larger (sub)systems.

Over the last decades such analyses became possible using an average workstation computer, a situation that could only be dreamed of in the 1970s. Additionally, the emergence of new noninvasive data collection methods such as fMRI and MEG that outperform more traditional methods like electroencephalography (EEG) in terms of spatial resolution (fMRI, MEG) or noise levels (MEG) made it possible to even obtain and analyze system-scale data of the human brain in vivo.

The goal of this chapter is to give a short introduction to the fundamentals of information theory and its application to data analysis problems in the neurosciences. And although information-theoretic analyses of neural systems were not often used in order to gain insight on or characterize neural dysfunction so far, this could prove to be a helpful tool in the future.

The chapter is organized as follows. We first talk a bit about the process of modeling in Sect. 5.2 that is fundamental for all what follows as it connects reality with theory. As information theory is fundamentally based on probability theory, following this we give an introduction to the mathematical notions of probabilities, probability distributions, and random variables in Sect. 5.3. If you are familiar with probability theory, you may well skim or skip this section. Section 5.4 deals with the main ideas of information theory. We first take a view on what we mean by information and introduce the core concept of information theory, namely, *entropy*. Starting from the concept of entropy, we will then continue to look at more complex notions such as *conditional entropy* and *mutual information* in Sect. 4.3. We will then consider a variant of *conditional mutual information* called *transfer entropy* in Sect. 4.5. We conclude the theoretical part by discussing methods used for the estimation of information-theoretic quantities from sampled data in Sect. 5.5. What follows will deal with the application of the theoretical measures to neural data. We then give a short overview of applications of the discussed theoretical methods in the neurosciences in Sect. 5.6, and last (but not least), Sect. 5.7 constrains a list of software packages that can be used to estimate information-theoretic quantities for some given data set.

5.2 Modeling

In order to analyze the dynamics and gain a theoretical understanding of a given complex system, one usually defines a model first, i.e., a simplified theoretical version of the system to be investigated. The rest of the analysis is then based on this model and can only capture aspects of the system that are also contained in the model. Thus, care has to be taken when creating the model as the following analysis crucially depends on the quality of the model.

When building a model based on measured data, there is an important thing we have to pay attention to, namely, that any data obtained by measurement of physical quantities is only accurate up to a certain degree and corrupted by noise. This naturally also holds for neural data (e.g., electrophysiological single- or multi-cell measurements, EEG, fMRI, or MEG data). Therefore, when observing the state of some system by measuring it, one can only deduce the true state of the system up to a certain error determined by the noise in the measurement (which may depend both on the measurement method and the system itself). In order to model this uncertainty in a mathematical way, one uses probabilistic models for the states of the measured quantities of a system. This makes probability theory a key ingredient to many mathematical models in the natural sciences.

5.3 Probabilities and Random Variables

The roots of the mathematical theory of probability lie in the works of Cardano, Fermat, Pascal, Bernoulli, and de Moivre in the sixteenth and seventeenth centuries, in which the authors attempted to analyze games of chance. Pascal and Bernoulli were the first to treat the subject as a branch of mathematics; see [106] for a historical overview. Mathematically speaking, probability theory is concerned with the analysis of random phenomena. Over the last centuries, it has become a well-established mathematical subject. For a more in-depth treatment of the subject see [47, 52, 98].

5.3.1 A First Approach to Probabilities via Relative Frequencies

Let us consider an experiment that can produce a certain fixed number of outcomes (say a coin toss, where the possible outcomes are heads or tails or the throw of a die where the die will show one of the numbers 1 to 6). The set of all possible outcomes is called the *sample space* of the experiment.

One possible result of an experimen t is called *outcome,* and a set of outcomes is called an *event* (for the mathematically adept: an event is a subset of the power set of all outcomes). Take, for example, the throw of a regular, six-sided die as an experiment. The set of results in this case would be the set of natural numbers $\{1, \ldots,6\}$, and examples of events are $\{1,3,5\}$ or $\{2,4,6\}$ corresponding to the events "an odd number was thrown" and "an even number was thrown," respectively.

The classical definition of the probability of an event is due to Laplace: "The probability of an event to occur is the number of cases favorable for the event divided by the number of total outcomes possible" [106].

We thus assign each possible outcome a *probability*, a real number between 0 and 1 that is thought of as to describe how "likely" it is that the given event will occur, where 0 means "the event does not ever occur" and 1 means "the event always occurs." The sum of all the assigned numbers is restricted to be 1 as we assume that one of our considered events always occurs. For the coin toss, the possible outcomes heads and tails thus each have probability $\frac{1}{2}$ (considering that the number of favorable outcomes is one and the number of possible outcomes is two), and for the throw of a die this number is $\frac{1}{6}$ for each digit. This assumes that we have a so-called *fair* coin or die, i.e., one that does not favor any particular outcome over the others.

The probability of a given event to occur is then just the sum of the probabilities of the outcomes the event is composed of, e.g., when considering the throw of a die, the probability of the event "an odd number is thrown" is $\frac{1}{6}+\frac{1}{6}+\frac{1}{6}=\frac{1}{2}$.

Such types of experiments in which all possible outcomes have the same probability (they are called *equiprobable*) are called *Laplacian experiments.* The simplest case of an experiment not having equiprobable outcomes is the so-called Bernoulli experiment. Here, two possible outcomes "success" and "failure" with probabilities $p \in [0,1]$ and $1 - p$ are considered. Let us now consider probabilities in the general setting.

5.3.2 An Axiomatic Description of Probabilities

The foundations of modern probability theory were laid by Kolmogorov [54] in the 1930s. He was the first to give an axiomatic description of probability theory based on measure theory, putting the field on a mathematically sound basis. We will state his axiomatic description of probabilities in the following. This rather technical approach might seem a little complicated and cumbersome at first, and we will try to give well-understandable explanations of the concepts and notions used as they are of general importance.

Kolmogorov's definition is based on what is known as measure theory, a field of mathematics that is concerned with measuring the (geometric) size of subsets of a given space. Measure theory gives an axiomatic description of a *measure* (as a function μ assigning a nonnegative number to each subset) that fulfills the usual properties of a geometric measure of length (in one-dimensional space), area (in two-dimensional space), volume (in three-dimensional space), and so on. For example, if we take the measure of two disjoint (i.e., non-overlapping) sets, we expect the measure of their union to be the sum of the measures of the two sets and so on.

One prior remark on the definition: When looking at sample spaces (remember, these are the sets of possible outcomes of a random experiment), we have to make a fundamental distinction between *discrete sample spaces* (i.e., ones in which the outcomes can be separated and counted, like in a pile of sand, where we think of each little sand particle representing one possible outcome) and *continuous sample spaces* (where the outcomes form a continuum and cannot be separated and counted, think of this sample space as some kind of dough in which the outcomes cannot be separated). Although in most cases the continuous setting can be treated as a straightforward generalization of the discrete case and we just have to replace sums by integrals in the formulas, some technical subtleties exist, that makes a distinction between the two cases necessary. This is why we separate the two cases in all of what follows.

Definition 3.1 Measure Space and Probability Space. *A measure space is a triple* $(\Omega, \mathrm{F}, \mu)$. *Here*

- The *base space* Ω denotes an arbitrary nonempty set.
- F denotes the set of *measurable sets* in Ω which has to be a so-called σ-algebra over Ω, i.e., it has to fulfill:
  - $\varnothing \in \mathrm{F}$
  - F is closed under complements: if $E \in \mathrm{F}$, then $(\Omega \setminus E) \in \mathrm{F}$.
  - F is closed under countable unions: if $E_i \in \mathrm{F}$ for $i = 1, 2, \ldots$, then $(\cup_i E_i) \in \mathrm{F}$.
- μ is the so-called measure: It is a function $\mu : \mathrm{F} \rightarrow \mathrm{R} \cup \{\infty\}$ with the following properties
  - $\mu(\varnothing) = 0$ and $\mu \geq 0$ (non-negativity).
  - μ is countably additive: if $E_i \in \mathrm{F}$, $i = 1, 2, \ldots$ is a collection of pairwise disjoint (i.e., non-overlapping) sets, then $\mu(\cup_i E_i) = \sum_i \mu(E_i)$.

Why this complicated definition of measurable sets, measures, etc.? Well, this is mathematically the probably (no pun intended) most simple way to formalize the notion of a "measure" (in terms of geometric volume) as we know it over the real numbers.

When defining a measure, we first have to fix the whole space in which we want to measure. This is the base space Ω. Ω can be any arbitrary set: the sample space of a random experiment, e.g., $\Omega = \{\text{heads, tails}\}$ when we look at a coin toss or $\Omega = \{1, \ldots, 6\}$ when we look at the throw of a die (these are two examples of discrete sets), the set of real numbers R, the real plane R^2 (these are two examples of continuous sets), or whatever you choose it to be. When modeling the spiking activity of a neuron, the two states could be "neuron spiked" or "neuron did not spike."

In a second step we choose a collection of subsets of Ω that we name F, the collection of subsets of Ω that we want to be measurable. Note that the measurable subsets of Ω are not given a priori, but that we determine those by choosing F. So, you may ask, why this complicated setup with F, why not make every possible subset of Ω measurable, i.e., make F the power set of Ω (the power set is the set of all subsets of Ω)? This is totally reasonable and can easily been done when the number of elements of Ω is finite. But as with many things in mathematics, things get complicated when we deal with the continuum: In many natural settings, e.g., when Ω is a continuous set, this is just not possible or desirable for technical reasons. That is why we choose only a subset of the power set (you might refer to its elements as the "privileged" subsets) and make only the contained subsets measurable. We want to choose this subset in a way that the usual constructions that we know from geometric measures still work in the usual way, though. This motivates the properties that we impose on F: We expect to be able to measure the complements of measurable sets, as well as the union and intersection of a finite number of measurable sets to again be measurable. These properties are motivated by the corresponding properties of geometric measures (i.e., the union, intersection and complement of intervals of certain lengths has a length and so on). So to sum up, the set F is a subset of the power set of Ω, and sets that are not in F are not measurable.

In a last step, we choose a function μ that assigns a measure (think of it as a generalized geometric volume) to each measurable set (i.e., each element of F), where the measure has to fulfill some basic properties that we know from geometric measures: The measure is nonnegative, the empty set (that is contained in every set) should have measure 0, and the measure is additive.

All together, this makes the triple (Ω, F, μ) a space in which we can measure events and use constructions that we know from basic geometry. Our definition makes sure that the measure μ behaves in the way we expect it to (mathematicians call this a natural construction). Take some time to think about it: Definition 3.1 above generalizes the notion of the geometric measure in terms of the length $l(I) = b - a$ of intervals $I = [a,b]$ over the real numbers.

In fact, when choosing the set $\Omega = R$, we can construct the so-called Borel σ-algebra B that contains all closed intervals $I = [a,b]$, $a < b$, and a measure μ_B that assigns each interval $I = [a,b] \in B$ its length $\mu_B(I) = b - a$. The measure μ_B is called *Borel measure*. It is the standard measure of length that we know from geometry and makes (R, B, μ_B) a measure space. This construction can easily be extended to arbitrary dimensions (using closed sets) resulting in the measure space $(\mathrm{R}^n, \mathrm{B}^n, \mu_{\mathrm{B}^n})$ that fulfills the properties of a n-dimensional geometric measure of volume.

Let us look at some examples of measure spaces now:

1. Let $\Omega = \{0,1\}$, $\mathrm{F} = \{\emptyset,\{0\},\{1\},\Omega\}$, and P with $P(0) = P(1) = 0.5$. This makes (Ω,F,P) a measure space for our coin toss experiment. Note that in this simple case, F equals the full power set of Ω.
2. Let $\Omega = \{a,b,c,d\}$ and let $\mathrm{F} = \{\emptyset,\{a,b\},\{c,d\},\Omega\}$ with $P(\{a,b\}) = p$ and $P(\{c,d\}) = 1 - p$, where p denotes an arbitrary number between 0 and 1. This makes (Ω,F,P) a measure space.

Having understood the general case of a measure space, defining a probability space and a probability distribution is easy.

Definition 3.2 Probability Space, Probability Distribution. *A probability space is a measure space* (Ω,F,μ) *for which the measure* μ *is normed,* i.e., $\mu : \Omega \to [0,1]$ *with* $\mu(\Omega) = 1$. *The measure* μ *is called probability distribution and is often also denoted by P (for probability).* Ω *is called the sample space, elements of* Ω *are called outcomes and* F *is the set of events.*

Note that again, we make the distinction between discrete and continuous sample spaces here. In the course of history, a probability distribution on a discrete sample space came to be called *probability mass function* (or *pmf*), and a probability distribution defined on a continuous sample space came to be called *probability density function* (or *pdf*).

Let us look at a few examples, where the probability spaces in the following are given by the triple (Ω,F,P):

1. Let $\Omega = \{\text{heads,tails}\}$ and let $\mathrm{F} = \{\emptyset,\{\text{head}\},\{\text{tails}\},\Omega\}$. This is a probability space for our coin toss experiment, where $\emptyset$ relates to the event "neither heads nor tails" and Ω to the event "either heads or tails." Note that in this simple case, F equals the full power set of Ω.
2. Let $\Omega = \{1, \ldots,6\}$ and let F be the full power set of Ω (i.e., the set of all subsets of Ω, there are $6^2 = 36$, can you enumerate them all?). This is a probability for our experiment of dice throws, where we can distinguish all possible events.

5.3.3 Theory and Reality

It is important to stress that probabilities themselves are a mathematical and purely theoretical construct to help in understanding and analyzing random experiments, and per se they do not have to do anything with reality. They can be understood as an "underlying law" that generates the outcomes of a random experiment and *can never* be directly observed; see Fig. 5.1. But with some restrictions they can be estimated for a certain given experiment by looking at the outcomes of many repetitions of that experiment.

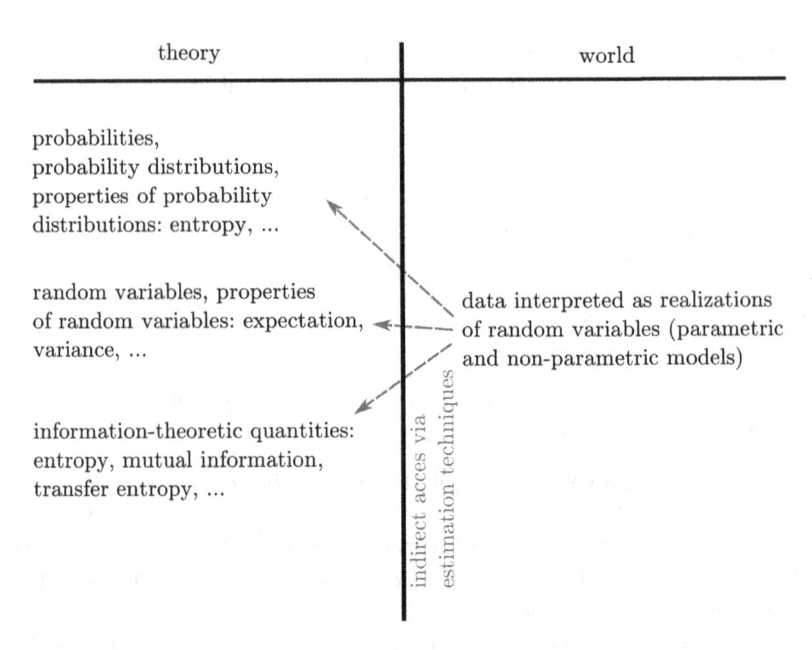

Fig. 5.1 Theoretical quantities and measurable quantities: The only things observable and accessible usually are data (measured or generated), all theoretical quantities are not directly accessible. They have to be estimated using statistical methods

Let us consider the following example. Assume that our experiment is the roll of a six-sided die. When repeating the experiment for ten times (also called *trials*), we will obtain frequencies for each of the numbers as given in Fig. 5.1. Repeating the experiment for 100 times, we will get frequencies that look similar to the ones given in Fig. 5.1. If we look at the relative frequencies (i.e., the frequency divided by the total number of trials), we see that these converge to the theoretically predicted value of $\frac{1}{6}$ as our number of trials grows larger.

This fundamental finding is also called the "Borel's law of large numbers."

Theorem 3.3 Borel's Law of Large Numbers. *Let Ω be a sample space of some experiment and let P be a probability mass function on Ω. Furthermore, let $N_n(E)$ be the number of occurrences of the event $E \subset \Omega$ when the experiment is repeated n times. Then the following holds:*

$$\frac{N_n(E)}{n} \to P(E) \quad \text{as } n \to \infty.$$

Borel's law of large numbers states that if an experiment is repeated many times (where the trials have to be independent and done under identical conditions), then the relative frequency of the outcomes converge to their probability as assigned by the probability mass function. The theorem thus establishes the notion of probability as the long run relative frequency of an event occurrence and thereby connects

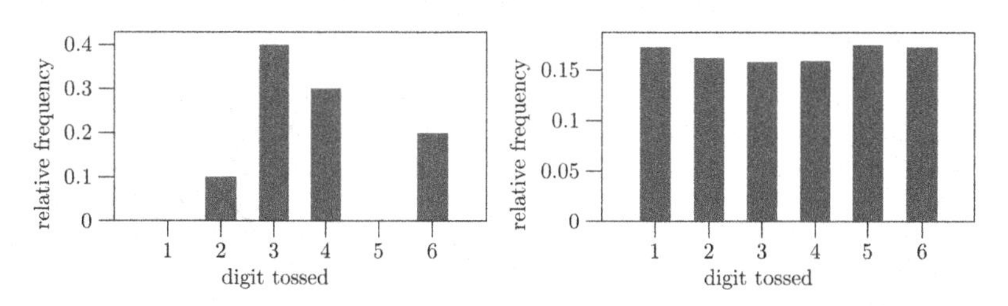

Fig. 5.2 Relative frequencies of tossed digits using a fair die: after 10 tosses (*left*) and after 1,000 tosses (*right*)

the theoretical side to the experimental side. Keep in mind though that we can never directly measure probabilities, and although relative frequencies will converge to the probability values, they will usually not be exactly equal (Fig. 5.2).

5.3.4 Independence of Events and Conditional Probabilities

A fundamental notion in probability theory is the idea of independence of events. Intuitively, we call two events independent if the occurrence of one does not affect the probability of occurrence of the other. Consider, for example, the events that it rains and the event that the current day of the week is Monday. These two are clearly independent, unless we lived in a world where there would be a correlation between the two, i.e., where the probability of rain would be different on Mondays compared to the other days of the week which is clearly not the case.

Similarly, we establish the notion of independence of two events in the sense of probability theory as follows.

Definition 3.4 Independent Events. *Let A and B be two events of some probability space* (Ω,Σ,P). *Then A and B are called independent if and only if*

$$P(A \cap B) = P(A)P(B). \tag{5.1}$$

The term $P(A \cap B)$ is referred to *joint probability* of A and B; see Fig. 5.3.

Another important concept is the notion of conditional probability, i.e., the probability of one event A occurring, given the fact that another event B occurred.

Definition 3.5 Conditional Probability. *Given two events A and B of some probability space* (Ω,F,P) *with* $P(B) > 0$ *we call*

$$P(A|B) = \frac{P(A \cap B)}{P(B)}$$

the *conditional probability of A given B.*

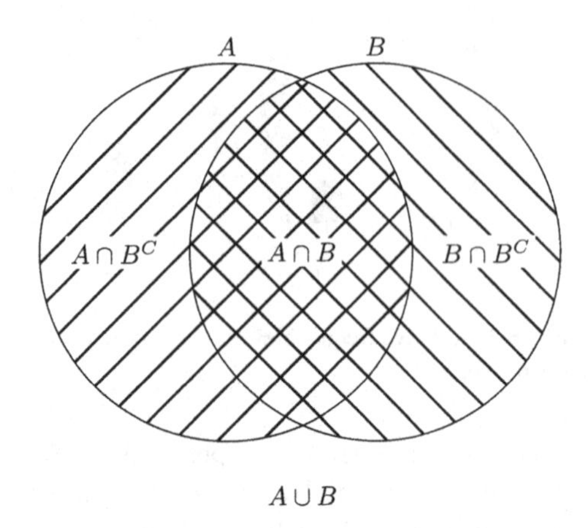

Fig. 5.3 Two events A and B, their union $A \cup B$, their intersection $A \cap B$ (i.e., common occurrence in terms of probability) and their exclusive occurrences $A \cap B^C$ (A and not B occurs), $B \cap A^C$ (B occurs and not A), where $\cdot^C$ denotes the complement in $A \cup B$

Note that for independent events A and B, we have $P(A \cap B) = P(A)P(B)$ and thus $P(A|B) = P(A)$ and $P(B|A) = P(B)$. We can thus write

$$\begin{aligned} P(A \cap B) &= P(A)P(B), \\ \Leftrightarrow P(A) &= \frac{P(A \cap B)}{P(B)} = P(A|B), \\ \Leftrightarrow P(B) &= \frac{P(A \cap B)}{P(A)} = P(B|A), \end{aligned}$$

and this means that the occurrence of A does not affect the conditional probability of B given A (and vice versa). This exactly reflects the intuitive definition of independence that we gave in the first paragraph of this section. Note that we could have also used the conditional probabilities to define independence in the first place. Nonetheless the definition of Eq. 5.1 is preferred, as it is shorter, symmetrical in A and B and more general as the conditional probabilities above are not defined in the case where $P(A) = 0$ or $P(B) = 0$.

5.3.5 Random Variables

In many cases the sample spaces of random experiments are a lot more complicated than the ones of the toy examples we looked at so far. Think, for example, of measurements of membrane potentials of certain neurons that we want to model mathematically, or the state of some complicated system, e.g., a network of neurons receiving some stimulus.

Thus mathematicians came up with a way to tame the sample spaces by looking at the events indirectly, namely, by first mapping the events to some better understood space, like the set of real numbers (or some higher dimensional real vector space), and then look at outcomes of the random experiment in the simplified space rather than in the complicated original space. Looking at spaces of numbers has many advantages: order relations exist (smaller, equal, larger), we can form averages, and much more. This leads to the concept of random variables.

A (real) *random variable* is a function that maps each outcome of a random experiment to some (real) number. Thus, a random variable can be thought of as a variable whose value is subject to variations due to chance. But keep in mind that a random variable is a mapping and not a variable in the usual sense.

Mathematically, a random variable is defined using what is called a *measurable function*. A measurable function is nothing more than a map from one measurable space to another for which the pre-image of each measurable set is again measurable (with respect to the two different measures in the two measure spaces involved). So a measurable map is nothing more than a "nice" map respecting the structures of the spaces involved (take as an example for such maps the continuous functions over R).

Definition 3.6 Random Variable. *Let* (Ω,Σ,P) *be a probability space and* (Ω',Σ') *a measure space. A* (Σ,Σ')*-measurable function* $X : \Omega \to \Omega'$ *is called* Ω'*-valued random variable (or just* Ω'*-random variable) on* Ω.

Commonly, a distinction between *continuous random variables* and *discrete random variables* is made, the former taking values on some continuum (in most cases R) and the latter on a discrete set (in most cases Z).

A type of random variable that plays an important role in modeling is the so-called Bernoulli random variable that only takes two distinct values 0 with probability p and 1 with probability $1 - p$ (i.e., it has a Bernoulli distribution as its underlying probability distribution). Spiking behavior of a neuron is often modeled that way, where 1 stands for "neuron spiked" and 0 for "neuron did not spike" (in some interval of time).

A real- or integer-valued random variable X thus assigns a number $X(E)$ to every event $E \in \Sigma$. A value $X(E)$ corresponds to the occurrence of the event E and is called a *realization of* X. Thus, random variables allow for the change of space in which outcomes of probabilistic processes are considered. Instead of considering an outcome directly in some complicated space, we first project it to a simpler space using our mapping (the random variable X) and interpret its outcome in that simpler space.

In terms of measure theory, a random variable $X : (\Omega,\Sigma,P) \to (\Omega',\Sigma')$ (again, considered as a measurable mapping here) induces a probability measure P_X on the measure space (Ω',Σ') via

$$P_X\left(S'\right) := P\left(X^{-1}\left(S'\right)\right),$$

where again $X^{-1}(S')$ denotes the pre-image of $S' \in \Sigma'$. This also justifies the restriction of X to be measurable: If it were not, such a construction would not be possible, but this is a technical detail. As a result, this makes (Ω',Σ',P_X) a probability space and we can think of the measure P_X as the "projection" of the measure P from Ω onto Ω' (via the measurable mapping X).

The measures P and P_X are probability densities for the probability distributions over Ω and Ω': They measure the likelihood of occurrence for each event (P) or value (P_X).

As a simple example of a random variable, consider again the example of the coin toss. Here, we have $\Omega = \{\text{heads,tails}\}$, $\mathrm{F} = \{\emptyset,\{\text{heads}\},\{\text{tails}\},\Omega\}$, and P that assigns to both heads and tails the probability $\frac{1}{2}$ forming the probability space. Consider as a random variable $X : \Omega \to \Omega'$ with $\Omega' = \{0,1\}$ that maps Ω to S such that $X(\text{heads}) = 0$ and $X(\text{tails}) = 1$. If we choose $\mathrm{F}' = \{\emptyset,\{0\},\{1\},\{0,1\}\}$ as a σ-algebra for Ω', this makes $M = (\Omega',\mathrm{F}')$ a measurable space and X induces a measure $P' = P_X$ on M with $P'(\{0\}) = P'(\{1\}) = \frac{1}{2}$. That makes (Ω',F',P') a measure space, and since P' is normed, it is a probability space.

5.3.5.1 Cumulative Distribution Function

Using random variables that take on values of whole or the real numbers, the natural total ordering of elements in these spaces enables us to define the so-called cumulative distribution function (or *cdf*) for a random variable.

Definition 3.7 Cumulative Distribution Function. *Let X be a* R*-valued or* Z*-valued random variable on some probability space (Ω,Σ,P). Then the function*

$$F(x) := P(X \leq x)$$

is called the *cumulative distribution function* of X.

The expression $P(X \leq x)$ evaluates to

$$P(X \leq x) = \int_{\tau \leq x} P(X = \tau)\ \mathrm{d}\tau,$$

in the continuous case and to

$$P(X \leq x) = \sum_{k \leq x} P(X = k)$$

in the discrete case.

In that sense, the measure P_X can be understood as the derivative of the cumulative distribution function F

$$P(x_1 \leq X \leq x_2) = F(x_2) - F(x_1),$$

and we also write $F(x) = \int_{\tau \leq x} P_X(\tau)\ \mathrm{d}\tau$ in the continuous case.

5.3.5.2 Independence of Random Variables

The definition of independent events directly transfers to random variables: Two random variables X, Y are called independent if the conditional probability distribution of $X(Y)$ given an observed value of $Y(X)$ does not differ from the probability distribution of $X(Y)$ alone.

Definition 3.8 Independent Random Variables. *Let X, Y be two random variables. Then X and Y are called independent, if the following holds for any observed values x of X and y of Y:*

$$P(X|Y=y) = P(X) \quad \text{and} \quad P(Y|X=x) = P(Y).$$

This notion can be generalized to the case of three or more random variables naturally.

5.3.5.3 Expectation and Variance

Two very important concepts of random variables are the so-called *expectation value* (or just *expectation*) and the *variance*. The expectation of a random variable X is the mean value of the random variable, where the weighting of the values corresponds to the probability density distribution. It thus tells us what value of X we should expect "on average."

Definition 3.9 Expectation Value. *Let X be a* R- *or* Z-*valued random variable. Then its expectation value (sometimes also denoted by μ) is given by*

$$E[X] := \int_{\mathbb{R}} x P_X(x) \ \mathrm{d}x = \int_{\mathbb{R}} x \ \mathrm{d}P_X,$$

for a real-valued random variable X and by

$$E[X] := \sum_{x \in \mathbb{Z}} x P_X(x)$$

if X is Z-valued.

Note that if confusion can be made as to which probability distribution the expectation value is taken, we will include the probability distribution to which the expectation value is taken in the index. Consider, for example, two random variables X and Y defined on the same base space but with different underlying probability distributions. In this case, we denote by $E_X[Y]$ the expectation value of Y taken with respect to the probability distribution of X.

Let us now look at an example. If we consider the throw of a fair die with $P(i) = \frac{1}{6}$ for each digit $i = 1, \ldots, 6$ and take X as the random variable that just assigns each digit its integer value $X(i) = i$, we get $E[X] = \frac{1}{6}(1 + \ldots + 6) = 3.5$.

Another important concept is the so-called *variance* of a random variable. The variance is a measure for how far the values of the random variable are spread around its expected value. It is defined as follows.

Definition 3.10 Variance. *Let X be a* R- *or* Z-*valued random variable. Then its variance is given as*

$$\mathrm{var}[X] := E\left[(E[X] - X)^2\right] = E[X^2] - (E[X])^2$$

sometimes also denoted as σ^2.

The variance is thus the expected squared distance of the values of the random variable to its expected value. Another commonly used measure is the so-called standard deviation $\sigma(X) = \sqrt{\mathrm{var}(X)}$, a measure for the average deviation of realizations of X from the mean value.

Often one also talks about the expectation value as "first-order moment" of the random variable, the variance as a "second-order moment." Higher-order moments can be constructed by iteration, but will not be of interest to us in the following.

Note again that the concepts of expectation and variance live on the theoretical side of the world, i.e., we cannot measure these quantities directly. The only thing that we can do is try to estimate them from a set of measurements (i.e., realizations of the involved random variables); see Fig. 5.1. The statistical discipline of estimation theory deals with question regarding the estimation of theoretical quantities from real data. We will talk about estimation in more detail in Sect. 5.5 and just give two examples here.

For estimating the expected value we can use what is called the *sample mean.*

Definition 3.11 Sample Mean. *Let X be a* R- *or* Z-*valued random variable with n realizations* $x_1, \ldots, x_n$. *Then the sample mean* $\hat{\mu}$ *of the realizations is given as*

$$\hat{\mu}(x_1, \ldots, x_n) := \frac{1}{n}\sum_{i=1}^{n} x_i$$

As we will see below, this sample mean provides a good estimation of the expected value if the number n of samples is large enough. Similarly, we can estimate the variance as follows.

Definition 3.12 Sample Variance. *Let X be a* R- *or* Z-*valued random variable with n realizations* $x_1, \ldots, x_n$. *Then the population variance* $\hat{\sigma}$ *of the realizations is given as*

$$\hat{\sigma}^2(x_1, \ldots, x_n) := \frac{1}{n}\sum_{i=1}^{n}(x_i - \hat{\mu}(x_1, \ldots, x_n))^2,$$

where $\hat{\mu}$ denotes the sample mean.

Before going on let us calculate some examples of expectations and variances of random variables. Take the coin toss example from above. Here, the expected value of X is $E[X] = \frac{1}{2} \cdot 0 + \frac{1}{2} \cdot 1 = \frac{1}{2}$, the variance $var(X) = E\left[(E[X] - X)^2\right] = \frac{1}{2} \cdot (0 - \frac{1}{2})^2 + \frac{1}{2} \cdot (1 - \frac{1}{2})^2 = \frac{1}{4}$. For the example of the dice roll (where the random variable X takes the value of the number thrown) we get $E[X] = \frac{1+2+3+4+5+6}{6} = \frac{7}{2}$ $= 3.5$ and $var(X) = E[X^2] - (E[X])^2 = \frac{91}{6} - \frac{49}{4} = \frac{35}{12} \approx 2.92$.

5.3.6 Laws of Large Numbers

The laws of large numbers (there exist two versions as we will see below) state that the sample average of a set of realizations of a random variable "almost certainly" converges the random variable's expected value when the number of realizations grows to infinity.

Theorem 3.13 Law of Large Numbers. *Let $X_1, X_2, \ldots$ be an infinite sequence of independent, identically distributed random variables with expected values $E(X_1) = E(X_2) = \ldots = \mu$. Let $\overline{X}_n = \frac{1}{n}(X_1 + \cdots + X_n)$ be the sample average.*

a. **Weak law of large numbers**. The sample average converges in probability towards the expected value, i.e., for any $\varepsilon > 0$

$$\lim_{n \to \infty} P\left(\left|\overline{X}_n - \mu\right| > \varepsilon\right) = 0.$$

   This is sometimes also expressed as

$$\overline{X}_n \xrightarrow{P} \mu \quad \text{when } n \to \infty.$$

b. **Strong law of large numbers**. The sample average converges almost surely towards the expected value, i.e.,

$$P\left(\lim_{n \to \infty} \overline{X}_n = \mu\right) = 1.$$

   This is sometimes also expressed as

$$\overline{X}_n \xrightarrow{a.s.} \mu \quad \text{when } n \to \infty.$$

The weak version of the law states that the sample average $\overline{X}_n$ is likely to be close to μ for some large value of n. But this does not exclude the possibility of $\left|\overline{X}_n - \mu\right| > \varepsilon$ occurring an infinite number of times.

The strong law says that this "almost surely" will not be the case: With probability 1, the inequality $\left|\overline{X}_n - \mu\right| < \varepsilon$ holds for all $\varepsilon > 0$ and all large enough n.

5.3.7 Some Parametrized Probability Distributions

Certain probability distributions often occur naturally when looking at typical random experiments. In the course of history, these were thus put (mathematicians like doing such things) into families or classes, and the members of one class are distinguished by a set of parameters (a parameter is just a number than can be chosen freely in some specified range). To specify a certain probability distribution we simply have to specify in which class it lies and which parameter values it exhibits, which is more convenient than specifying the probability distribution explicitly every time. This also allows proving (and reusing) results for whole classes of probability distributions and facilitates communication with other scientists.

Note that we will only give a concise version of the most important distributions relevant in neuroscientific applications here and point the reader to [47, 52, 98] for a more in-depth treatment of the subject.

The *normal distribution* $N(\mu,\sigma^2)$ is a family of continuous probability distributions parametrized by two real-valued parameters $\mu \in \mathrm{R}$ and $\sigma^2 \in \mathrm{R}^+$, called *mean* and *variance*. Its probability density function is given as

$$\begin{aligned} f(x;\mu,\sigma) : \mathrm{R} &\rightarrow \mathrm{R}_0^+ \\ x \mapsto & \frac{1}{\sigma\sqrt{2\pi}} e^{-\frac{1}{2}\left(\frac{x-\mu}{\sigma}\right)^2} . \end{aligned}$$

The family is closed under linear combinations, i.e., linear combinations of normally distributed random variables are again normally distributed. It is the most important and often used probability distribution in probability theory and statistics as many other probability distributions can be approximated by a normal distribution when the sample size is large enough (this fact is called the *central limit theorem*). See Fig. 5.4 for examples of the pdf and cdf for normally distributed random variables.

The *Bernoulli probability distribution* $Ber(p)$ describes the two possible outcomes of a Bernoulli experiment with the probability of success and failure being p and $1 - p$, respectively. It is thus a discrete probability distribution on two elements and it is parametrized by one parameter $p \in [0,1] \subset \mathrm{R}$. Its probability mass function is given by the two values $P(\text{success}) = p$ and $P(\text{failure}) = 1 - p$.

The *binomial probability distribution* $B(n,p)$ is a discrete probability distribution parametrized by two parameters $n \in \mathbf{N}$ and $p \in [0,1] \subset \mathrm{R}$. Its probability mass function is

$$f(k;n,p) = \binom{n}{k} p^k (1-p)^{n-k}, \tag{5.2}$$

and it can be thought of as a model for the probability of k successful outcomes in a trial with n independent Bernoulli experiments, each having success probability p, see Fig. 5.5.

The *Poisson distribution* $Poiss(\lambda)$ is a family of discrete probability distributions parametrized by one real parameter $\lambda \in \mathrm{R}^+$. Its probability mass function is given by

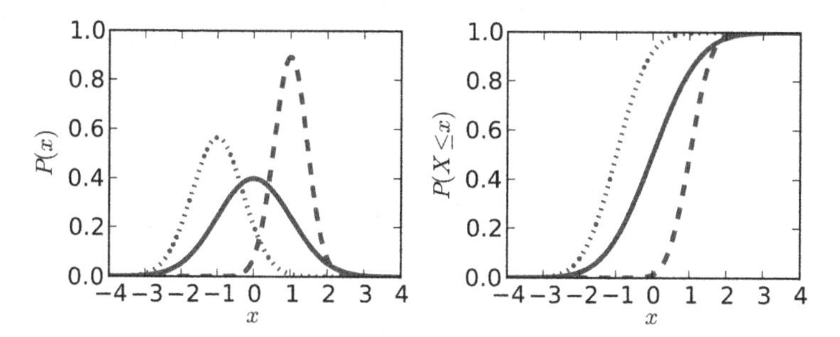

Fig. 5.4 Normal distribution: probability density function (*left*) and cumulative density function (*right*) for selected parameter values of μ and σ. *Solid line*: $\mu = 0$, $\sigma = 1$; *dashed line*: $\mu = 1$, $\sigma^2 = 0.2$; *dotted line*: $\mu = -1$, $\sigma^2 = 0.5$

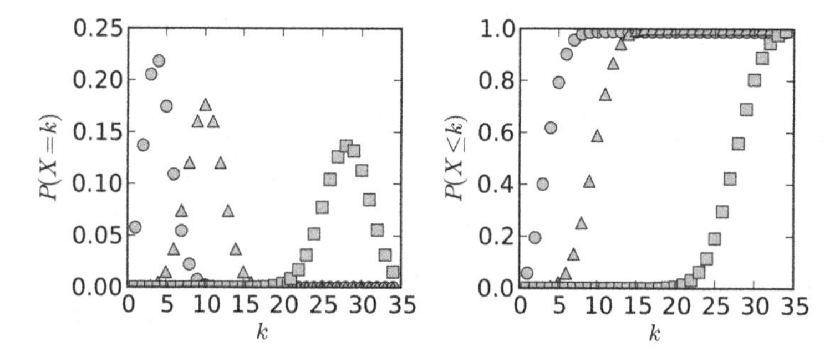

Fig. 5.5 Binomial distribution: probability mass function (*left*) and cumulative density function (*right*) for selected parameter values of p and n. *Circle*: $p = 0.2$, $n = 20$; *triangle*: $p = 0.5$, $n = 20$; *square*: $p = 0.7$, $n = 40$

$$\begin{aligned} f(k;\lambda) : \mathbf{N} &\rightarrow \mathbf{R}_0^+ \\ k &\mapsto \frac{\lambda^k e^{-\lambda}}{k!}. \end{aligned}$$

The Poisson distribution plays an important role in the modeling of neuroscience data. This is the case because the firing statistics of cortical neurons (and also other kinds of neurons) can often be well fit by a Poisson process, where λ is considered the mean firing rate of a given neuron; see [24, 74, 101].

This fact comes at no surprise if we invest some thought. The Poisson distribution can be seen as a special case of the binomial distribution. A theorem known as Poisson limit theorem (sometimes also called "law of rare events") now tells us that in the limit $p \rightarrow 0$ and $n \rightarrow \infty$ the binomial distribution converges to the Poisson distribution with $\lambda = np$. Consider, for example, the spiking activity of our neuron that we could model via a Binomial distribution. We discretize time and consider time bins of say 2 ms and assume a mean firing rate of the neuron denoted by λ (measured in Hertz). Clearly, in most time bins the neuron does not spike (corresponding to a small value of p), and the number of bins is large (corresponding

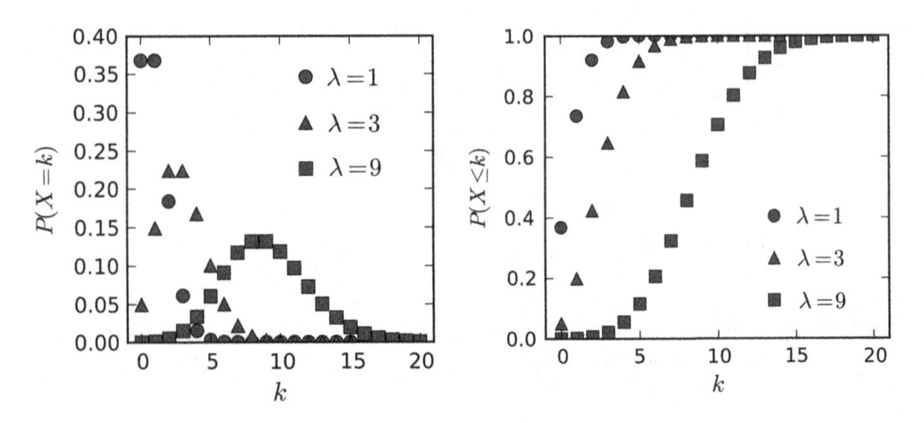

Fig. 5.6 Poisson distribution: probability mass function (*left*) and cumulative density function (*right*) for selected parameter values of λ

to a large n). The Poisson limit theorem tells us that in this case the probability distribution concerning spike emission is well matched by a Poisson distribution.

See Fig. 5.6 for examples of the pmf and cdf for Poisson-distributed random variables for a selection of parameters λ.

The so-called exponential distribution $Exp(\lambda)$ is a continuous probability distribution parametrized by one real parameter $\lambda \in \mathbb{R}^+$. Its probability density function is given by

$$\begin{aligned} f(x;\lambda) : \mathbb{R} &\rightarrow \mathbb{R}_0^+ \\ x &\mapsto \begin{cases} \lambda e^{-\lambda x} & \text{for } x > 0 \\ 0 & \text{for } x \leq 0 \end{cases}. \end{aligned}$$

The exponential distribution with parameter λ can be interpreted as the probability distribution describing the time between two events in a Poisson process with parameter λ; see the next section.

See Fig. 5.7 for examples of the pdf and cdf for exponentially distributed random variables for a selection of parameters λ.

We want to conclude our view on families on probability distributions at this point and point the interested reader to [47, 52, 98] regarding further examples and details of families of probability distributions.

5.3.8 Stochastic Processes

A *stochastic process* (sometimes also called *random process*) is a collection of random variables indexed by a totally ordered set, which is usually taken as time. Stochastic processes are commonly used to model the evolution of some random variable over time. We will only look at discrete-time processes in the following,

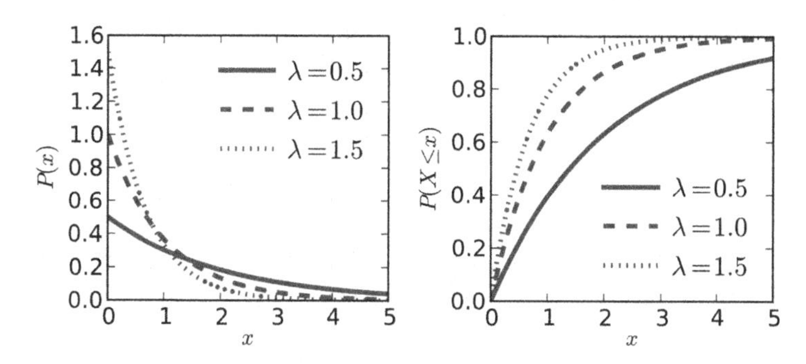

Fig. 5.7 Exponential distribution: probability density function (*left*) and cumulative density function (*right*) for selected parameter values of λ

i.e., stochastic processes that are indexed by a discrete set. The extension to the continuous case is straightforward; see [18] for an introduction to the subject.

Mathematically, a stochastic process is defined as follows.

Definition 3.14. *Let* $(\Omega,\mathbf{F},P)$ *be a probability space and let* $(S,\mathbf{S})$ *be a measure space. Let furthermore* $X_t : \mathbf{F} \rightarrow \mathbf{S}$ *be a set of random variables, where* $t \in T$*. Then an S-valued stochastic process* P *is given by*

$$\mathbf{P} := \{X_t : t \in T\},$$

where T is some totally ordered set, commonly interpreted as time. The space S is referred to as the *sample space of the process* P.

If the distribution underlying the random variables X_t does not vary over time, the process is called *homogeneous*, in the case where the probability distributions P_{X_t} depend on the time t, it is called *inhomogeneous*.

A special kind and well-studied type of stochastic process is the so-called Markov process. A discrete Markov process of order $k \in \mathbf{N}$ is a inhomogeneous stochastic process subject to the restriction that for any time $t = 0, 1, \ldots,$ the probability distribution underlying X_t only depends on the preceding k probability distributions of $X_{t-1}, \ldots, X_{t-k}$, i.e., that for any t and any set of realizations x_i of X_i $(0 \leq i \leq t)$, we have

$$P(X_t = x_t | X_{t-1} = x_{t-1}, \ldots, X_{t-k} = x_{t-k}) = P(X_t = x_t | X_{t-1} = x_{t-1}, \ldots, X_0 = x_0).$$

Another process often considered in neuroscientific applications is the *Poisson process*. It is a discrete-time stochastic process P for which the random variables are Poisson distributed with some parameter $\lambda(t)$ (in the inhomogeneous case, for the homogeneous case, we have $\lambda(t) = \lambda =$ constant). As can be shown, the time delay between each pair of consecutive events of a Poisson process is exponentially distributed. See Fig. 5.8 for examples of the number of instantaneous (occurring during one time slice) and the number of cumulated events (over all preceding time slices) of Poisson processes for a selection of parameters λ.

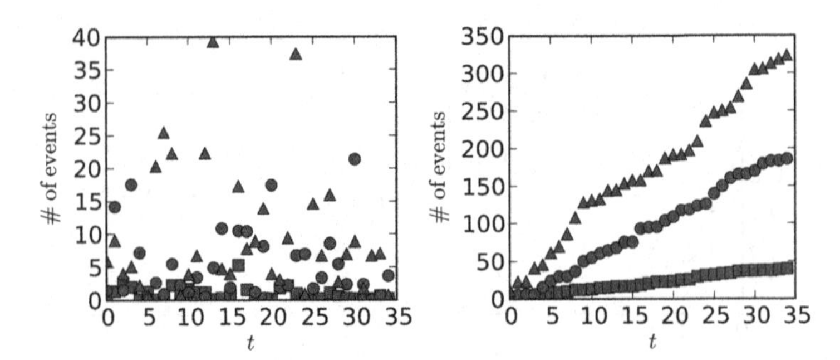

Fig. 5.8 Examples of the number of events in one time window of size $\Delta t = 1$ (*left*) and the number of accumulated events since $t = 0$ (*right*) for Poisson processes with certain rates $\lambda = 1$ (*circle*), $\lambda = 5$ (*triangle*) and $\lambda = 10$ (*square*)

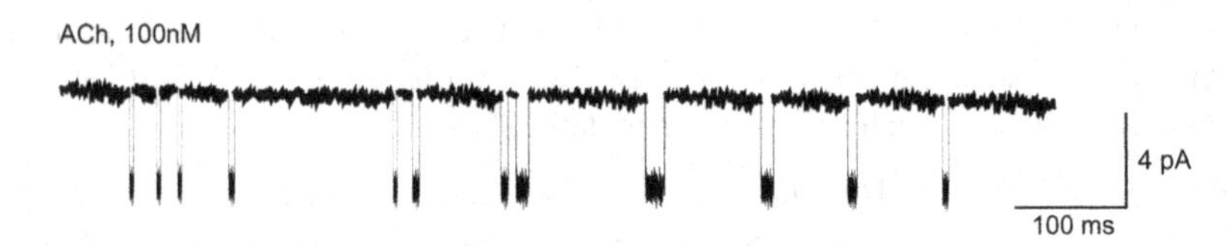

Fig. 5.9 Random opening and closing of ion channels (Modified from [23], Fig. 12)

Poisson processes have proven to be a good model for many natural as well as man-made processes such as radioactive decay, telephone calls and queues, and also for modeling neural data. An influential paper in the neurosciences was [23], showing the random nature of the closing and opening of single ion channels in certain neurons. Using as a model a Poisson process with the right parameter provides a good fit to the measured data here Fig. 5.9.

Another prominent example of neuroscientific models employing a Poisson process is the commonly used model for the sparse and highly irregular firing patterns of cortical neurons in vivo [24, 74, 101]. The firing patterns of such cells are usually modeled using inhomogeneous Poisson processes (with $\lambda(t)$ modeling the average firing rate of a cell).

5.4 Information Theory

Information theory was introduced by Shannon [97] as a mathematically rigid theory to describe the process of transmission of information over some channel of communication. His goal was to quantitatively measure the "information content" of a "message" sent over some "channel"; see Fig. 5.10. In what follows we will not go into detail regarding all aspects of Shannon's theory, but we will mainly focus on his idea of measuring "information content" of a message. For a more in-depth treatment of the subject, the interested reader is pointed to the excellent book [26].

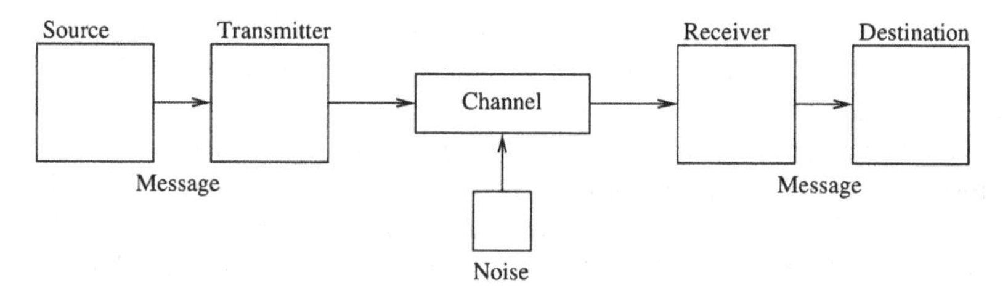

Fig. 5.10 The setting of Shannon's information theory: information is transferred from a source to a destination via a message that is first encoded and then subsequently sent over a noisy channel to be decoded by the receiver

The central elements of Shannon's theory are depicted in Fig. 5.10. In the standard setting considered in information theory, an *information source* produces *messages* that are subsequently encoded using *symbols* from an *alphabet* and sent over a noisy *channel* to be received by a *receiver* that decodes the message and attempts to reconstruct the original message.

A communication channel (or just channel) in Shannon's model transmits the encoded message from the sender to the receiver. Due to noise present in the channel, the receiver does not receive the original message dispatched by the sender but rather some noisy version of it.

The whole theory is set in the field of probability theory (hence our introduction to the concepts in the last section) and in this context, the messages emitted by the source are modeled as a random variable X with some underlying probability distribution P_X. For each message x (a realization of X), the receiver sees a corrupted version y of x and this fact is modeled by interpreting the received messages as realizations of a random variable Y with some probability distribution P_Y (that depends both on P_X and the channel properties). The transmission characteristics of the channel itself are characterized by the stochastic correspondence of the signals transmitted by the sender to the ones received by the receiver, i.e., by modeling the channel as a conditional probability distribution $P_{Y|X}$.

Being based upon probability theory, keep in mind that all the information-theoretic quantities that we will look at in the following such as "entropy" or "mutual information" are just properties of the random variables involved, i.e., properties of the probability distributions underlying these random variables.

Information-theoretic analyses have proven to be a valuable tool in many areas of science such as physics, biology, chemistry, finance, and linguistics and generally in the study of complex systems [62, 88]. We will have a look at applications in the neurosciences in Sect. 5.6.

Note that a vast number of works was published in the field of information theory and its applications since its first presentation in the 1950s. We will focus on the core concepts in the following and point the reader to [26] for a more in-depth treatment of the subject.

In the following we will start by looking at a notion of information and using this proceed to define *entropy* (sometimes also called *Shannon entropy*), a core concept in information theory. As all further information-theoretic concepts are based on the idea of entropy, it is of vital importance to understand this concept well. We will then look at mutual information, the information shared by two or more random variables. Furthermore, we will look at a measure of distance for probability distributions called Kullback–Leibler divergence and give an interpretation of mutual information in terms of Kullback–Leibler divergence. After a quick look at the multivariate case of mutual information between more than two variables and the relation between mutual information and channel capacity, we will then proceed to an information-theoretic measure called transfer entropy. Transfer entropy is based on mutual information but in contrast to mutual information is of directed nature.

5.4.1 A Notion of Information

Before defining entropy, let us try to give an axiomatic definition of the concept of "information", see [115]. The entropy of a random variable will then be nothing more than the expected (i.e., average) amount of information contained in a realization of that random variable.

We want to consider a probabilistic model in what follows, i.e., we have a set of events, each occurring with a given probability. The goal is to assess how informative the occurrence of a given event is. What would we intuitively expect from a measure of information h that maps the set of the events to the set of nonnegative real number, i.e., when we restrict h to be a non-negative real number?

First of all, it should certainly be additive for independent events and sub-additive for non-independent events. This is easily justified: If you read two newspaper articles about totally unrelated subjects, the total amount of information you obtain consists of both the information in the first and the second article. When you read articles about related subjects on the other hand, they often have some common information.

Furthermore, events that occur regularly and unsurprisingly are not considered informative and the more seldom or surprising an event occurs, the more informative it is. Think about an article about your favorite sports team winning a match that usually wins all matches. You will consider this not very informative. But when the local newspaper reports about an earthquake with its epicenter being in the part of town where you live, this will certainly be informative to you (unless you were at home during the time the earthquake happened), assuming that earthquakes do not occur on a regular basis where you live.

We thus have the following axioms for the information content h of an event, where we look at the information content of events contained in some probability space (Ω,Σ,P):

1. h is nonnegative: $h : \Sigma \rightarrow \mathbf{R}^{+}$.
2. h is sub-additive: For any two messages $\omega_1, \omega_2 \in \Sigma$, we have $h(\omega_1 \cap \omega_2) \leq h(\omega_1) + h(\omega_2)$, where equality holds if and only if ω_1 and ω_2 are independent.

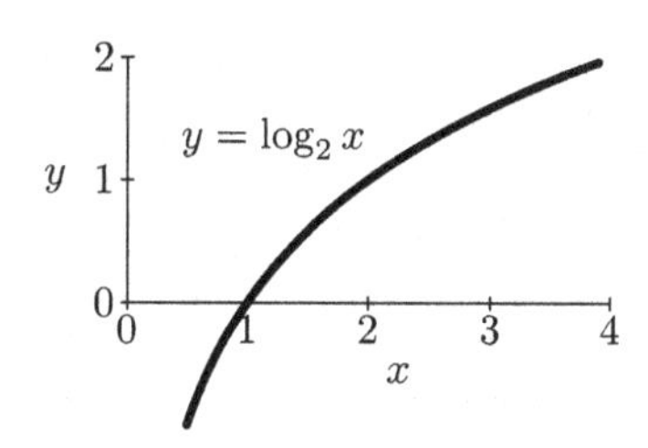

Fig. 5.11 The logarithm to the basis of 2

3. h is continuous and monotonic with respect to the probability measure P.
4. Events with probability 1 are not informative: $h(\omega) = 0$ for $\omega \in \Sigma$ with $P(\omega) = 1$.

Now calculus tells us (this is not hard to show – you paid attention in the mathematics class at school, did you not?) that these four requirements leave only one possible function that fulfills all these requirements: the logarithm (Fig. 5.11). This leads us to the following natural definition.

Definition 4.1 Information. *Let* (Ω,Σ,P) *be a probability space. Then the information* h *of an event* $\sigma \in \Sigma$ *is defined as*

$$h(\sigma) := h(P(\sigma)) = -\log_b(P(\sigma)),$$

where b denotes the basis of the logarithm.

For the basis of the logarithm, usually $b = 2$ or $b = e$ is chosen, fixing the unit of h as "bit" or "nat," respectively. We resort to using $b = 2$ for the rest of this chapter and write *log* for the logarithm to the basis of two. The natural logarithm will be denoted by *ln*.

Note that the information content in our definition only depends on the probability of the occurrence of the event and not the event itself. It is thus a property of the probability distribution P.

Let us give some examples in order to illustrate this idea of information content.

Consider a toss of a fair coin, where the possible outcomes are heads (H) or tails (T), each occurring with probability $\frac{1}{2}$. What is the information contained in a coin toss? As the information solely depends on the probability, we have $h(H) = h(T)$, which comes at no surprise. Furthermore we have $h(H) = h(T) = -\log\frac{1}{2} = -(\log(1) - \log_2(2)) = \log 2 = 1$ bit, when we apply the fundamental logarithmic identity $\log(a \cdot b) = \log(a) + \log(b)$. Thus one toss of a fair coin gives us one bit of information. This fact also lets us explain the unit attached to h. If measured in bit (i.e., with $b = 2$), this is the amount of bits needed to store that information. For the toss of a coin we need one bit, assigning each outcome to either 0 or 1.

Repeating the same game for the roll of a fair die where each digit has probability $\frac{1}{6}$, we again have the same amount of information for each digit $E \in \{1, \ldots,6\}$, namely, $h(E) = \log(6) \approx 2.58$ bit. This means that in this case we need three bits to store the information associated to each outcome, namely, the number shown.

Looking at the two examples above, we can give another (hopefully intuitive) characterization of the term information content: It is the minimal number of yes-no questions that we have to ask until we know which event occurred, assuming that we have a knowledge of the underlying probability distribution. Consider the example of the coin toss above. We have to ask exactly one question and we know the outcome ("Was it heads?" "Was it tails?").

Things get more interesting when we look at the case of the die throw. Here, several question asking strategies are possible and you can freely choose your favorite – we will give one example below.

Say a digit d was thrown. The first question could be, "Was the digit less or equal to 3?" (other strategies, "Was the digit greater than 3?" "Was the digit even?" "Was the digit odd?"). We then go on depending on the answer and cut off at least half of the remaining probability mass in each step, leaving us with a single possibility after at most 3 steps. From the information content, we know that on average we have to ask 2.58 times on average.

The two examples above were both cases with uniform probability distributions but in principle the same applies to arbitrary probability distributions.

5.4.2 *Entropy as Expected Information Content*

The term entropy is at the heart of Shannon's information theory [97]. Using the notion of the information as discussed in Sect. 4.1, we can readily define the entropy of a discrete random variable as its expected information.

Definition 4.2 Entropy. *Let X be a random variable on some probability space (Ω,Σ,P) with values in the integer or the real numbers. Then its entropy*[1] *(sometimes also called Shannon entropy or self-information) $H(X)$ is defined as the expected amount of information of X,*

$$H(X) := E[h(X)]. \tag{5.3}$$

If X is a random variable that takes integer values (i.e., a discrete random variable), Eq. 5.3 evaluates to

$$H(X) = \sum_{x\in Z} P(X=x)h(P(X=x)) = -\sum_{x\in Z} P(X=x)\log(P(X=x)),$$

in the case of a real-valued, continuous random variable, we get

---

[1] Shannon chose the letter H for denoting entropy after Boltzmann's H-theorem in classical statistical mechanics.

$$H(X) = \int_{\mathrm{R}} P(X = x)h(P(X = x)) \, \mathrm{d}x$$

and the resulting quantities is called *differential entropy* [26].

As the information content is a function solely dependent on the probability of the events one also speaks of the entropy of a probability distribution.

Looking at the definition in Eq. 5.3, we see that entropy is a measure for the average amount of information that we expect to obtain when looking at realizations of a given random variable X. An equivalent characterization would be to interpret it as the average information one is missing when one would not know the value of the random variable (i.e., its realization), and a third one would be to interpret it as the average reduction of uncertainty about the possible values of a random variable having observed one or more realizations.

Akin to the information content h, entropy H is a dimensionless number and usually measured in bits (i.e., the expected number of binary digits needed to store the information), taking a logarithm to the base of 2.

Shannon entropy has many applications as we will see in the following and constitutes the core of all things labeled "information theory." Let us thus look a bit closer at this quantity.

Lemma 4.3. *Let X be some discrete random variable. Then its entropy $H(X)$ satisfies the two inequalities*

$$0 \leq H(X) \leq \log(n).$$

Note that the first inequality is a direct consequence of the properties of the information content, and the second follows from Gibbs' inequality [26].

With regard to entropy, probability distributions having maximal entropy are often of interest in applications as they can be seen as the least restricted ones (i.e., having the least a priori assumptions), given the model parameters. The *principle of maximum entropy* states that when choosing among a set of probability distributions with certain fixed properties, the preference should be given to distributions that have the maximal entropy among all considered distributions. This choice is justified as the one making the fewest assumptions on the shape of the distribution apart from the prescribed properties.

For discrete probability distributions, the uniform distribution is the one with the highest entropy among all other distributions on the same base set. This can be well seen in the example in Fig. 5.12: The entropy of a Bernoulli distribution takes its maximum at $p = 1/2$, the parameter value for which it corresponds to the uniform probability distribution on the two elements 0 and 1, each occurring with probability 1/2.

For continuous, real-valued random variables with a given finite mean μ and variance σ^2, the normal distribution with mean μ and variance σ^2 has highest entropy. Demanding non-negativity and a non-vanishing probability on the positive real numbers (i.e., an infinite support) with positive given mean μ yields the exponential distribution with parameter $\lambda = 1/\mu$ as a maximum entropy distribution.

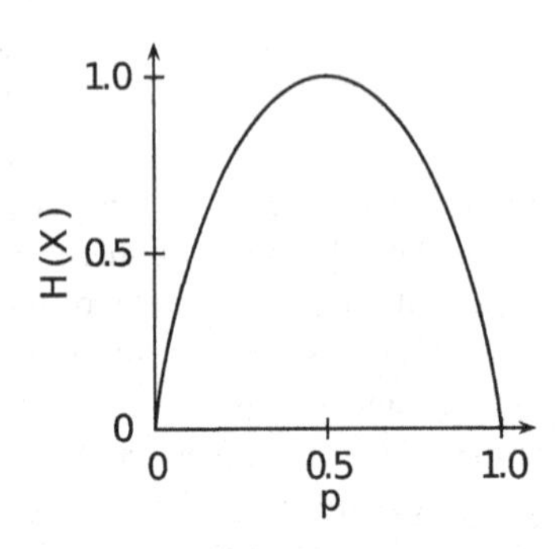

Fig. 5.12 Entropy $H(X)$ of a Bernoulli random variable X as a function of success probability $p = P(X = 1)$. The maximum is attained at $p = 1/2$

Examples

Before continuing, let us now compute some more entropies in order to get a feeling for this quantity.

For a uniform probability distribution P on n events $\Omega = \{\omega_1, \ldots, \omega_n\}$ each event has probability $P(\omega_i) = 1/n$ and we obtain

$$H(P) = -\sum_{i=1}^{n} \frac{1}{n} \log \frac{1}{n} = \log n,$$

as the maximal entropy for all discrete probability distributions on the set Ω.

Let us now compute the entropy of a Bernoulli random variable X, i.e., a binary random variable X taking values 0 and 1 with probability p and $1 - p$, respectively. For the entropy of X we get

$$H(X) = -(p \ \log \ p + (1-p) \ \log(1-p)).$$

See Fig. 5.12 for a plot of the entropy seen as a function of the success probability p. As expected, the maximum is attained at $p = 1/2$, corresponding to the case of the uniform distribution.

Computing the differential entropy of a normal distribution $N(\mu,\sigma^2)$ with mean μ and variance σ^2 yields

$$H\big(\big(N\big(\mu, \sigma^2\big)\big) = \frac{1}{2} \log\big(2\pi e \sigma^2\big),$$

and we see that the entropy does not depend on the mean value of the distribution but just its variance. This is not surprising, as the shape of the probability distribution is only changed by σ^2 and not μ.

For an example of how to compute the entropy of spike trains, see Sect. 5.6.

5.4.2.1 Joint Entropy

Generalizing the notion of entropy to two or more variables, we can define the so-called joint entropy to quantify the expected uncertainty (or expected information) in a joint distribution of random variables.

Definition 4.4 Joint Entropy. *Let X and Y be discrete random variables on some probability spaces. Then the joint entropy of X and Y is given by*

$$H(X, Y) = -E_{X,Y}[\log\ P(x, y)] = -\sum_{x,y} P(x, y)\ \log P(x, y), \qquad (5.4)$$

where $P_{X,Y}$ denotes the joint probability distribution of X and Y and the sum runs over all possible values x and y of X and Y, respectively.

This definition allows a straightforward extension to the case of more than two random variables.

The conditional entropy $H(X|Y)$ of two random variables X and Y quantifies the expected uncertainty (respectively expected information) remaining in a random variable X under the condition that a second variable Y was observed or equivalently as the reduction of the expected uncertainty in X upon the knowledge of Y.

Definition 4.5 Conditional Entropy. *Let X and Y be discrete random variables on some probability spaces. Then the conditional entropy of X given Y is given by*

$$H(X|Y) = -E_{X,Y}[\log\ P(x|y)] = -\sum_{x,y} P(x, y)\ \log P(x|y),$$

where $P_{X,Y}$ denotes the joint probability distribution of X and Y.

5.4.3 *Mutual Information*

In this section we will introduce the notion of mutual information, an entropy-based measure for the information shared between two (or more) random variables. Mutual information can be thought of as a measure for the mutual dependence of random variables, i.e., as a measure for how far they are from being independent.

We will give two different approaches to this concept in the following: a direct one based on the point-wise mutual information i and one using the idea of conditional entropy. Note that in essence, these are just different approaches to defining the same object. We give the two approaches in the following, hoping that they help in understanding the concept better. In Sect. 4.4 we will see yet another characterization in terms of the Kullback–Leibler divergence.

5.4.3.1 Point-Wise Mutual Information

In terms of information content, the case of considering two events that are independent is straightforward: One of the axioms tells us that the information content of the two events occurring together is the sum of the information contents of the single events. But what about the case where the events are non-independent? In this case we certainly have to consider the conditional probabilities of the two events occurring: If one event often occurs given that the other one occurs (think of the two events "It is snowing" and "It is winter"), the information overlap is higher than when the occurrence of one given the other is rare (think of "It is snowing" and "It is summer").

Using the notion of information from Sect. 4.1, let us express this in a mathematical way by defining the *mutual information* (i.e., shared information content) of two events. We call this the *point-wise mutual information* or *pmi*.

Definition 4.6 Point-Wise Mutual Information. *Let x and y be two events of a probability space* (Ω,Σ,P). *Then their point-wise mutual information (pmi) is given as*

$$\begin{aligned} i(x;y) :&= -\log \frac{P(x,y)}{P(x)P(y)} \\ &= -\log \frac{P(x|y)}{P(x)} \\ &= -\log \frac{P(y|x)}{P(y)}. \end{aligned} \tag{5.5}$$

Note that we used joint probability distribution of x and y is for the definition of $i(x;y)$ to avoid the ambiguities introduced by the conditional distributions. Yet, the latter are probably the easier way to gain a first understanding of this quantity.

Let us note that this measure of shared information is symmetric ($i(x;y) = i(y;x)$) and that it can take any real value, particularly also negative values. Such negative values of point-wise mutual information are commonly referred to as *misinformation* [64]. Point-wise mutual information is zero if the two events x and y are independent and it is bounded above by the information content of x and y. More generally, the following inequality holds:

$$-\infty \leq i(x;y) \leq \min\left\{ \underbrace{-\log\ P(x)}_{=h(x)}, \underbrace{-\log\ P(y)}_{=h(y)} \right\}.$$

Defining the information content of the co-occurrence of x and y as

$$i(x,y) := -\log\ P(x,y),$$

another way of writing the point-wise mutual information is

$$\begin{aligned} i(x;y) &= i(x) + i(y) - i(x,y), \\ &= i(x) - i(x|y), \\ &= i(y) - i(y|x), \end{aligned} \tag{5.6}$$

where the first identity above is readily obtained from Eq. 5.5 by just expanding the logarithmic term, and in the second and third line the formula for the conditional probability was used.

Table 5.1 Table of joint probabilities $P(\omega_a,\omega_b)$ of two events ω_a and ω_b

ω_a	ω_b	P(x,y)
a_1	b_1	0.2
a_1	b_2	0.5
a_2	b_1	0.25
a_2	b_2	0.05

Before considering mutual information of random variables as a straightforward generalization of the above, let us look at an example.

Say we have two probability spaces (Ω_a,Σ_a,P_a) and (Ω_b,Σ_b,P_b), with $\Omega_a = \{a_1,a_2\}$ and $\Omega_b = \{b_1,b_2\}$. We want to compute the point-wise mutual information of two events $\omega_a \in \Omega_a$ and $\omega_b \in \Omega_b$ subject to the joint probability distributions of ω_a and ω_b as given in Table 5.1. Note that the joint probability distribution can also be written as matrix

$$P(\omega_a, \omega_b) = \begin{pmatrix} 0.2 & 0.5 \\ 0.25 & 0.05 \end{pmatrix},$$

if we label rows by possible outcomes of ω_a and columns by possible outcomes of ω_b. The marginal distributions $P(\omega_a)$ and $P(\omega_b)$ are now obtained as row, respectively, column sums as $P(\omega_a = a_1) = 0.7$, $P(\omega_a = a_2) = 0.3$, $P(\omega_b = b_1) = 0.45$, and $P(\omega_b = b_2) = 0.55$.

We can now calculate the point-wise mutual information of, for example,

$$i(a_2; b_2) = -\log \frac{0.05}{0.3 \cdot 0.55} \approx 1.7 \text{ bits},$$

and

$$i(a_1; b_1) = -\log \frac{0.2}{0.7 \cdot 0.45} \approx -0.65 \text{ bits}.$$

Note again that in contrast to mutual information (that we will discuss in the next section), point-wise mutual information can take negative values called; see [64].

5.4.3.2 Mutual Information as Expected Point-Wise Mutual Information

Using point-wise mutual information, the definition of mutual information of two random variables is straightforward: Mutual information of two random variables is the expected value of the point-wise mutual information of all realizations.

Definition 4.7 Mutual Information. *Let X and Y be two discrete random variables. Then the mutual information $I(X;Y)$ is given as the expected point-wise mutual information:*

$$I(X;Y): = E_{X,Y}[i(x;y)] = \sum_y \sum_x P(x,y) i(x;y) = -\sum_y \sum_x P(x,y) log\left(\frac{P(x,y)}{P(x)P(y)}\right), \quad (5.7)$$

where the sums are taken over all possible values x of X and y of Y.

Remember again that the joint probability $P(x,y)$ is just a two-dimensional matrix where the rows are indexed by X-values and the columns by Y-values and that each row (column) tells us how likely each possible value of $Y(X)$ is, given the value x of X (y of Y) determined by the row (column) index. The rows (columns) sum to the marginal probability distributions $P(x)$ ($P(y)$), that can be written as vectors.

If X and Y are continuous random variables we just replace the sums by integrals and obtain what is known as *differential mutual information*:

$$I(X;Y) := \int_R \int_R P(x,y) \log\left(\frac{P(x,y)}{P(x)P(y)}\right) dx dy. \quad (5.8)$$

Here $P(x,y)$ denotes the joint probability distribution function of X and Y, and $P(x)$ and $P(y)$ the marginal probability distribution functions of X and Y, respectively.

As we can see, mutual information can be interpreted as the information (i.e., entropy) shared by the two variables, hence its name. Like point-wise mutual information, it is a symmetric quantity $I(X;Y) = I(Y;X)$ and in contrast to point-wise mutual information it is nonnegative, $I(X;Y) \geq 0$. Note though that it is not a metric, as in the general case it does not satisfy the triangle inequality. Furthermore, we have $I(X;X) = H(X)$, and this identity is the reason why entropy is sometimes also called *self-information*.

Taking the expected value of Eq. 5.5 and using the notion of conditional entropy, we can define mutual information between two random variables as follows:

$$I(X;Y): = H(X) + H(Y) - H(X,Y), = H(X) - H(X|Y), = H(Y) - H(Y|X), \quad (5.9)$$

where in the last two steps the identity $H(X,Y) = H(X) + H(Y|X) = H(Y) + H(X|Y)$ was used. Note that Eq. 5.9 is the generalization of Eq. 5.6 to the case of random variables. See Fig. 5.13 for an illustration of the relation between the different entropies and mutual information.

A possible interpretation of mutual information of two random variables X and Y is to consider it as a measure for the shared entropy between the two variables.

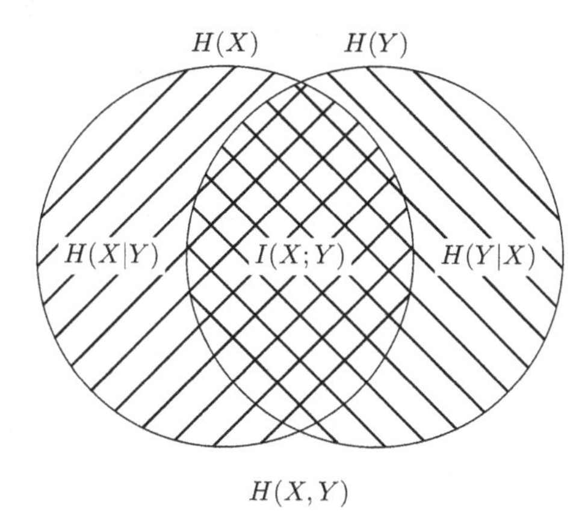

Fig. 5.13 Venn diagram showing the relation between the entropies $H(X)$ and $H(Y)$, the joint entropy $H(X,Y)$, the conditional entropies $H(X|Y)$ and $H(Y|X)$, and mutual information $I(X;Y)$

5.4.3.3 Mutual Information and Channel Capacities

We will look at channels in Shannon's sense of communication in the following and relate mutual information to channel capacity. But rather than looking at the subject in its full generality, we restrict ourselves to discrete, memoryless channels. The interested reader is pointed to [26] for a more thorough treatment of the subject.

Let us take as usual X and Y for the signal transmitted by some sender and received by some receiver, respectively. In terms of information transmission, we can interpret mutual information $I(X;Y)$ as the average amount of information the received signal constrains about the transmitted signal, where the averaging is done over the probability distribution of the source signal P_X. This makes mutual information a function of P_X and $P_{Y|X}$ and as we know, it is a symmetric quantity.

Shannon defines the *capacity* C of some channel as the maximum amount of information that a signal Y received by the receiver can contain about the signal X transmitted through the channel by the source.

In terms of mutual information $I(X;Y)$ we can define the channel capacity as the maximum mutual information $I(X;Y)$ among all realizations of the signal X. Channel capacity is thus not dependent on the distribution of P_X of X but rather a property of the channel itself, i.e., a property of the conditional distribution $P_{Y|X}$ and as such asymmetric and causal [85, 112].

Note that channel capacity is bound from below by 0 and from above by the entropy $H(X)$ of X, with the maximal capacity being attained by a noise-free channel. In the presence of noise the capacity is lower.

We will have a look at channels again when dealing with applications of the theory in Section 6.

5.4.3.4 Normalized Measures of Mutual Information

In many applications one is often interested in making values of mutual information comparable by employing a suitable normalization. Consequently, there exists a variety of proposed normalized measures of mutual information, most based on the simple idea of normalizing by one of the entropies that appear in the upper bounds of the mutual information. Using the entropy of one variable as a normalization factor, there are two possible choices and both were proposed: The so-called *coefficient of constraint* $C(X|Y)$ [25]

$$C(X|Y) := \frac{I(X;Y)}{H(Y)}$$

and the *uncertainty coefficient* $U(X|Y)$ [105]

$$U(X|Y) := \frac{I(X;Y)}{H(X)}.$$

These two quantities are obviously nonsymmetric but can easily be symmetrized, for example, by setting

$$U(I,J) := \frac{H(I)U(I|J) + H(J)U(J|I)}{H(I) + H(J)}.$$

Another symmetric normalized measure for mutual information, usually referred to as *redundancy measure*, is obtained when normalizing using the sum of the entropy of the variables

$$R = \frac{I(X;Y)}{H(X) + H(Y)}.$$

Note that R takes its minimum of 0 when the two variables are independent and its maximum when one variable is completely redundant knowing the other.

Note that the list of normalized variants of mutual information given here is far from complete. But as said earlier, the principle behind most normalizations is to use one or a combination of the entropies of the involved random variables as a normalizing factor.

5.4.3.5 Multivariate Case

What if we want to calculate the mutual information between not only two random variables but rather three or more? A natural generalization of mutual information to this so-called *multivariate* case is given by the following definition using conditional entropies and is also called *multi-information* or *integration* [107].

The mutual information of three random variables $X_1, \ldots, X_3$ is given by

$$I(X_1;X_2;X_3) := I(X_1;X_2) - I(X_1;X_2|X_3),$$

where the last term is defined as

$$I(X_1;X_2|X_3) := E_{X_3}[I(X_1;X_2)|X_3],$$

the latter being called the *conditional mutual information* of X_1 and X_2 given X_3. The conditional mutual information $I(X_1; X_2|X_3)$ can also be interpreted as the average common information shared by X_1 and X_2 that is not already contained in X_3.

Inductively, the generalization to the case of n random variables $X_1, \ldots, X_n$ is straightforward:

$$I(X_1;\ldots;X_n) := I(X_1;\ldots;X_{n-1}) - I(X_1;\ldots;X_{n-1}|X_n),$$

where the last term is again the conditional mutual information

$$I(X_1;\ldots;X_{n-1}|X_n) := E_{X_n}[I(X_1;\ldots;X_{n-1})|X_n].$$

Beware that while the interpretations of mutual information directly generalize from the bivariate case $I(X;Y)$ to the multivariate case $I(X_1; \ldots;X_n)$, there is an important difference between the bivariate and the multivariate measure. Whereas mutual information $I(X;Y)$ is a nonnegative quantity, multivariate mutual information (MMI for short) behaves a bit differently than the usual mutual information in the aspect that it can also take negative values which makes this information-theoretic quantity sometimes difficult to interpret.

Let us first look at an example of three variables with positive MMI. To make things a bit more hands on, let us look at three binary random variables, one telling us whether it is cloudy, the other whether it is raining, and the third one whether it is sunny. We want to compute $I(rain;\ no\ sun;\ cloud)$. In our model, clouds can cause rain and can block the sun, and so we have

$$I(rain; \text{no}\ \ sun|cloud) \leq I(rain; \text{no}\ \ sun),$$

as it is more likely that it is raining and there is no sun visible when it is cloudy than when there are no clouds visible. This results in positive MMI for $I(rain;\ no\ sun;\ cloud)$, a typical situation for a common-cause structure in the variables: here, the fact that the sun is not shining can partly be due to the fact that it is raining and partly due to the fact that there are clouds visible.

In a sense the inverse is the situation where we have two causes with a common effect: This situation can lead to negative values for the MMI; see [67]. In this situation, observing a common effect induces a dependency between the causes that did not exist before. This fact is called "explaining away" in the context of Bayesian networks; see [84]. Pearl [84] also gives a car-related example where the three (binary) variables are "engine fails to start" (X), "battery dead" (Y), and "fuel pump broken" (y). Clearly, both Y and Z can cause X and are uncorrelated if we have no knowledge of the value of X. But fixing the common effect X, namely, observing that the engine did not start, induces a dependency between Y and Z that can lead to negative values of the MMI.

Another problem with the n-variate case to keep in mind is the combinatorial explosion of the degrees of freedom regarding their interactions. As a priori every nonempty subset of the variables could interact in an information-theoretic sense, this yields 2^{n-1} degrees of freedom.

5.4.4 A Distance Measure for Probability Distributions: The Kullback–Leibler Divergence

The *Kullback–Leibler divergence* [57] (or *KL divergence* for short) is a kind of "distance measure" on the space of probability distributions: Given two probability distributions on the same base space Ω interpreted as two points in the space of all probability distributions over the base set Ω, it tells us how far they are "apart."

We again use the usual expectation-value construction as used for the entropy before.

Definition 4.8 Kullback–Leibler Divergence. *Let P and Q be two discrete probability distributions over the same base space Ω. Then the Kullback–Leibler divergence of P and Q is given by*

$$D_{\mathrm{KL}}(P||Q) := \sum_{\omega \in \Omega} P(\omega) \log \frac{P(\omega)}{Q(\omega)}. \tag{5.10}$$

The Kullback–Leibler divergence is nonnegative $D_{\mathrm{KL}}(P||Q) \geq 0$ (and it is zero if P equals Q almost everywhere), but it is not a metric in the mathematical sense as in general it is nonsymmetric $D_{\mathrm{KL}}(P||Q) \neq D_{\mathrm{KL}}(Q||P)$, and it does not fulfill the triangle inequality. Note that in their original work, Kullback and Leibler [57] defined the divergence via the sum

$$D_{\mathrm{KL}}(P||Q) + D_{\mathrm{KL}}(Q||P),$$

making it a symmetric measure. $D_{\mathrm{KL}}(P||Q)$ is additive for independent distributions, namely,

$$D_{\mathrm{KL}}(P||Q) = D_{\mathrm{KL}}(P_1||Q_1) + D_{\mathrm{KL}}(P_2||Q_2),$$

where the two pairs P_1, P_2 and Q_1, Q_2 are independent probability distributions with the joint distributions $P = P_1P_2$ and $Q = Q_1Q_2$, respectively.

Note that the expression in Eq. 5.10 is nothing else than the expected value $E_P[\log P - \log Q]$ with the expectation value taken with respect to P, which in term can be interpreted as "expected distance of P and Q," measured in terms of the information content. Another interpretation can be given in the language of codes: $D_{\mathrm{KL}}(P||Q)$ is the average number of extra bits needed to code samples from P using a code book based on Q.

Analogous to previous examples, the KL divergence can also be defined for continuous random variables in a straightforward way via

$$D_{\mathrm{KL}}(P||Q) = \int_{\mathrm{R}} p(x) \log \left(\frac{p(x)}{q(x)} \right) \, \mathrm{d}x,$$

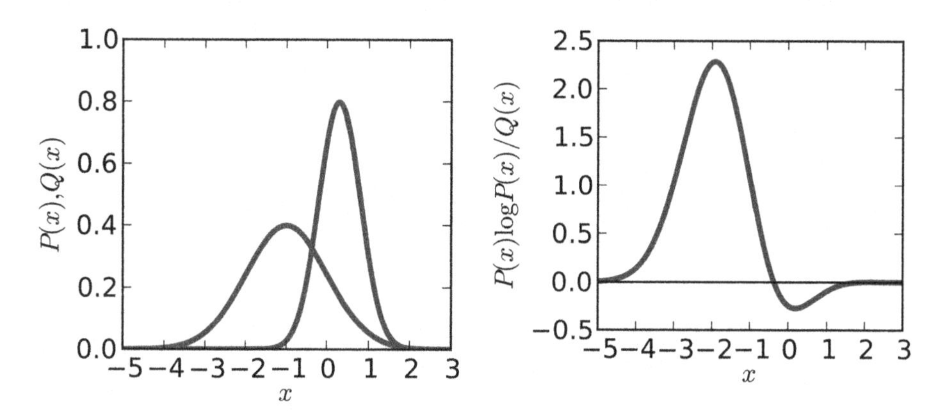

Fig. 5.14 The probability densities of two Gaussian probability distributions (*left*) and the quantity $P(x) log P(x)/Q(x)$ that yields the KL divergence when integrated (*right*)

where P_1, P_2 and Q_1, Q_2 denote the pdf of two continuous probability distributions $P = P_1P_2$ and $Q = Q_1Q_2$.

Expanding the logarithm in Eq. 5.10 we can write the Kullback–Leibler divergence between two probability distributions P and Q in terms of entropies as

$$D_{\mathrm{KL}}(P||Q) = -E_P(\log\ q(x)) + E_P(\log\ p(x)) = H^{\mathrm{cross}}(P,Q) - H(P),$$

where p and q denote the pdf or pmf of the distributions P and Q and $H(P,Q)^{\mathrm{cross}}$ is the so-called *cross-entropy* of P and Q given by

$$\mathrm{H}^{\mathrm{cross}}(P,Q) := -E_P(\log\ Q).$$

This relation lets us easily compute a closed form of the KL divergence for many common families of probability distributions. Let us, for example, look at the value of the KL divergence between two normal distributions $P : N(\mu_1,{\sigma_1}^2)$ and $Q : N(\mu_2,{\sigma_2}^2)$; see Fig. 5.14. This can be calculated as

$$D_{\mathrm{KL}}(P||Q) = \frac{(\mu_1 - \mu_2)^2}{2\sigma_2^2} + \frac{1}{2}\left(\frac{\sigma_1^2}{\sigma_2^2} - \log\frac{\sigma_1^2}{\sigma_2^2} - 1\right).$$

Another example: The KL divergence between two exponential distributions $P : Exp(\lambda_1)$ and $Q : Exp(\lambda_2)$ is

$$D_{\mathrm{KL}}(P||Q) = \log(\lambda_1) - \log(\lambda_2) + \frac{\lambda_2}{\lambda_1} - 1.$$

Using the Kullback–Leibler divergence, we can give yet another characterization of mutual information: It is a measure of how far two measured variables are from being independent, this time in terms of the Kullback–Leibler divergence.

$$
\begin{aligned}
I(X;Y) &= H(X) - H(X|Y) \\
&= \underbrace{-\sum_{x} P(x)\log(P(x))}_{=-\sum_{x,y} P(x,y) log(P(x))} \quad + \sum_{x,y} P(x,y)\log(P(x|y)) \\
&= \sum_{x,y} P(x,y)\log\left(\frac{P(x|y)}{P(x)}\right) \\
&= \sum_{x,y} P(x,y)\log\left(\frac{P(x,y)}{P(x)P(y)}\right) \\
&= D_{\mathrm{KL}}(P(x,y)||P(x)P(y))
\end{aligned}
\tag{5.11}
$$

Thus, mutual information of two random variables can be seen as the KL divergence of their underlying joint probability distribution from the products of their marginal probability distributions, i.e., as a measure for how far the two variables are from being independent.

5.4.5 Transfer Entropy: Conditional Mutual Information

In the past, mutual information was often used as a measure of information transfer between units (modeled as random variables) in some system. This approach faces the problem that mutual information is a symmetric measure and does not have an inherent directionality. In some applications this symmetry is not desired though, namely, whenever we want to explicitly obtain information about the "direction of flow" of information, for example, to measure causality in an information-theoretic setting; see Sect. 6.5.

In order to make mutual information a directed measure, a variant called *time-lagged mutual information* was proposed, calculating mutual information for two variables including a previous state of the source variable and a next state of the destination variable (where discrete time is assumed).

Yet, as Schreiber [94] points out, while time-lagged mutual information provides a directed measure of information transfer, it does not allow for a time-dynamic aspect as it measures the statically shared information between the two elements. With a suitable conditioning on the part of the variables, the introduction of a time-dynamic aspect is possible though. The resulting quantity is commonly referred to as *transfer entropy* [94]. Its common definition is the following.

Definition 4.9 Transfer Entropy. *Let X and Y be discrete random variables given on a discrete-time scale, and let $k, l \geq 1$ be two natural numbers. Then the transfer entropy from Y to X with k memory steps in X and l memory steps in Y is defined as*

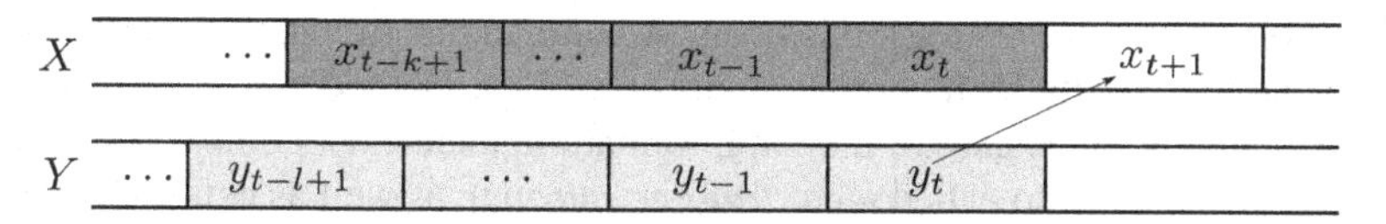

Fig. 5.15 Computing transfer entropy $TE_{Y \to X}$ from source Y to target X at time t as a measure of the average information present in y_t about the future state x_{t+1}. The memory vectors x_n^k and y_n^k are shown in *gray*

$$TE_{Y \to X} : \sum_{x_{n+1}, x_n^k, y_n^l} P\left(x_{n+1}, x_n^k, y_n^l\right) \log \frac{P\left(x_{n+1} \middle| x_n^k, y_n^l\right)}{P\left(x_{n+1} \middle| x_n^k\right)},$$

where we denoted by x_n, y_n the value of X and Y at time n and by y the past k values of X, counted from time n on $x_n^k := (x_n, x_{n-1}, \ldots, x_{n-k+1})$ and analogously $y_n^l := (y_n, y_{n-1}, \ldots, y_{n-l+1})$, Fig. 5.15.

Although this definition might look complicated at first, the idea behind it is quite simple. It is merely the Kullback–Leibler divergence between the two conditional probability distributions $P(x_{n+1}|x_n^k)$ and $P(x_{n+1}|x_n^k, y_n^l)$,

$$TE_{Y \to X} = D_{\mathrm{KL}}\left(P\left(x_{n+1} \middle| x_n^k\right) || P\left(x_{n+1} \middle| x_n^k, y_n^l\right)\right),$$

i.e., a measure of how far the two distributions are from fulfilling the generalized Markov property (see Sect. 3.8)

$$P\left(x_{n+1} \middle| x_n^k\right) = P\left(x_{n+1} \middle| x_n^k, y_n^l\right). \tag{5.12}$$

Note that for small values of transfer entropy, we can say that Y has little influence on X at time t, whereas we can say that information is transferred from Y to X at time t when the value is large. Yet, keep in mind that transfer entropy is just a measure of statistical correlation; see Sect. 6.5.

Another interpretation of transfer entropy is seeing it as a conditional mutual information $I(Y^{(l)}; X'|X^{(k)})$, measuring the average information the source Y constrains about the next state X' of the destination X that was not contained in the destination's past $X^{(k)}$ (see [62]) or alternatively as the average information provided by the source about the state transition in the destination; see [51, 62].

As so often before, the concept can be generalized to the continuous case [51], although the continuous setting introduces some subtleties that have to be addressed.

Concerning the memory parameters k and l of the source and the destination, although arbitrary choices are possible, the values chosen fundamentally influence the nature of the questions asked. In order to get correct measures for systems being far from Markovian (i.e., systems which states are not influenced by more than a certain fixed number of preceding system states), high values of k have to be used, and for non-Markovian systems, the case $k \to \infty$ has to be considered. On the other hand, commonly just one previous state of the source variable is considered in applications, setting $l = 1$ [62], this being also due to the growing data intensity in k and l and the usually high computational cost of the method.

Note that akin to the case of mutual information, there exist point-wise versions of transfer entropy (also called *local transfer entropy*), as well as extensions to the multivariate case; see [62].

5.5 Estimation of Information-Theoretic Quantities

As we have seen in the preceding sections, one needs to know the full sample spaces and probability distributions of the random variables involved in order to precisely calculate information-theoretic quantities such as the entropy, mutual information, or transfer entropy. But obtaining this data is in most cases impossible in reality, as the spaces are usually high dimensional and sparsely sampled, rendering the direct methods for the calculation of such quantities impossible to carry out. A way around this problem is to come up with estimation techniques that estimate entropies and derived quantities such as mutual information from the data. Over the last decades a large body of research was published concerning the estimation of entropies and related quantities, leading to a whole zoo of estimation techniques, each class having its own advantages and drawbacks. So rather than a full overview, we will give a sketch of some central ideas here and give references to further literature. The reader is also pointed to the review articles [10, 79].

Before looking at estimation techniques for neural (and other) data, let us first give a swift and painless review of some important theoretical concepts regarding statistical estimation.

5.5.1 A Bit of Theory Regarding Estimations

From a statistical point of view, the process of estimation in its most general form can be regarded in the following setting: We have some data (say measurements or data obtained via simulations) that is believed to be generated by some stochastic process with an underlying non-autonomous, i.e., time dependent or autonomous probability distribution. We then want to estimate either the value of some function defined on that probability distribution (e.g., the entropy) or the shape of this probability distribution as a whole (from which we can then obtain an estimate of a derived quantity). This process is called *estimation* and a function mapping the data to an estimated quantities *estimator*. In this section we will first look at estimators and their desired properties and then look at what is called maximum likelihood estimation, the most commonly used method for the estimation of parameters in the field of statistics.

5.5.1.1 Estimators

Let $x = (x_1, \ldots, x_n)$ be a set of realizations of the random variable X that is believed to have a probability distribution that comes from a family of probability distributions P_θ parametrized by a parameter θ and assume that the underlying probability distribution of X is $P_{\theta_{\text{true}}}$.

Let $T : x \mapsto \hat{\theta}_{\text{true}}$ be an estimator for the parameter θ with the true value θ_{true}. For the value of the estimated parameter, we usually write $\hat{\theta}_{\text{true}} := T(x)$. The *bias* of $T(x)$ is the expected difference between $\hat{\theta}_{\text{true}}$ and θ_{true}:

$$bias(T) := E_X\left[\hat{\theta}_{\text{true}} - \theta_{\text{true}}\right],$$

and an estimator with vanishing bias is called *unbiased.*

One usually strives to obtain unbiased estimators that are also *consistent*, i.e., for which the estimated value $\hat{\theta}_{\text{true}}$ converges to the value of the true parameter θ_{true} in probability as the sample x increases in size, i.e., as $n \to \infty$:

$$\lim_{n\to\infty} P(|T(X) - \theta_{\text{true}}| > \varepsilon) = 0.$$

Another important property of an estimator is its variance *var*(T), and an unbiased estimator having the minimal variance among all unbiased estimators of the same parameter is called *efficient.*

Yet another measure often used when assessing the quality of an estimator T is its mean squared error

$$MSE(T) = (bias(T))^2 + var(T),$$

and as we can see, any unbiased estimator with minimal variance minimizes the mean squared error.

Without further going into detail here, it is noted that there exists a theoretical lower bound to the minimal variance obtainable by an unbiased estimator, the *Cramér-Rao bound.* The Cramér-Rao bound sets the variance of the estimator in relation to the so-called Fisher information (that can be set into relation with mutual information, see [17, 113]). The interested reader is pointed to [2, 58].

5.5.1.2 Estimating Parameters: The Maximum Likelihood Estimator

Maximum likelihood estimation is the most widely used estimation technique in statistics and, as we will see in the next few paragraphs, a straightforward procedure that in essence tells us what the most likely parameter value in an assumed family of probability distributions is, given a set of realizations of a random variable that is believed to have an underlying probability distribution from the family considered.

In statistical applications one often faces the following situation: We have a finite set of realizations $\{x_i\}_i$ of a random variable X. We assume X to have a probability distribution $f(x,\theta_{\text{true}})$ in a certain parametrized class of probability distributions $\{f(x,\theta)\}_\theta$, where the true parameter θ_{true} is unknown. The goal is to get an estimate $\hat{\theta}_{\text{true}}$ of θ_{true} using the realizations $\{x_i\}_i$, i.e., to do statistical inference of the parameter θ. Let us consider the so-called likelihood function

$$L(\theta|x) = P_\theta(X = x) = f(x|\theta)$$

as a function of θ. It is a measure of how likely it is that the parameter of the probability distribution has the value θ, given the observed realization x of X. In maximum likelihood estimation, we look for the parameter that maximizes the likelihood function. This is $\hat{\theta}_{\text{true}}$:

$$\hat{\theta}_{\text{true}} = argmax_\theta L(\theta|x).$$

Choosing a value of $\theta = \hat{\theta}_{\text{true}}$ minimizes the KL divergence between P_θ and $P_{\theta_{\text{true}}}$ for all possible values of θ. The value $\hat{\theta}_{\text{true}}$, often written as $\hat{\theta}_{\text{MLE}}$, is called the *maximum likelihood estimate* (MLE for short) of θ_{true}.

In this setting, one often not uses the likelihood function directly, but works with the *log* of the likelihood function (this is referred to as log-likelihood). Why? The likelihood functions are often very complicated and situated in high dimensions, making it impossible to find a maximum of the function analytically. Thus, numerical methods (such as Newton's method and variants or the simplex method) have to be employed in order to find a solution. These numerical methods work best (and can be shown to converge to a unique solution) if the function they operate on is concave (bowl-shaped, where the closed end is on the top). The log function has the property to make the likelihood function concave in many cases, that being the reason why one considers the log-likelihood function, rather than the likelihood function directly; see also [80].

5.5.2 Regularization

Having looked at some core theoretical concepts regarding the estimation of quantities depending on probability distributions, let us now come back to dealing with real data.

As in real-world data, the involved probability distributions are often continuous and infinite-dimensional, the resulting estimation problem is very difficult (if not impossible) to solve in its original setting. As a remedy, the problem is often *regularized*, i.e., mapped to a discrete, more easily solvable problem. This of course introduces errors and often makes a direct estimation of the information-theoretic quantities impossible, but even in that simplified model we can estimate lower bounds of the quantities that we are interested in.

By using Shannon's *information-processing inequality* [26]

$$I(X;Y) \geq I(S(X);T(Y)),$$

where X and Y are (discrete) random variables and S and T are measurable maps and choosing the mappings S and T as our regularization mappings (you might also regard them as parameters) we can change the coarseness of the regularization. The regularization can be chosen arbitrarily coarse, i.e., choosing S and T as constant functions, but this of course comes with a price. For example, in the latter case of constant S and T, the mutual information $I(S(X);\ S(Y))$ would be equal to 0, clearly not a very useful estimate. This means that a trade-off between complexity reduction and the quality of the estimation has to be made. In general, there exists no all-purpose recipe for this, each problem requiring an appropriate regularization.

As this discretization technique has become the standard method in many fields, we will solely consider the regularized, discrete case in the following and point the reader to the review article [10] concerning the continuous case.

In the neurosciences, such a regularization technique was also proposed and is known as the "direct method" [19, 104]. Here, spike trains of recorded neurons are discretized into time bins of a given fixed width, and the neuronal spiking activity is interpreted as a series of symbols from an alphabet defined via the observed spiking pattern in the time bins.

5.5.3 Nonparametric Estimation Techniques

Commonly, two different classes of estimation techniques regarding the shape of probability distributions are distinguished. Parametric estimation techniques assume that the probability distribution is contained in some family of probability distributions having some prescribed shape (see Sect. 3.7). Here, one estimates the value of the parameter from the data observed, whereas nonparametric estimation techniques make no assumptions about the shape of the underlying distribution. We will solely look at nonparametric estimation techniques in the following as in many cases one tries to not assume prior information about the shape of the distribution.

Histogram-based estimation is the most popular and most widely used estimation technique. As the name implies, this method uses a histogram obtained from the data to estimate the probability distribution of the underlying random generation mechanism.

For the following, assume that we obtained a finite set of N samples $x = \{x_i\}_i$ of some real random variable X defined over some probability space (Ω,Σ,P). We then divide the domain of X into $m \in \mathrm{N}$ equally sized bins $\{b_i\}_i$ and subsequently count the number of realizations x_i in our data set contained in each bin. Here, the number m of bins can be freely chosen. It controls the coarseness of our discretization,

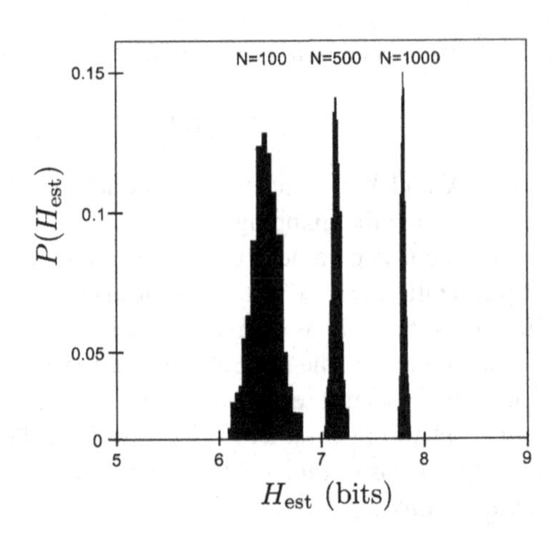

Fig. 5.16 Estimation bias for a non-bias-corrected histogram-based maximum likelihood estimator H_{est} of the entropy of a given distribution with true entropy $H = 8$ bits. Estimated values are shown for three different sample sizes N (Adapted from [79], Fig. 1)

where the limit $m \to \infty$ is the continuous case. This allows us to define relative frequencies of occurrences for X with respect to each bin that we interpret as estimations $\hat{p}_i^m$ (note that we make the dependence on the number of bins m explicit in the notation) of the probability of X taking a value in bin b_i which we denote by $p_i^m = P(X \in b_i)$. The law of large numbers then tells us that our estimated probability values converge to the real probabilities as $N \to \infty$.

Note that although histogram-based estimations are usually called nonparametric as they do not assume a certain shape of the underlying probability distribution, they do have parameters, namely, one parameter for each bin and the estimated probability value $\hat{p}_i^m$. These estimates $\hat{p}_i^m$ can also be interpreted as maximum likelihood estimates of p_i^m.

The following defines an estimator of the entropy based on the histogram. It is often called "plug-in" estimator:

$$\hat{H}_{\text{MLE}}(x) := -\sum_{i=1}^{m} \hat{p}_i^m \log p_i^m. \tag{5.13}$$

The are some problems with this estimator $\hat{H}_{\text{MLE}}(X)$, though. Its convergence to the true value $H(X)$ can be slow and it is negatively biased, i.e., its value is almost always below the true value $H(X)$; see [4, 79, 82, 83]. This shift can be quite significant even for large N; see Fig. 5.16 and [79]. More specifically, one can show that the expected value of the estimated entropy is always smaller than the true value of the entropy

$$E_X\left[\hat{H}_{\text{MLE}}(x)\right] \leq H(X),$$

where the expectation value is taken with respect to the true probability distribution P.

Bias generally is a problem for history-based estimation techniques [14, 82, 95], and although we can correct for the bias, this may not always be a feasible solution [4]. Nonetheless we will have a look at a bias-corrected version of the estimator given in Eq. 5.13 below.

As a remedy to the bias problem, Miller and Madow [71] calculated the bias of the estimator of Eq. 5.13 and came up with a bias-corrected version of the maximum likelihood estimator for the entropy, referred to as *Miller-Madow* estimator:

$$\hat{H}_{\mathrm{MM}}(x) := \hat{H}_{\mathrm{MLE}}(x) + \frac{\hat{m} - 1}{2N},$$

where $\hat{m}$ is an estimate of the number of bins with nonzero probability. We will not go into the detail of the method here; the interested reader is referred to [71].

Another way of bias-correction $\hat{H}_{\mathrm{MLE}}(X)$ is the so-called "jack-knifed" version of the maximum likelihood estimator by Efron and Stein [34]:

$$\hat{H}_{\mathrm{JK}}(x) := N \cdot \hat{H}_{\mathrm{MLE}}(x) + \frac{N-1}{N} \sum_{j=1}^{N} \hat{H}_{\mathrm{MLE}}(x \backslash \{x_i\}),$$

Yet another bias-corrected variant of the MLE estimator based on polynomial approximation is presented in [79], for which also bounds on the maximal estimation error were derived.

In an effort to overcome the problems faced by histogram-based estimation, many new and more powerful estimation techniques have emerged over the last years, both for entropy and other information-theoretic quantities. As our focus here is to give an introduction to the field, we will not review all of those methods here but rather point the interested reader to the literature where a variety of approaches are discussed. There exist methods based on the idea of adaptive partitioning of sample space [21], ones using entropy production rates and allowing for confidence intervals [99], ones using Bayesian methods [75, 90, 99], and ones based on density estimation using nearest neighbors [55], along with many more. See [46] for an overview concerning several estimation techniques for entropy and mutual information. We note here that in contrast to estimations of entropy, estimators of mutual information are usually positively biased, i.e., tend to overestimate mutual information.

5.6 Information-Theoretic Analyses of Neural Systems

Some time after its discovery by Shannon, neuroscientists started to recognize information theory as a valuable mathematical tool to assess information processing in neural systems. Using information theory, several questions regarding information processing and the neural code can be addressed in a quantitative way, among those:

- How much information single cells or populations carry about a stimulus and how this information is coded.
- What aspects of a stimulus are encoded in the neural system.

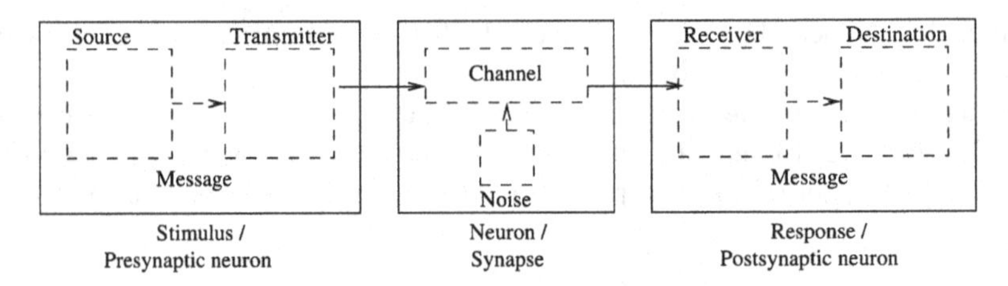

Fig. 5.17 An information-theoretic view on neural systems. Neurons can either act as channels in the information-theoretic sense, relaying information about some stimulus or as senders and receivers with channels being synapses

- How "effective connectivity" [40] in neural systems can be defined via causal relationships between units in the system.

See Fig. 5.17 for an illustration of how Shannon's theory can be used in a neural setting.

Attneave [6] and Barlow [9] were the first to consider information processing in neural systems from an information-theoretic point of view. Subsequently, Eckhorn and Pöpel [32, 33] applied information-theoretic methods to electrophysiologically recorded data of neurons in a cat. But being data intensive in nature, these methods faced some quite strong restrictions during that time, namely, the limited amount of computing power (and computer memory) and the limited amount (and often low quality) of data obtainable via measurements at that time.

But over the last decades, available computing became more and more available, and classical measurement techniques were improved, along with new ones emerging such as fMRI, MEG, and calcium imaging. This made information-theoretic analyses of neural systems more and more feasible, and through the invention of recording techniques such as MEG and fMRI, it is nowadays even possible to perform such analyses on a system scale for the human brain in vivo. Yet, even with the newly available recording techniques today, there are some conceptual difficulties with information-theoretic analyses as it is often a challenge to obtain enough data in order to get good estimates of information-theoretic quantities. Special attention has to be paid to using the data efficiently, and the validity of such analyses has to be assessed to their statistical significance.

In the following we will discuss some conceptual questions relevant when regarding information-theoretic analyses of neural systems. More detailed reviews can be found in [15, 35, 93, 111].

5.6.1 The Question of Coding

Marr described "three levels at which any machine carrying out an information-processing task must be understood" [68] [Chap. 1.2]. They are:

1. Computational theory: What is the goal of the computation, why is it appropriate, and what is the logic of the strategy by which it can be carried out?

2. Representation and algorithm: How can this computational theory be implemented? In particular, what is the representation for the input and output, and what is the algorithm for the transformation?
3. Hardware implementation: How can the representation and algorithm be realized physically?

When performing an information-theoretic analysis of a system, one naturally faces the fundamental problem related to the coding of the information: In order to calculate (i.e., estimate) information-theoretic quantities, one has to define a family of probability distributions over the state space of the system, each member of that family describing one system state that is to be considered. As we know, all information-theoretic quantities such as entropy and mutual information (between the system state and the state of some external quantity) are determined by the probability distributions involved. The big question now is how to define the system state in the first point, a question which is especially difficult to answer in the case of neural systems on all scales.

One possible way to construct such a probabilistic model for a sensory neurophysiological experiment involving just one neuron is the following. Typically, the experiment consists of many trials, where per trial $i = 1, \ldots, n$ in some defined time window a stimulus S_i is presented eliciting a neural response $R(S_i)$ consisting of a sequence of action potentials. Presenting the same stimulus S many times allows for the definition of a probability distribution of responses $R(S)$ of the neuron to a stimulus S. This is modeled as a conditional probability distribution $P_{R|S}$. As noted earlier, we usually have no direct access to $P_{R|S}$ but rather have to find an estimate $\hat{P}_{R|S}$ from the available data. Note that in practice, usually the joint probability distribution $P(R,S)$ is estimated and estimates of conditional probability distributions are subsequently obtained from the estimate of the joint distribution.

Let us now assume that the stimuli are drawn from the set of stimuli $S = \{S_1, \ldots, S_k\}$ according to some probability distribution P_S (that can be freely chosen by the experimenter). We can then compute the mutual information between the stimulus ensemble S and its elicited response $R(S)$

$$I(S; R(S)) := H(R(S)|S) - H(S) = H(S|R(S)) - H(R(S))$$

using the probability distributions P_S and $\hat{P}_{R|S}$; see Sect. 4.3.

As usual, by mutual information we assess the expected shared information between the stimulus and its elicited response averaged over all stimuli and responses. In order to break this down to the level of single stimuli, we can either consider the point-wise mutual information or employ one of the proposed decompositions of mutual information such as *stimulus-specific information* or *stimulus-specific surprise*; see [20] for a review.

Having sketched the general setting, let us come back to the question of coding of information by the neurons involved. This is important as we have to adjust our model of the neural responses accordingly, the goal being to capture all relevant features of the neural response in the model.

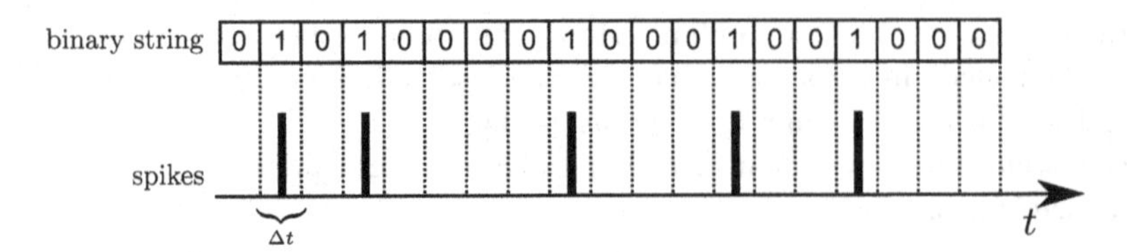

Fig. 5.18 Model of a spike train. The binary string is obtained through a binning of time

Regarding neural coding, there are two main hypotheses of how single neurons might code information: Neurons could use a *rate code*, i.e., encode the information via their mean firing rates, neglecting the timing patterns of spikes, or they could employ a *temporal code*, i.e., a code where the precise timing of single spikes plays an important role. Yet another hypothesis would be that neurons code information in bursts of spikes, i.e., groups of spikes emitted in a small time window, which is a variant of the time code. For questions regarding coding in populations, see the review [89].

Note that the question of neural coding is a highly debated one in the neurosciences as of today (see [42, 96]), and we do not want to favor one view point over the other in the following. As with many things in nature, there does not seem to be a clear black and white picture regarding neuronal coding. Rather it seems that a gradient of different coding schemes is employed depending on which sensory system is considered and at which stage of neuronal processing; see [19, 22, 42, 81, 93].

5.6.2 Computing Entropies of Spike Trains

Let us now compute the entropy of spike trains and subsequently single spikes, assuming that the neurons we model employ either a rate or a time code. We are especially interested in the maximal entropy attainable by our model spike trains as these can give us upper bounds for the amount of information such trains and even single spikes can carry in theory. The following examples here are adapted from [108]. Concerning the topics of spike trains and their analysis, the interested reader is also pointed to [92].

First, we define a model for the spike train emitted by a neuron measured for some fixed time interval of length T. We can consider two different models for the spike train, a continuous and a discrete one. In the continuous case, we model each spike by a Dirac delta function and the whole spike train as a combination of such functions. The discrete model is obtained from the continuous one by introducing small time bins of size Δt in a way that one bin can at most contain one spike, say $\Delta t = 2$ ms. We then assign to each bin in which no spike occurred a value of 0 and ones in which a spike occurred a value of 1; see Fig. 5.18.

Let us use this discrete model for the spike train of a neuron, representing a spike train as a binary string S in the following. Fixing the time span to be T and the bin

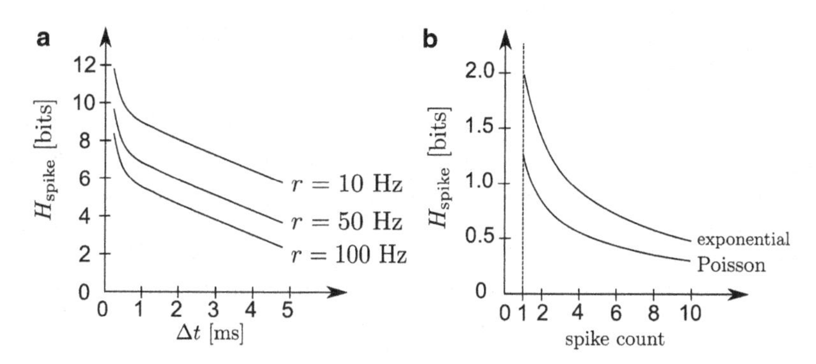

Fig. 5.19 Maximum entropy per spike for spike trains. (**a**) Time code with different rates r as a function of the size Δt of the time bins. (**b**) Rate code using Poisson and exponential spiking statistics (Figure adapted from [108] Fig. D.4)

width to be Δt, each spike train S has length $N = T/\Delta t$. We want to calculate the maximal entropy among all such spike trains n, subject to the condition that the number of spikes in S is a fixed number $r \leq N$ which we call the spike rate of S.

Let us now calculate the entropy in the firing pattern of a neuron of which we assume that spike timing carries important information, i.e., a neuron employing a time code. In order to keep the model simple, let us further assume that the spiking behavior is not restricted in any way, i.e., that all possible binary strings S are equiprobable. Then we can calculate the entropy of this uniform probability distribution P as

$$H(P) = \log \binom{N}{r}, \tag{5.14}$$

where $\binom{N}{r}$ denotes the binomial coefficient $\binom{N}{r} = \frac{N!(N-r)!}{r!}$, the number of all distinct binary strings of length N having exactly r nonzero entries. The entropy in Eq. 5.14 can be approximated by

$$H(P) \approx -\frac{N}{\ln 2}\left(\frac{N}{r}\ln\frac{N}{r} + \left(1 - \frac{N}{r}\right)\ln\left(1 - \frac{N}{r}\right)\right), \tag{5.15}$$

where ln denotes the natural logarithm to the base e. The expression in Eq. 5.15 is obtained by using the approximation formula

$$\log \binom{n}{k} \approx n\left(\frac{k}{n}\log\left(\frac{k}{n}\right) - \left(1 - \frac{k}{n}\right)\log\left(1 - \frac{k}{n}\right)\right)$$

which is valid for large n and k and in turn based on Stirling's approximation formula for $\ln n!$.

See Fig. 5.19a for the maximum entropy attainable by the time code as a function of bin size Δt for different firing rates r.

On the other hand, modeling a neuron that reacts to different stimuli with a graded response in its firing rate is usually done using a rate code. Assuming a rate code where the timing of spikes does not play any role yields different results, as we will see in the following; see Fig. 5.19b. In the rate code only the number of spikes N occurring in a given time interval of length T matters, i.e., we consider probability distributions $P_{N,T}$ parametrized by N and T describing how likely the occurrence of N spikes in a time window of length T is. Being well-backed with experimental data [24, 74, 101], a popular choice of $P_{N,T}$ is taking a Poisson distribution with some fixed mean $N = r \cdot T$, where r is thought of as the mean firing rate of the neuron.

The probability $P_{N,T}(n)$ of observing n spikes in an interval of length T now is given by the pmf of the Poisson distribution

$$P_{N,T}(n) = \frac{N^n e^{-N}}{n!}$$

and the entropy of $P_{N,T}$ computes as

$$H(P_{N,T}) = -\sum_n P_{N,T}(n) \log P_{N,T}(n).$$

Again using Stirling's formula this can be written as

$$H(P_{N,T}) \approx \frac{1}{2}(\log N - \log 2\pi). \tag{5.16}$$

Dividing the entropy $H(P_{N,T})$ by the number of spikes that occurred yields the entropy per spike. See Fig. 5.19b for a plot of the entropy per spike as a function of the number of observed spikes.

An interesting question is to ask for the maximal information (i.e., entropy) that spike trains can carry, assuming a rate code. Assuming continuous time and prescribing mean and variance of the firing rate, this leaves the exponential distribution P_{exp} as the one with the maximal entropy. The entropy of an exponentially distributed spike train with mean rate $r = 1/T(e^{\lambda} - 1)$ is

$$H(P_{\text{exp}}) \approx \log(1 + N) + N \log\left(1 + \frac{1}{N}\right),$$

see also Fig. 5.19b.

Note that while it was possible to compute the exact entropies in the preceding as we assumed full knowledge of the underlying probability distributions. This is of course not the case for data obtained by recordings. Here the estimation of entropies faces the bias-related problems of sparsely sampled probability distributions as discussed earlier. Concerning entropy estimation in spike trains, the reader is also pointed to [82].

5.6.3 *Efficient Coding?*

The principle of efficient coding [6, 9, 100] (also called *Infomax principle*) was first proposed by Attneave and Barlow. It views the early sensory pathway as a channel in Shannon's sense and postulates that early sensory systems try to maximize information transmission under the constraint of an efficient code, i.e., that neurons maximize mutual information between a stimulus and their output spike train, using as few spikes as possible. This minimization of spikes for a given stimulus results in a maximal compression of the stimulus data, minimizing redundancies between different neurons on a population level. One key prediction of this optimality principle is that neurons involved in the processing of stimulus data (and ultimately the whole brain) is adapted to natural stimuli, i.e., some form of natural (and structured) sensory input such as sounds or images rather than noise. For some sensory systems it could be shown that there is strong evidence that early stages of processing indeed perform an optimal coding; see, e.g., [77]. While first mainly the visual system was studied and it was shown that the Infomax principle holds here [9], other sensory modalities were also considered in the following years [13, 59–61, 109, 114].

But whereas the Infomax principle could explain certain experimental findings in the early sensory processing stages, the picture becomes less clear the more upstream the information processing in neural networks is considered. Here, other principles were also argued for; see, for example, [43].

On the system level, Friston et al. [36, 38] proposed an information-theoretic measure of free energy in the brain that can be understood as generalization of the concept of efficient coding. Also arguing for optimal information transfer, Norwich [76] gave a theory of perception based on information-theoretic principles. He argues that the information present in some stimulus is relayed to the brain by the sensory system with negligible loss. Many empirical equations of psychophysics can be derived from this model.

5.6.4 *Scales*

There are many scales at which information-theoretic analyses of neural systems can be performed. From the level of a single synapse [30, 65] over the level of single neurons [29, 93] over the population level [27, 35, 50, 87, 89] up to the system level [78, 110]. In the former cases the analyses are usually carried out on electrophysiologically recorded data of single cells, whereas on the system level data is usually obtained by EEG, fMRI, or MEG measurements.

Notice that most of the information-theoretic analyses of neural systems were done for early stages of sensory systems, focusing on the assessment of the amount of mutual information between some stimulus and its neural response. Here different questions can be answered about the nature and efficiency of the neural code and the information conveyed by neural representations of stimuli; see [12, 15, 91, 93]. This stimulus–response-based approach has already provided a lot of insight into the processing of information in early sensory systems, but things get more and more complicated the more downstream an analysis is performed [22, 93].

On the systems level, the abilities of neural systems to process and store information are due to interactions of neurons, populations of neurons, and sub-networks. As these interactions are highly nonlinear and in contrast to the early sensory systems neural activity is mainly driven by the internal network dynamics (see [5, 110]), stimulus–response-type models often are not very useful here. Here, transfer entropy has proven to be a valuable tool, making analyses of information transfer in the human brain in vivo possible [78, 110]. Transfer entropy can also be used as a measure for causality, as we will discuss in the next section.

5.6.5 Causality in the Neurosciences

The idea of *causality*, namely, the question of what are the causes resulting in the observable state and dynamics of complex systems of physical, biological, or social nature is a deep, philosophical question that has been driving scientists in all fields ever since. In a sense this question lies at the heart of science itself and as such is often notoriously difficult to answer.

In the neurosciences, this principle is related to one of the core questions of neural coding and subsequently neural information processing: What stimuli make neurons spike (or change their membrane potential for non-spiking neurons)? For many years now, neuroscientists have investigated neurophysiological correlates of information presented to a sensory system in form of stimuli.

While considerable progress has been made regarding the answer to this question in the early stages of sensory processing (see the preceding sections), where often a clear correlation between a stimulus and the resulting neuronal activity could be found, things get less and less clear the further downstream this question is addressed. In the latter case, neuronal activity is subject to higher and higher degrees of internal dynamics and a clear stimulus–response relation is often lacking.

Considering early sensory systems, even though merely a correlation between a stimulus and neural activity can be measured, it is justified to speak of causality here, as it is possible to actively influence the stimulus and observe the change in neural activity. Note that the idea of intervention is crucial here; see [7, 85].

Looking at more downstream systems or at the cognitive level, an active intervention albeit possible (but often not as directly as for sensory systems) may not have the same easy to detect effects on system dynamics. Here, often just statistical correlations can be observed, and in most cases, it is very hard if not impossible to show that the principle of causality in its purest form holds. Yet, one can still make some statements regarding what one might call "statistical causality" in this case, as we will see.

In an attempt to give a statistical characterization of the notion of causality, the mathematician Wiener [112] came up with the following probabilistic framing of this concept that came to be known as *Wiener causality*: Consider two stochastic processes $X = (X_t)_{t \in \mathbb{N}}$ and $Y = (Y_t)_{t \in \mathbb{N}}$. Then Y is said to Wiener-cause X if the knowledge of past values of Y diminishes uncertainty about the future values of X. Note that Wiener causality is a measure of predictive information transfer and not one of causality, and thus the naming is a bit unfortunate; see [63].

The economist Granger employed Wiener's principle of causality and developed the notion of what is nowadays called *Wiener-Granger causality* [16, 44]. Subsequently, the linear Wiener-Granger causality and its generalizaticons were often employed as measure of statistical causality in the neurosciences; see [16, 46]. Another model for causality in the neurosciences is *dynamic causal modeling* [37, 41, 102].

In contrast to dynamic causal modeling, causality measures based on information-theoretic concepts are usually purely data-driven and thus inherently model-free [46, 110]. This fact can be of advantage in some cases but we do not want to make a judgment here, calling one method better per se, as each has its advantages and drawbacks [39].

The directional and time-dynamic nature of transfer entropy allows using it as a measure of Wiener causality, as was proposed in the field of neurosciences recently [110]. As such, transfer entropy can be seen as a nonlinear extension of the concept of Wiener-Granger causality; see [66] for a comparison of transfer entropy to other measures.

Note again that transfer entropy still essentially is a measure of conditional correlation rather than one of direct effect (i.e., causality) and that correlation is not causation. Thus it is a philosophical question to which extent transfer entropy can be used to infer some form of causality, a question that we will not further pursue here, rather pointing the reader to [7, 46, 66, 85].

In any case the statistical significance of the inferred causality (remember that transfer entropy just measures conditional correlation) has to be verified. For trial-based data sets as often found in the neurosciences, this testing is usually done against the null hypothesis H_0 of average transfer entropy obtained by random shuffling of the data.

5.6.6 *Information-Theoretic Aspects of Neural Dysfunction*

Given the fact that information-theoretic analyses can provide insights about the functioning of neural systems, the next logical step is to ask how these might help in better understanding neural dysfunction and neural diseases.

The field one might call "computational neuroscience of disease" is an emerging field of research within the neurosciences; see the special issue of

Neural Networks [28]. The discipline faces some hard questions as in many cases dysfunction is observed on the cognitive (i.e., systems) level but has causes on many scales of neural function (subcellular, cellular, population, system).

Over the last years, different theoretical models regarding neural dysfunction and disease were proposed, among them computational models applicable to the field of psychiatry [48, 72], models for brain lesions [1], models of epilepsy [3], models for deep brain stimulation [70, 86], and models for aspects of Parkinson's [45, 73] and Alzheimer's [11, 56] disease, of abnormal auditory processing [31, 56], and for congenital prosopagnosia (a deficit in face identification) [103].

Some of these models employ information-theoretic ideas in order to assess differences between the healthy and dysfunctional states [8, 103]. For example, information-theoretic analyses of cognitive and systems-level processes in the prefrontal cortex were carried out recently [8, 53], and differences in information processing could be assessed between the healthy and dysfunctional system by means of information theory [8].

Yet, computational neuroscience of disease is a very young field of research, and it remains to be elucidated if and in what way analyses of neural systems employing information-theoretic principles could be of help in medicine on a broader scale.

5.7 Software

There exist several open source software packages that can be used to estimate information-theoretic quantities of neural data. The list below is by no means complete, but should give a good overview of things; see also [49]:

- Entropy: Entropy and mutual information estimation
  - URL: http://cran.r-project.org/web/packages/entropy.
  - Authors: Jean Hausser and Korbinian Strimmer.
  - Type: R package.
  - From the website: This package implements various estimators of entropy, such as the shrinkage estimator by Hausser and Strimmer, the maximum likelihood and the Millow-Madow estimator, various Bayesian estimators, and the Chao-Shen estimator. It also offers an R interface to the NSB estimator. Furthermore, it provides functions for estimating mutual information.
- Information-dynamics tool kit
  - URL: http://code.google.com/p/information-dynamics-toolkit.
  - Author: Joseph Lizier.
  - Type: standalone Java software.
  - From the website: Provides a Java implementation of information-theoretic measures of distributed computation in complex systems: i.e., information

storage, transfer, and modification. Includes implementations for both discrete and continuous-valued variables for entropy, entropy rate, mutual information, conditional mutual information, transfer entropy, conditional/complete transfer entropy, active information storage, excess entropy/predictive information, and separable information.

- ITE (information-theoretical estimators)
  - URL: https://bitbucket.org/szzoli/ite/.
  - Author: Zoltan Szabo.
  - Type: Matlab/Octave plug-in.
  - From the website: ITE is capable of estimating many different variants of entropy, mutual information, and divergence measures. Thanks to its highly modular design, ITE supports additionally the combinations of the estimation techniques, the easy construction and embedding of novel information-theoretical estimators, and their immediate application in information-theoretical optimization problems. ITE can estimate Shannon and Rényi entropy; generalized variance, kernel canonical correlation analysis, kernel generalized variance, Hilbert-Schmidt independence criterion, mutual information (Shannon, L2, Rényi, Tsallis), copula-based kernel dependency, and multivariate version of Hoeffding's Phi; complex variants of entropy and mutual information; and divergence (L2, Rényi, Tsallis), maximum mean discrepancy, and J-distance. ITE offers solution methods for Independent Subspace Analysis (ISA) and its extensions to different linear-, controlled-, post nonlinear-, complex-valued, partially observed systems, as well as to systems with nonparametric source dynamics.
- PyEntropy
  - URL: http://code.google.com/p/pyentropy.
  - Authors: Robin Ince, Rasmus Petersen, Daniel Swan, and Stefano Panzeri.
  - Type: Python module.
  - From the website: PyEntropy is a Python module for estimating entropy and information-theoretic quantities using a range of bias-correction methods.
- Spike train analysis tool kit
  - URL: http://neuroanalysis.org/toolkit.
  - Authors: Michael Repucci, David Goldberg, Jonathan Victor, and Daniel Gardner.
  - Type: Matlab/Octave plug-in.
  - From the website: Information-theoretic methods are now widely used for the analysis of spike train data. However, developing robust implementations of these methods can be tedious and time-consuming. In order to facilitate further adoption of these methods, we have developed the Spike Train Analysis Toolkit, a software package which implements several information-theoretic spike train analysis techniques.

- TRENTOOL
  - URL: http://trentool.de.
  - Authors: Michael Lindner, Raul Vicente, Michael Wibral, Nicu Pampu, and Patricia Wollstadt.
  - Type: Matlab plug-in.
  - From the website: TRENTOOL uses the data format of the open source MATLAB toolbox Fieldtrip that is popular for electrophysiology data (EEG/MEG/LFP). Parameters for delay embedding are automatically obtained from the data. TE values are estimated by the Kraskov-Stögbauer-Grassberger estimator and subjected to a statistical test against suitable surrogate data. Experimental effects can then be tested on a second level. Results can be plotted using Fieldtrip layout formats.

Acknowledgements The author would like to thank *Nihat Ay*, *Yuri Campbell*, *Aleena Garner*, *Jörg Lehnert*, *Timm Lochmann*, *Wiktor Młynarski*, and *Carolin Stier* for their useful comments on the manuscript.

References

1. J. Alstott, M. Breakspear, P. Hagmann, L. Cammoun, and O. Sporns. Modeling the impact of lesions in the human brain. *PLoS computational biology*, 5(6):e1000408, June 2009.
2. S-I. Amari, H. Nagaoka, and D. Harada. *Methods of information geometry*, volume 191 of *Translations of Mathematical Monographs*. American Mathematical Society, Providence, RI, 2000.
3. I. S. And and K. Staley, editors. *Computational Neuroscience in Epilepsy*. Academic Press, 2011.
4. A. Antós and I. Kontoyiannis. Convergence properties of functional estimates for discrete distributions. *Random Structures and Algorithms*, 19(3–4):163–193, 2001.
5. M. M. Arnold, J. Szczepanski, N. Montejo, J. M. Amigó, E. Wajnryb, and M. V. Sanchez-Vives. Information content in cortical spike trains during brain state transitions. *J Sleep Res*, 22(1):13–21, 2013.
6. F. Attneave. Some informational aspects of visual perception. *Psychol Rev*, 61(3):183–193, 1954.
7. N. Ay and D. Polani. Information Flows in Causal Networks. *Advances in Complex Systems*, 11(01):17–41, 2008.
8. F. Barcelo and R. T. Knight. An information-theoretical approach to contextual processing in the human brain: evidence from prefrontal lesions. *Cerebral cortex*, 17 Suppl 1:51–60, 2007.
9. H. B. Barlow. *Sensory Communication*, chapter Possible principles underlying the transformation of sensory messages, pages 217–234. MIT Press, 1961.
10. J. Beirlant and E. J. Dudewicz. Nonparametric entropy estimation: An overview. *Intern J Math Stat Sci*, 6(1):1–14, 1997.
11. B. S. Bhattacharya, D. Coyle, and L. P. Maguire. A thalamo-cortico-thalamic neural mass model to study alpha rhythms in Alzheimer's disease. *Neural Networks*, 24(6):631–645, 2011.
12. W. Bialek, F. Rieke, R. de Ruyter van Steveninck, and D. Warland. Reading a neural code. *Science*, 252(5014):1854–1857, 1991.
13. W. Bialek, R. Scalettar, and A. Zee. Optimal performance of a feed-forward network at statistical discrimination tasks. *Journal of Statistical Physics*, 57(1–2):141–156, 1989.

14. C. R. Blyth. Note on Estimating Information Author. *The Annals of Mathematical Statistics*, 30(1):71–79, 1959.
15. A. Borst and F. E. Theunissen. Information theory and neural coding. *Nat Neurosci*, 2(11):947–957, 1999.
16. S. L. Bressler and A. K. Seth. Wiener-Granger causality: a well established methodology. *Neuroimage*, 58(2):323–329, 2011.
17. N. Brunel and J. P. Nadal. Mutual information, Fisher information, and population coding. *Neural Comput*, 10(7):1731–1757, 1998.
18. Z Brzeniak and T. J Zastawniak. *Basic Stochastic Processes: A Course Through Exercises*. Springer, 1999.
19. G. T. Buracas, A. M. Zador, M. R. DeWeese, and T. D. Albright. Efficient discrimination of temporal patterns by motion-sensitive neurons in primate visual cortex. *Neuron*, 20(5):959–969, 1998.
20. D. A. Butts. How much information is associated with a particular stimulus? *Network*, 14(2):177–187, 2003.
21. C. Cellucci, A. Albano, and P. Rapp. Statistical validation of mutual information calculations: Comparison of alternative numerical algorithms. *Physical Rev E*, 71(6): 066208, 2005.
22. G. Chechik, M. J. Anderson, O. Bar-Yosef, E. D. Young, N. Tishby, and I. Nelken. Reduction of information redundancy in the ascending auditory pathway. *Neuron*, 51(3):359–368, 2006.
23. D. Colquhoun and B. Sakmann. Fast events in single-channel currents activated by acetylcholine and its analogues at the frog muscle end-plate. *The Journal of Physiology*, 369:501–557, 1985.
24. A. Compte, C. Constantinidis, J. Tegner, S. Raghavachari, M. V. Chafee, P. S. Goldman-Rakic, and X-J. Wang. Temporally irregular mnemonic persistent activity in prefrontal neurons of monkeys during a delayed response task. *J Neurophysiol*, 90(5):3441–3454, 2003.
25. C. H. Coombs, R. M. Dawes, and A. Tversky. *Mathematical psychology: an elementary introduction*. Prentice-Hall, 1970.
26. T. M. Cover and J. A. Thomas. *Elements of Information Theory*, volume 2012. John Wiley & Sons, 1991.
27. M. Crumiller, B. Knight, Y. Yu, and E. Kaplan. Estimating the amount of information conveyed by a population of neurons. *Frontiers in Neurosci*, 5(July):90, 2011.
28. V. Cutsuridis, T. Heida, W. Duch, and K. Doya. Neurocomputational models of brain disorders. *Neural Networks*, 24(6):513–514, 2011.
29. R. de Ruyter van Steveninck and W. Bialek. Real-time performance of a movement-sensitive neuron in the blowfly visual system: coding and information transfer in short spike sequences. *Proc. R. Soc. Lond. B*, 234(1277):379–414, 1988.
30. R. de Ruyter van Steveninck and S. B. Laughlin. The rate of information transfer at graded-potential synapses. *Nature*, 379:642–645, 1996.
31. X. Du and B. H. Jansen. A neural network model of normal and abnormal auditory information processing. *Neural Networks*, 24(6):568–574, 2011.
32. R. Eckhorn and B. Pöpel. Rigorous and extended application of information theory to the afferent visual system of the cat. I. Basic concepts. *Kybernetik*, 16(4):191–200, 1974.
33. R. Eckhorn and B. Pöpel. Rigorous and extended application of information theory to the afferent visual system of the cat. II. Experimental results. *Biol Cybern*, 17(1):71–77, 1975.
34. B. Efron and C. Stein. The jackknife estimate of variance. *The Annals of Statistics*, 9(3):586–596, 1981.
35. A. Fairhall, E. Shea-Brown, and A. Barreiro. Information theoretic approaches to understanding circuit function. *Curr Opin Neurobiol*, 22(4):653–659, 2012.
36. K. Friston. The free-energy principle: a unified brain theory? *Nat Rev Neurosci*, 11(2):127–138, 2010.
37. K. Friston. Dynamic causal modeling and Granger causality Comments on: the identification of interacting networks in the brain using fMRI: model selection, causality and deconvolution. *Neuroimage*, 58(2):303–310, 2011.

38. K. Friston, J. Kilner, and L. Harrison. A free energy principle for the brain. *J Physiol Paris*, 100(1–3):70–87, 2006.
39. K. Friston, R. Moran, and A. K. Seth. Analysing connectivity with Granger causality and dynamic causal modelling. *Current opinion in neurobiology*, pages 1–7, December 2012.
40. K. J Friston. Functional and effective connectivity in neuroimaging: A synthesis. *Human Brain Mapping*, 2(1–2):56–78, October 1994.
41. K. J. Friston, L. Harrison, and W. Penny. Dynamic causal modelling. *Neuroimage*, 19(4):1273–1302, 2003.
42. W. Gerstner, A. K. Kreiter, H. Markram, and A. V. Herz. Neural codes: firing rates and beyond. *Proc Natl Acad Sci U S A*, 94(24):12740–12741, 1997.
43. A. Globerson, E. Stark, D. C. Anthony, R. Nicola, B. G. Davis, E. Vaadia, and N. Tishby. The minimum information principle and its application to neural code analysis. *Proc Natl Acad Sci U S A*, 106(9):3490–3495, 2009.
44. C. W. J. Granger. Investigating causal relations by econometric models and cross-spectral methods. *Econometrica: Journal of the Econometric Society*, 37:424–438, 1969.
45. M. Haeri, Y. Sarbaz, and S. Gharibzadeh. Modeling the Parkinson's tremor and its treatments. *J Theor Biol*, 236(3):311–322, 2005.
46. K. Hlavackovaschindler, M. Palus, M. Vejmelka, and J. Bhattacharya. Causality detection based on information-theoretic approaches in time series analysis. *Physics Reports*, 441(1):1–46, 2007.
47. P. G. Hoel, S. C. Port, and C. J. Stone. *Introduction to probability theory*. Houghton Mifflin Co., Boston, Mass., 1971.
48. Q. J. M. Huys, M. Moutoussis, and J. Williams. Are computational models of any use to psychiatry? *Neural Networks*, 24(6):544–551, 2011.
49. R. A. A. Ince, A. Mazzoni, R. S. Petersen, and S. Panzeri. Open source tools for the information theoretic analysis of neural data. *Frontiers in Neurosci*, 4(1):62–70, 2010.
50. R. A. A. Ince, R. Senatore, E. Arabzadeh, F. Montani, M. E. Diamond, and S. Panzeri. Information-theoretic methods for studying population codes. *Neural Networks*, 23(6):713–727, 2010.
51. A. Kaiser and T. Schreiber. Information transfer in continuous processes. *Physica D*, 166(March):43–62, 2002.
52. A. Klenke. *Probability Theory*. Universitext. Springer London, London, 2008.
53. E. Koechlin and C. Summerfield. An information theoretical approach to prefrontal executive function. *Trends in Cognitive Sciences*, 11(6):229–235, 2007.
54. A. Kolmogoroff. *Grundbegriffe der Wahrscheinlichkeitsrechnung*. Springer-Verlag, Berlin, 1973.
55. A. Kraskov, H. Stögbauer, and P. Grassberger. Estimating mutual information. *Physical Rev E*, 69(6):066138, 2004.
56. S. Krishnamurti, L. Drake, and J. King. Neural network modeling of central auditory dysfunction in Alzheimer's disease. *Neural Networks*, 24(6):646–651, 2011.
57. S. Kullback and R. A. Leibler. On information and sufficiency. *The Annals of Mathematical Statistics*, 22(1):79–86, 1951.
58. E. L. Lehmann and G. Casella. *Theory of Point Estimation*. Springer, 1998.
59. R. Linsker. Self-organization in a perceptual network. *Computer*, 21(3):105–117, 1988.
60. R. Linsker. Perceptual neural organization: some approaches based on network models and information theory. *Annu Rev Neurosci*, 13:257–281, 1990.
61. R. Linsker. Local synaptic learning rules suffice to maximize mutual information in a linear network. *Neural Comput*, 702(1):691–702, 1992.
62. J. T. Lizier. *The Local Information Dynamics of Distributed Computation in Complex Systems*. Number October. Springer, springer edition, 2013.
63. J. T. Lizier and M. Prokopenko. Differentiating information transfer and causal effect. *The European Physical Journal B*, 73(4):605–615, January 2010.
64. J.T. Lizier, M. Prokopenko, and A.Y. Zomaya. The information dynamics of phase transitions in random Boolean networks. In *Proc Eleventh Intern Conf on the Simulation and Synthesis of Living Systems (ALife XI)*, pages 374–381. MIT Press, 2008.

65. M. London, A. Schreibman, M. Häusser, M. E. Larkum, and I. Segev. The information efficacy of a synapse. *Nat Neurosci*, 5(4):332–340, 2002.
66. M. Lungarella, K. Ishiguro, Y. Kuniyoshi, and N. Otsu. Methods for Quantifying the Causal Structure of Bivariate Time Series. *International Journal of Bifurcation and Chaos*, 17(03):903–921, 2007.
67. D. J. C. MacKay. *Information theory, inference and learning algorithms*. Cambridge University Press, 2003.
68. David Marr. *Vision: A Computational Investigation into the Human Representation and Processing of Visual Information*. MIT Press, 1982.
69. R. Marschinski and H. Kantz. Analysing the information flow between financial time series. *The European Physical Journal B*, 30(2):275–281, 2002.
70. C. C. McIntyre, S. Miocinovic, and C. R. Butson. Computational analysis of deep brain stimulation. *Expert Rev Med Devices*, 4(5):615–622, 2007.
71. G. A. Miller. *Information Theory in Psychology: Problems and Methods*, chapter Note on the bias of information estimates, pages 95–100. Free Press, 1955.
72. P. R. Montague, R. J. Dolan, K. J. Friston, and P. Dayan. Computational psychiatry. *Trends in Cognitive Sciences*, 16(1):72–80, 2012.
73. A. A. Moustafa and M. A. Gluck. Computational cognitive models of prefrontal-striatal-hippocampal interactions in Parkinson's disease and schizophrenia. *Neural Networks*, 24(6):575–591, 2011.
74. M. P. Nawrot, C. Boucsein, V. Rodriguez Molina, A. Riehle, A. Aertsen, and S. Rotter. Measurement of variability dynamics in cortical spike trains. *J Neurosci Methods*, 169(2):374–390, 2008.
75. I. Nemenman, W. Bialek, and R. R. de Ruyter van Steveninck. Entropy and information in neural spike trains: Progress on the sampling problem. *Physical Rev E*, 69(5):056111, 2004.
76. K. H. Norwich. *Information, Sensation, and Perception*. Academic Press, 1993.
77. B. A. Olshausen and D. J. Field. Sparse coding with an overcomplete basis set: A strategy employed by VI? *Vision Res*, 37(23):3311–3325, 1997.
78. D. Ostwald and A. P. Bagshaw. Information theoretic approaches to functional neuroimaging. *Magn Reson Imaging*, 29(10):1417–1428, 2011.
79. L. Paninski. Estimation of entropy and mutual information. *Neural Comput*, 15(6):1191–1254, 2003.
80. L. Paninski. Maximum likelihood estimation of cascade point-process neural encoding models. *Network: Computation in Neural Systems*, 15(4):243–262, November 2004.
81. S. Panzeri, R. S. Petersen, S. R. Schultz, M. Lebedev, and M. E. Diamond. The role of spike timing in the coding of stimulus location in rat somatosensory cortex. *Neuron*, 29(3):769–777, 2001.
82. S. Panzeri, R. Senatore, M. A. Montemurro, and R. S. Petersen. Correcting for the sampling bias problem in spike train information measures. *Journal of neurophysiology*, 98(3):1064–72, 2007.
83. S. Panzeri and A. Treves. Analytical estimates of limited sampling biases in different information measures. *Network*, 7:87–107, 1995.
84. J. Pearl. *Probabilistic Reasoning in Intelligent Systems: Networks of Plausible Inference*. Kaufmann, M, 1988.
85. J. Pearl. *Causality: Models, Reasoning, and Inference*. Cambridge University Press, 2000.
86. M. Pirini, L. Rocchi, M. Sensi, and L. Chiari. A computational modelling approach to investigate different targets in deep brain stimulation for Parkinson's disease. *J Comput Neurosci*, 26(1):91–107, 2009.
87. A. Pouget, P. Dayan, and R. Zemel. Information processing with population codes. *Nat Rev Neurosci*, 1(2):125–132, 2000.
88. M. Prokopenko, F. Boschetti, and A. J. Ryan. An information-theoretic primer on complexity, self-organization, and emergence. *Complexity*, 15(1):11–28, 2009.

89. R. Q. Quiroga and S. Panzeri. Extracting information from neuronal populations: information theory and decoding approaches. *Nat Rev Neurosci*, 10(3):173–195, 2009.
90. K. R. Rad and L. Paninski. Information Rates and Optimal Decoding in Large Neural Populations. In *NIPS 2011: Granada, Spain*, pages 1–9, 2011.
91. F. Rieke, D. Warland, and W. Bialek. Coding efficiency and information rates in sensory neurons. *EPL (Europhysics Letters)*, 22(2):151–156, 1993.
92. F. Rieke, D. Warland, R. de Ruyter van Steveninck, and W. Bialek. *Spikes: Exploring the Neural Code (Computational Neuroscience)*. A Bradford Book, 1999.
93. E. T. Rolls and A. Treves. The neuronal encoding of information in the brain. *Prog Neurobiol*, 95(3):448–490, 2011.
94. T. Schreiber. Measuring Information Transfer. *Phys Rev Lett*, 85(2):461–464, 2000.
95. T. Schürmann. Bias analysis in entropy estimation. *Journal of Physics A: Mathematical and General*, 37(27):L295–L301, 2004.
96. T. J. Sejnowski. Time for a new neural code? *Nature*, 376(July):21–22, 1995.
97. C. E. Shannon. A Mathematical Theory of Communication. *The Bell System Technical Journal*, 27(July, October 1948):379–423, 623–656, 1948.
98. A. N. Shiryayev. *Probability*, volume 95 of *Graduate Texts in Mathematics*. Springer-Verlag, New York, 1984.
99. J. Shlens, M. B. Kennel, H. D. I. Abarbanel, and E. J. Chichilnisky. Estimating information rates with confidence intervals in neural spike trains. *Neural Comput*, 19(7):1683–1719, 2007.
100. E. P. Simoncelli and B. A. Olshausen. Natural image statistics and neural representation. *Annu Rev Neurosci*, 24:1193–1216, 2001.
101. W. R. Softky and C. Koch. The highly irregular firing of cortical cells is inconsistent with temporal integration of random EPSPs. *The Journal of Neuroscience*, 13(1):334–350, 1993.
102. K. E. Stephan, L. M. Harrison, S. J. Kiebel, O. David, W. D. Penny, and K. J. Friston. Dynamic causal models of neural system dynamics:current state and future extensions. *Journal of biosciences*, 32(1):129–144, 2007.
103. R. Stollhoff, I. Kennerknecht, T. Elze, and J. Jost. A computational model of dysfunctional facial encoding in congenital prosopagnosia. *Neural Networks*, 24(6):652–664, 2011.
104. S. Strong, R. Koberle, R. de Ruyter van Steveninck, and W. Bialek. Entropy and Information in Neural Spike Trains. *Phys Rev Lett*, 80(1):197–200, 1998.
105. H. Theil. *Henri Theil's Contributions to Economics and Econometrics: Econometric Theory and Methodology*. Springer, 1992.
106. I. Todhunter. *A History of the Mathematical Theory of Probability from the Time of Pascal to that of Laplace*. Elibron Classics, 1865.
107. G. Tononi, O. Sporns, and G. M. Edelman. A measure for brain complexity: relating functional segregation and integration in the nervous system. *Proc Natl Acad Sci U S A*, 91(11):5033–5037, 1994.
108. T. Trappenberg. *Fundamentals of Computational Neuroscience*. Oxford University Press, 2010.
109. J. H. van Hateren. A theory of maximizing sensory information. *Biol Cybern*, 29:23–29, 1992.
110. R. Vicente, M. Wibral, M. Lindner, and G. Pipa. Transfer entropy–a model-free measure of effective connectivity for the neurosciences. *J Comput Neurosci*, 30(1):45–67, 2011.
111. J. D. Victor. Approaches to information-theoretic analysis of neural activity. *Biological theory*, 1(3):302–316, 2006.
112. N Wiener. The theory of prediction. In E. Beckenbach, editor, *Modern mathematics for engineers*. McGraw-Hill, New-York, 1956.
113. S. Yarrow, E. Challis, and P. Seriès. Fisher and shannon information in finite neural populations. *Neural Comput*, 1780:1740–1780, 2012.
114. L. Zhaoping. Theoretical understanding of the early visual processes by data compression and data selection. *Network*, 17(4):301–334, 2006.
115. I. Csiszár. Axiomatic characterizations of information measures. *Entropy*, 10(3):261–273, 2008.

Chapter 6
Machine Learning-Based Imputation of Missing SNP Genotypes in SNP Genotype Arrays

Aleksandar R. Mihajlovic

6.1 Introduction

As organisms age, health is increasingly threatened. Health problems are the most common cause of death in humans. Poor health is attributed to disease, a structural or functional disorder(s) of any organ(s) in the living body. One of mankind's constant endeavors is to eradicate diseases, engaging tens of thousands of scientists, engineers, economists, politicians, and business people worldwide. The medical discoveries of the nineteenth century gave birth to a new science that would change the way we think and view diseases: genetics. With the help of new gene research procedures, many diseases were shown to be genetic in nature. Since the first discovery of heritable diseases, along with the observation that genes may influence certain diseases, genetics found its way into the heart of almost all medical disciplines. With the explosive growth in use and development of computer-based information systems, high-throughput technology has aggressively penetrated the field of medicine. The discoveries of new genes and their phenotypes, made with the aid of advanced state-of-the-art computing technology, painted a completely new research domain for doctors and researchers. This new research domain, the science of genes and how they influence bodily functions, was coined *genomics*. One of the most important genomic discoveries was the existence of tight associations between *alleles* or editions of genes and disease [1].

A.R. Mihajlovic (✉)
Mathematical Institute of the Serbian Academy of Sciences and Art (MISANU), Belgrade, Serbia
e-mail: mihajlovic@mi.sanu.ac.rs

G. Rakocevic et al. (eds.), *Computational Medicine in Data Mining and Modeling*,
DOI 10.1007/978-1-4614-8785-2_6,

6.2 The Missing Genotype Problem

By comparing the genetic sequences of large groups of healthy and diseased individuals, researchers are able to create a rough map of the faulty genes causing the disease [2]. Such a data-mining procedure is known as a *Genome Wide Association Study* or GWAS for short. Relative gene locations, i.e., which chromosome the gene is located on and which part of the chromosome, gives researchers a starting point to localize genes, understand, and cure diseases. However, researchers are presented with a problem. Due to the volatility and relative chemical sensitivity of genetic material and the accuracy of genetic data-reading equipment, the encoding or translation process responsible for digitizing the sampled genetic data to discrete alphanumeric encoded data is frequently compromised, resulting in missing values of genetic reads. This is an instance of the age-old problem of frequent occurrence in massive data sets, the *missing value problem*. In order to cost-effectively solve this problem, various machine learning-based algorithms relying on probabilistic imputation are used to estimate the missing genetic data [3, 4]. Massaging the data set using machine learning algorithms prior to the GWAS data-mining-based analysis is critical. In GWAS, imputation usually refers to the substitution of missing *single-nucleotide polymorphism* or SNP values with one of two possible SNP values. Missing SNP data is fairly common in association studies, sometimes with rates as high as 5–10 % [4]. In association studies, missing values can lead to very poor analysis results. SNP genotype imputation methods are used to improve data set quality and to complete the data set, which is necessary for accurate data analysis results.

6.3 The Biological Problem Domain

In GWAS, SNPs are collected from disease-relevant *cohorts* and healthy cohorts or patient groups. The SNPs correspond to physical locations of particular bases on the DNA chain of the chromosomes of the cohorts. The SNP is of a dual nature. The first nature of the SNP corresponds to the location of the base along the DNA chain. SNP locations are fixed locations for all human beings. Their exact positions along the chromosome are uniform for all humans. Since their nature is so static, in research they serve and are referred to as *markers*, milestones along the DNA chain. The second nature of the SNP corresponds to the value of the particular base on the DNA chain, at the marker position. Marker value data, for statistical measurements, is organized into a tabular data set. Typical analysis of this data is the comparison of sequences of markers between the two cohorts. This is made possible by the mentioned tabular organization, where all the markers belonging to a specific patient are located on a single row of the table. The columns represent the values present at the particular marker location, i.e., each column is a specific marker location within the genome. Given that 99 % of the genome will have the same marker values, especially in healthy individuals, specific and differing marker sequences belonging to the diseased patient group stand out and are easily located. When the differing SNP sequences are localized, the physical locations of these differences would be recorded. This location is where the suspected gene, causing the disease, is most probably located. Due to the analytical sensitivity of GWAS, such studies are not possible without high quality, i.e., complete data sets [5]. Unfortunately, SNP genotyping technology is still too sensitive and causes missing values during the digitization process, resulting in instances of *the missing SNP genotype problem*. Bad quality data, with missing SNP values, have a negative impact on the quality of the final data analysis results. The need for *data completeness* and *correctness* is of utmost importance for reliable results. Eliminating missing SNP values by accurate substitution, i.e., imputation in a cost-effective manner, is of essential importance. An in-depth view of GWAS and a detailed description of the missing value problem are provided in Sects. 6.4 and 6.5. In order to better understand the problem, first a short introduction to the genetic concepts behind GWAS is provided in following subsections of Sect. 6.3.

6.3.1 Chromosomes

Chromosomes are long linear chains of bundled DNA present in all cell nuclei. Each chromosome is essentially a single long strand of DNA. The DNA of a chromosome contains heritable trait segments of DNA known as *genes*. Most of the DNA in chromosomes serves no purpose and is eminently useless. The ~28,000 segments of DNA, the genes, on the chromosomes, however are essential for human life. Genes in a DNA context are the scribed instructions for assembling proteins from available amino acids within the cytoplasm of a cell. Thus, the main purpose of genes is to instruct the creation of proteins which account for almost all of our body's dry weight, i.e., weight exclusive of water. Within a nondividing human somatic cell nucleus, 23 pairs of unbundled chromosomes also referred to as *chromatin* are found. Chromatin assumes the shaft- or bar-like shape only upon cell division (see Fig. 6.1). The chromosome pairs are labeled 1–22, with the 23rd pair, the sex chromosomes, labeled as XX or XY. The two *chromatids* of a chromosome pair are homologous in nature. One chromatid is inherited form one parent while the other from the other parent. The locations of the genes, their positions along the chromosomes, are the same. The homologous set of genes located on homologous chromosomes work together in expressing a trait. Each homologous gene-pair alleles make a tuple. This tuple is referred to as the *genotype*.

Each homologous gene can have its own version, or *allele*. When two of the same alleles are present in a homologous pair, the genotype is termed *homozygous*, and when two different alleles are present in a homologous pair, the genotype is termed *heterozygous*. The traits that the genotypes control are classified as being either dominant, recessive, or codominant. In the first case, one of the homologous gene alleles is exclusively expressed, while the other homolog is silenced.

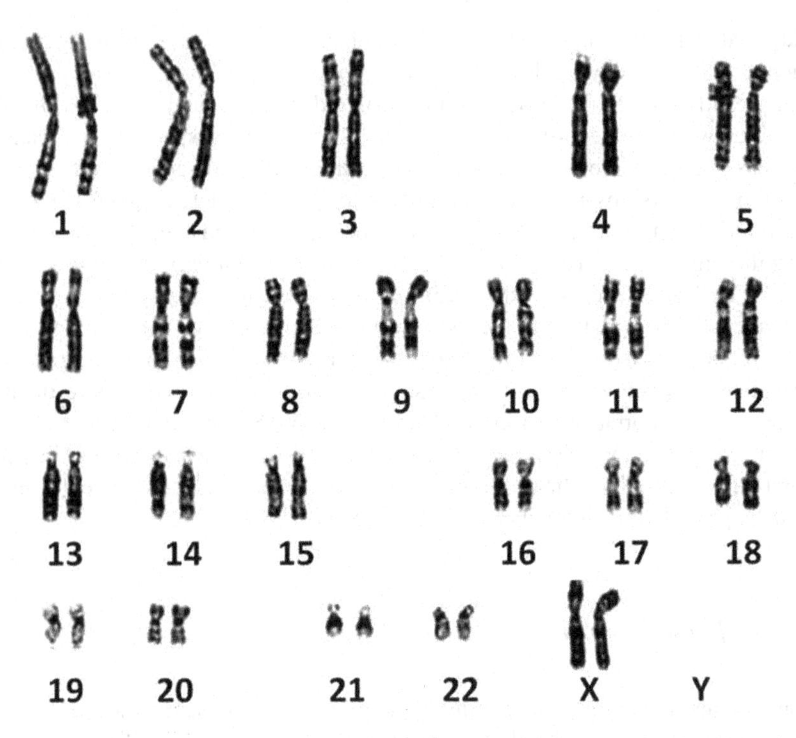

Fig. 6.1 Human female karyotype with the Y-chromosome missing and each pair of homologous chromosomes labeled with 1–22, *X* and *Y*. Between the two homologous chromosomes of chromosome pairs, recombination events take place [6]

This usually requires at least one allele to be dominant. In the second case, a trait is recessive if both of the homologous gene alleles are recessive. In the third case, both alleles are dominant and codominate the expression of the trait.

6.3.2 DNA

Deoxyribonucleic acid or DNA is one of two nucleic acids. It has a helical ladderlike molecular three-dimensional structure where each rail of the ladder is made of a chain of *nucleotides*. Nucleotides are the fundamental building blocks of DNA. Two nucleotides joined by a hydrogen bond at complementary nitrogenous bases form a base pair unit of DNA. The two bases can be viewed as the rungs of the ladder. The nucleotide in order to reinforce the hydrogen bond between the two nitrogenous bases has a pentose sugar and phosphate group chained together, forming the rails of the ladder.

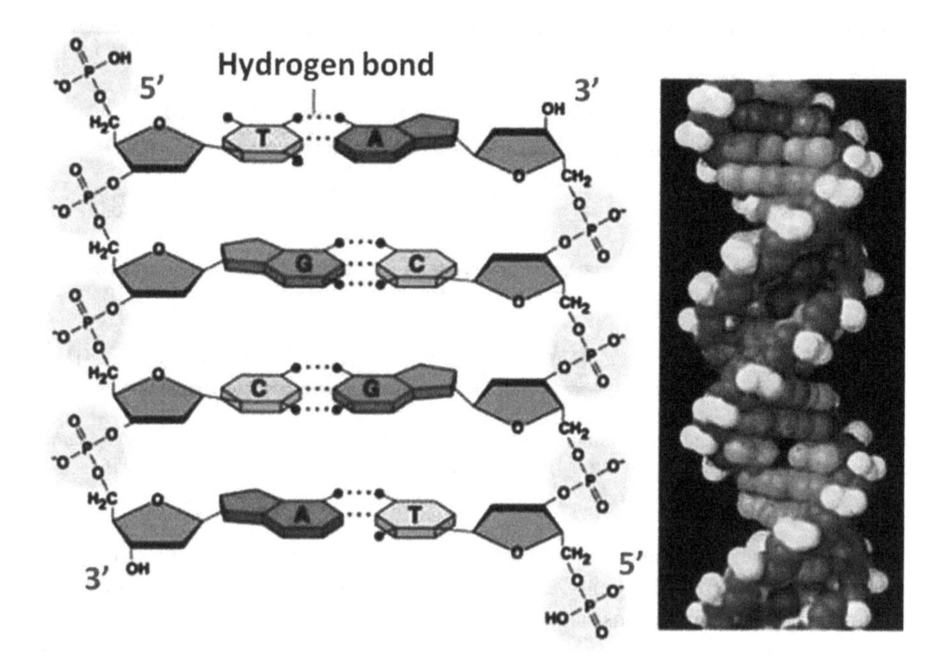

Fig. 6.2 DNA chain with clearly indicated nitrogenous base relationships

Four of DNA nitrogenous bases are *adenine*, *guanine*, *thymine*, and *cytosine*. The character of the four bases is complementary in nature, meaning only one of two possible base pair combinations can be present at each rung of the ladder (see Fig. 6.2). Hydrogen bonds hold adenine and thymine together and cytosine and guanine together. Adenine with cytosine or guanine cannot bond together, nor can thymine with guanine or cytosine. The bonds themselves between the bases are initiated in a zipper-like fashion, bonding the two complementary bases together and forming the rungs of the ladder of Fig. 6.2. Given their complementary nature, knowledge of one linear sequence of bases along one rail of the DNA chain is enough to describe the DNA chain as a whole. A gene is a specific linear sequence of these bases on a DNA chain. The sequence is read by climbing the raillike structure.

6.3.3 SNPs and Point Mutations

Benign mutations of gene coding regions are the culprits of hereditary variation. These benign mutations correspond to the 0.01 % difference between all humans. One study on genetic variations between different species of Drosophila suggests that if a mutation changes a protein produced by a gene, the result is likely to be

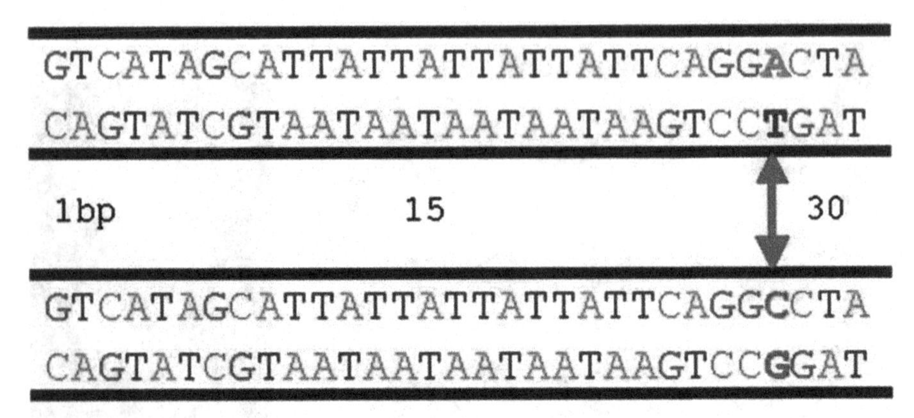

Fig. 6.3 An example of a point mutation, i.e., an SNP. Both DNA sequences are exactly the same except for one location, pointed out by the *red* double-headed *arrow* (Adapted from [11])

harmful, with an estimated 70 % of amino acid polymorphisms having damaging effects, and the remainder being either neutral or weakly beneficial [7]. The most common type of mutations are *point mutations*. Point mutations are single-nucleotide differences, i.e., single-base differences between humans referred to as SNPs. They are in fact the culprits of variation both in a phenotype, i.e., physical sense, and thus in a genotype or gene sense, i.e., in terms of alleles. The minor differences in gene coding sequences, gene base sequences, resulting from point mutations, give rise to many different editions of genes, i.e., gene alleles. When any two human genomes are compared side by side, they are 99.9 % identical [8]. However, with a 3.2×10^9 base pair genome, each person harbors roughly 3.2×10^6 differences (see Fig. 6.3).

These differences are attributed to SNPs. While the majority of the SNPs are of no significant biological consequence, since they cover irrelevant regions of the DNA, a fraction of the substitutions have functional importance and are the basis for the diversity such as one found among humans [9]. SNPs usually occur in noncoding regions more frequently than in coding regions, i.e., regions coded for protein synthesis or transcription. SNPs are biallelic, base pairs have only two possible combinations, and thus they are easily assayed [10].

Given the complementary nature of bases, we have only two possible alleles or base bonds for any single-base change, A-T and C-G, where A-T is the same as T-A and C-G is the same as G-C. SNPs, which make up about 90 % of all human genetic variation, occur every 100–300 bases along the 3×10^9 base human *genome*, the set of all 46 DNA chains within a cell nucleus. Two of every three SNPs involve the replacement of cytosine (C) with thymine (T) [12]. SNPs are also evolutionary stable, i.e., are not changing much, from generation to generation, making them easier to follow in population studies [12].

6.3.4 SNP Haplotypes and Haplogroups

A *haplotype* in genetics is a combination of SNP alleles within the genome that are inherited together from generation to generation. Recombination events during cellular meiosis in sex cells rarely can separate these inheritable allelic chunks. *Association* in this case refers to strong linkage between the SNPs under allelic conditions, which means under a particular configuration of SNP marker values, the markers and their values are inherited together regardless of any recombination events where homologous chromosomes are spliced and mixed up for sex cell gamete variation. In molecular evolution, a *haplogroup* is a group of similar haplotypes that share a common ancestor having the same SNP mutation in both haplotypes [13]. Since haplogroups contain similar haplotypes, it is possible to predict a haplogroup from haplotypes. An SNP test confirms a haplogroup with 100 % accuracy, but the haplogroup can be estimated statistically with average certainty. The haplogroup is best determined through SNP gene-chip testing. Haplogroups are assigned code names in a form of an alphanumeric character string, e.g., R1b1. Y-chromosome and mitochondrial DNA haplogroups have different haplogroup designations. Haplogroups pertain to deep ancestral origins that may date back thousands of years [13].

6.3.5 GWAS in Detail

Genome-wide association studies are used to identify common genetic factors that influence health and disease [14]. It is an examination of many common genetic/SNP variants in different individuals to see if any variant is associated with a trait. GWAS typically focus on associations between SNPs and traits, especially those expressed by major diseases. These studies normally compare the SNPs on homologous DNA chains of two groups of participants or cohorts. Cohort commonly refers here to a group of people with the disease and similar people without it. The two test groups are carefully selected. The selection process for the diseased cohorts involves setting a standard *anamnesis test*. Both subgroups or groups of cohorts, the healthy and the diseased groups, must belong to the same anamnesis group; the only thing setting their anamnesis apart is the disease. This eliminates the role any environmental factors might play on disease onset or even development of disease. All individuals in each group are genotyped for the majority of common known SNPs. The exact number of SNPs depends on the study, but typically around a million. The SNP genotype information from these two groups is analyzed and if one type of SNP allele or group of alleles, i.e., haplotype is more frequent in people with the disease than in those without the disease, the SNP or haplotype is said to be "associated" with the disease. The associated SNPs are then considered to mark or span a region or regions of the human genome which influences the expression of the disease.

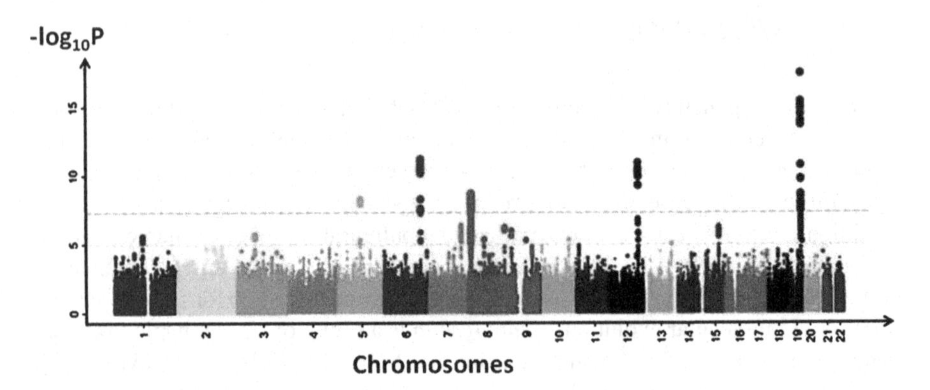

Fig. 6.4 Image of Manhattan plot [15]

In contrast to methods which specifically test one or a few genetic regions, the GWA studies investigate the entire genome. The approach is therefore said to be *noncandidate driven* in contrast to gene-specific candidate-driven studies, where the chromosomal location of the disease-associated gene culprit is roughly known. GWA studies identify SNPs and other variants in DNA which are associated with a disease but cannot on their own specify which genes are causal, rather they do specify the active location(s) on the chromosome(s) that are expressive in diseased individuals [16–18]. After necessary measures such as odds ratios and P-values have been calculated for all SNPs, a *Manhattan plot* is created. A Manhattan plot is a type of scatter plot. It is used to display data with a large number of data points. In GWAS Manhattan plots, genomic coordinates are displayed along the X-axis, with the negative logarithm of the association P-value for each SNP displayed on the Y-axis [19].

Because the strongest associations have the smallest P-values, such as Y values 10–15, their negative logarithms will be the greatest (see Fig. 6.4). Each point in the plot is an SNP laid out across the human chromosomes from left to right, and the heights correspond to the strength of the association to the particular disease being studied by the GWAS. The strongest associations form neat peaks where nearby correlated SNPs all show the same signal (see Fig. 6.4) [20]. Any Manhattan plot with points all over the place should be viewed as highly suspicious.

6.3.6 Missing SNP Genotypes

Genotyping is one of the primary procedures in beginning a GWAS. It is the process of determining the allelic composition of homologous genetic data for an individual or a data set of individuals. Genotypes exist for SNPs. SNP genotypes are of primary concern in such studies. Genotyped SNPs are presented in tabular form for data mining, i.e., analysis. Each row of the table represents one patient and each

Table 6.1 Sample of a large genome-wide, SNP genotype map. Once the SNPs have been called, they are typed into a genotype matrix shown here. The "?" resembles the typing of a "No Call" value, i.e., a missing SNP genotype. The column titles of the matrix shown represent the names of the SNPs typed

Selectin	FASPII	FGB_1	FGB_2	FGB_3	Factor_VII	GplB	HL	ICAM
A/A	G/G	C/C	G/G	C/C	C/C	C/C	C/T	A/A
C/A	A/G	C/C	G/G	C/C	C/T	C/C	C/C	A/G
A/A	?	?	?	C/C	C/C	?	?	A/G
A/A	A/G	C/C	G/G	C/C	C/C	C/C	C/T	A/G
A/A	A/G	C/C	G/G	C/T	C/C	C/C	?	A/G
A/A	G/G	C/C	G/G	C/C	C/C	C/T	?	A/A
A/A	A/G	C/T	A/G	C/T	C/C	C/C	?	A/G
A/A	A/G	C/C	G/G	C/C	T/T	C/T	?	A/G
A/A	A/A	C/T	A/G	C/T	C/C	C/C	?	A/G
A/A	G/G	C/T	A/G	C/T	C/C	C/C	C/C	A/A
A/A	A/G	C/C	G/G	C/C	C/C	C/C	?	G/G
C/A	A/G	C/C	G/G	C/C	C/C	C/C	?	G/G
A/A	A/G	C/T	A/G	C/T	C/C	C/T	?	G/G

column represents a specific marker. At each element of the table, we see the genotypes of the homologous SNPs at the specific marker locations for each patient.

In Table 6.1, a genotyped table is presented with missing values represented by the "?" character. In this table the syntax for two homologous SNP alleles, i.e., the SNP genotype, is presented in "A/B" form, where A represents the allele of the SNP for that particular marker (column) inherited from one parent and B the allele inherited from the other parent of the patient. Genotyping errors, i.e., missing genotypes, occur when the observed genotype does not correspond to the true underlying genetic information, as a result of a mistake in data entry [21]. Genotyping errors result from diverse, complex, and sometimes cryptic origins; thus grouping errors into discrete categories according to their causes is challenging because different causes sometimes interact to generate an error. Errors such as missing data from "No Calls" can be caused by low-quality DNA samples, poorly performing SNP assays, and/or errors in sample preparation. For clarity, Pompanon and associates [22] have proposed categories for these "pre-calling" errors. They have grouped errors into four categories:

- Errors that are linked to the DNA sequence itself
- Errors that are due to the low quality or quantity of the DNA
- Biochemical artifacts
- Human factors

In the few studies designed to analyze the precise causes of genotyping error, the main cause was related to human factors. Many times missing data is unavoidable, even when measures are taken to correct them prior to SNP calling. Nonetheless, these errors must be corrected. Missing data create imbalanced and complicated calculations required for statistical analyses [4]. Erroneous genotypes such as missing values can drastically affect linkage and association studies (data-mining procedures for measuring the differences between common and uncommon SNP genotypes in data sets), the main studies involved in GWAS, by masking the true segregation of alleles. For example, in linkage studies genotyping errors can affect haplotype frequencies [23]. Error rates as low as 3 % can have serious effects on linkage disequilibrium analysis [24, 25], i.e., analysis where associated SNPs are distinguished from unassociated ones, and a 1 % error rate can generate a loss of 53–58 % of the linkage information for a trait locus [26]. Errors, therefore, decrease the power for detecting associations. Missing SNP data are fairly common in these association studies, sometimes with a rate of 5–10 %.

The importance of resolving missing data is evident. Re-genotyping the missing genotypes is the optimal solution for correcting errors such as missing values; however, it is often not practical due to its high cost [4]. Even though the cost of SNP genotyping per SNP for GWAS decreased from $1.00 in 2001 to 0.1¢ in 2007, considering that the average number of SNPs genotyped is between 500,000 and 10,000,000 per individual and the standard number of individuals taken into account is 2,000, i.e., the cost is still in the millions, it is still too expensive [17, 27]. Older solutions included exclusion of rows with missing values from data sets. This approach also isn't effective since it can decrease the data set size drastically. Imputation algorithms are therefore more commonly used than any other alternative method [4, 28]. Genotype imputation is a machine learning approach that infers missing genotypes. It is a cost-effective and statistically accurate enough solution for solving the missing value problem concerning SNP genotypes. The most accurate imputation algorithms used today rely on probabilistic modeling of data sets and efficient estimation of inferred genotype values.

6.4 The Mathematical Problem Domain

From the mathematical perspective, missing values in data sets such as those of GWAS set are imputed or inferred probabilistically. Imputation is an algorithmic substitution of blanks, open, unknown missing values within a given set of data with an estimated value based on a probabilistic model of the data set. Substitution of one data point is referred to as "unit imputation," whereas substituting a component of a data point is known as "item imputation." Upon imputation, association analysis of the data set can be performed; the necessary data analysis steps of GWAS can be resumed. The data analysis must take into account that there exists a degree of uncertainty due to imputation than if the imputed values had actually been observed; the values were not missing in the first place. It is the goal of modern imputation methods to eliminate this uncertainty as best as possible. For genotype data, numerous imputation techniques are available.

Prior to the invention of the personal computer and high, user-friendly computing technology, the punch-card and punch-tape systems were used. An earlier method of imputation used on the punch-card system was *hot-deck imputation* where a missing value was imputed from a randomly selected similar record from within the same data set, i.e., the same punch card. The term hot was coined in order to emphasize the local nature of utilizing a single data set. Hence, if similar data records exist within a data set, then they resemble hot cards from a deck in a poker game [29]. The opposite of hot-deck imputation is *cold-deck imputation* where, by contrast, randomly selected similar records are provided by another data set, data sets that are not currently being processed. According to later discussed classifications of imputation algorithms, the early cold-deck imputation method has become the fundamental building block of the knowledge-based imputation algorithms. Since computing power has advanced rapidly and punch cards are no longer in use, more sophisticated methods of imputation have generally superseded the original random and sorted hot-deck imputation techniques, such as the *nearest neighbor hot-deck* imputation or the *approximate Bayesian bootstrap* [29]. High-throughput, sophisticated imputation algorithms which rely on more sophisticated probabilistic models such as probabilistic graphic models of data sets are used more frequently today.

6.4.1 Probabilistic Graphic Models

Probabilistic models are fair representations of real-world phenomena and systems which involve uncertainties. Algorithms as step-by-step descriptions of how to go about solving a given problem are based on the underlined problem model which includes the relevant observed as well as hidden or missing data type models. As the number of random variables of the model and observed value samples increase, the semantics can easily become very complex and hard to comprehend. Fortunately, complex probabilistic models can be graphically represented by translation of mathematical expressions into the language of visual objects of graphs [30].

6.4.2 Single Random Variable Graphical Modeling

One particular sort of graphical representation of interdependencies, known simply as a *graph*, is of particular interest here.

Definition. In mathematics, a graph is a graphical expression of some complex mathematical statement using a combination of two elementary visual objects:

- A labeled node or vertex
- An edge, also known as an arc, line, or link, beginning and ending at one or beginning at one and ending at another node

Nodes are typically depicted as circles containing some label in it (see Fig. 6.5). The edges may be directed (asymmetric appearing as an arrow) or undirected

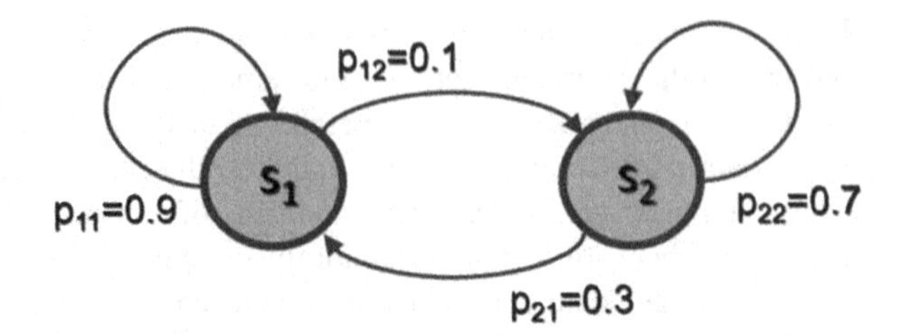

Fig. 6.5 Graph of a single binary random variable X (with two possible values or states, S_1 and S_2), showing the transitions between the two states depicted by the labeled edges

(symmetric appearing as a simple curve segment without any arrows). In addition, an edge may have a label or be unlabeled when its sole existence implies a value (e.g., relationship exists).

Definition. Formally, a graph G is a pair of sets (V, E), where V is the set of vertices and E is the set of edges.

Tables or matrices easily may be represented by graphs. The engineering community may refer to graphs as state machines or state transition diagrams. The term "graph" commonly used to represent a plot of some function against some base line variable is improperly applied. The mathematical discipline known as graph theory refers to graphs as defined above and not to the so-called plots of functions. The term "graph" was first used in this sense by Sylvester in 1878 [31, 32].

A graph as a diagram may represent a set of possible system activities by depicting the so-called state transitions, where each node represents one state or a value from the set of possible values. For instance, Fig. 6.5 illustrates one binary random variable model, where transition between two variable values S_1 and S_2 may take place with some conditional probability. For example,

$$P(X = S_2 | X = S_1) = p_{1,2} = 0.1. \tag{6.1}$$

As the dimensionality of the model increases, the growth of the set of possible values escalates the underlined complexity. For example, the graph modeling a two-dimensional variable (X_1,X_2) with binary coordinate X_1 and ternary coordinate X_2 represents 2×3 state transition matrix as $n = 6$ node graph having $n(n - 1)/2$ links. The complexity of the model exponentially grows with the complexity of observations.

We may use graphs to represent a set of conditional probabilities and large set of variable interdependencies. The graph may help analysts capture probability distributions over spaces with a large number of factors or associations.

6.4.3 Markov Models

Markov models are probabilistic models that follow the Markov property. The Markov property assumes that the distribution for the current variable depends only on the distribution of the previous variable. The two essential Markov models are the *Markov chain* and the *hidden Markov model* or *HMM* [33]. An extension of the

Markov chain models to practical situations with incomplete observed data, known as a hidden Markov model or HMM, is drawing the attention of researchers that are trying to model sequences of data. The underlying mathematical domain of the missing value problem in the context of SNP genotype data deals with the problem of modeling spatial sequence of DNA data. The sequential nature of SNP genotypes and the missing data imputation problem qualifies the HMM approach as one that is applicable.

6.4.3.1 Markov Chains

A Markov chain is an experiment that undergoes transitions from one state to another, between a countable number of possible, i.e., discrete states. A Markov chain is a random process that may be characterized as memoryless, where the next state depends only on the current state and not on the sequence of past states. This is known as the Markov property of the first order. A Markov chain is a sequence of random variables $X1$, $X2$, $X3$, ... with the *Markov property*, namely, that, given the present state, the future and past states are independent:

$$P(X_{n+1} = x | X_1 = x_1, X_2 = x_2, \ldots, X_n = x_n) = P(X_{n+1} = x | X_n = x_n). \quad (6.2)$$

The possible values of X_i are drawn from a countable se0074 S called the state space of the chain. Markov chains are graphically described by a directed graph, where the edges are labeled as the probabilities of going from one state to the other states. A graphical example of a Markov chain shown in Fig. 6.5 illustrates a simple Markov chain with two states S_1 and S_2. The transitions between the two states are depicted by the edges labeled by the respective transition probabilities. Missing SNP genotype value imputation may be approximated by a Markov chain, if we would make an assumption that the probability of the next marker value k, i.e., the next missing SNP genotype, depends only on the value observed at the previous location $k - 1$ in the chain, or inversely, in case of traversing the genotype sequence backwards, from the value observed at $k + 1$ location. Only one of the two alternative traversal directions may be applied at one time, in one chain. The forward-backward algorithm, presented in Sect. 6.4.4, for HMMs performs exactly this traversal in both directions in order to maximize the estimate for a missing value.

6.4.3.2 Hidden Markov Models

A hidden Markov model or HMM is used to model sequence or spatial data. HMMs are most similar to a mixture model. Mixture models are probabilistic models representing the presence of subpopulations within a grander population. Given that a super set population is comprised of sets of data, i.e., observable data and hidden data, the key correspondent feature of HMMs and mixture models is that probabilistic mixture model does not require that the observed data identify the

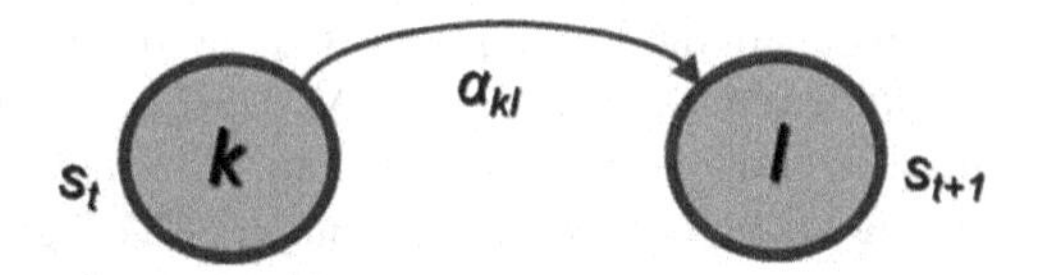

Fig. 6.6 Two-state sequence with transition probability $a_{k.\,l}$

subpopulation to which an observation belongs [34], hence resulting in the subpopulation being hidden. This population is later referred to as a hidden population or a hidden data set. Similarly in HMMs, the subpopulation is considered to be a hidden state, while the observed data set refers to the observed state. In HMMs as in mixture models, the observed states are generated by the hidden states, i.e., the observed state is one of sometimes numerous values that belong to the hidden state.

Since HMMs model sequential or spatial data, the underlying states follow a Markov chain behavior mentioned in the previous section, with the state-based transitional scheme, where the probabilities of a transition from one state to another are modeled graphically by edges. The probabilistic weights of all edges leaving an individual state node in the graph must sum to 1. Suppose that we have given set u of observed sequential data (e.g., a sequence of bits) $v = (v_1, v_2, v_3, \ldots v_t, \ldots, v_T)$. Every v_t, $t = 1, \ldots, T$, is generated by a hidden state, s_t. In this respect each hidden state can be viewed as a random variable and each observed state can be viewed as a value that the hidden variable, i.e., random variable s, can take on. The transition from state to state occurs between the hidden states. The observed states are the results of the hidden states and their transitions. The hidden states follow Markov chain behavior and Markov rule that "the future state only depends on the present state," or

$$P(s_{t+1}|s_t, s_{t-1}, \ldots, s_0) = P(s_{t+1}|s_t). \tag{6.3}$$

Since HMMs deal with sequences, the probabilities of these sequences occurring within a data set of many sequences correspond to the probability of transition from one value of a state to another value of a state. For instance, with a random bit sequence, what is the probability that the next value in the sequence will be a "1" given that the present value in the sequence is "0"? The transition probabilities are modeled as

$$a_{k.l} = P(s_{t+1} = l|s_t = k). \tag{6.4}$$

In this case $a_{k,l}$ is graphically represented by an edge from the current state $s_t = k$ to the next probable state $s_{t+1} = l$, (see Fig. 6.6).

In this graphical model, the state s_t takes on the value k, $s_t = k$, and the state s_{t+1} takes on the value l, $s_{t+1} = l$. Here $k, l = 1, 2, \ldots, M$, where M is the total number of possible state values from which any present state of value k, can transition to future state of value l. Each edge from k to any other possible next state l must collectively sum up to 1 (i.e., Markov process must take some value of a state). This means that there the next state s_{t+1} with probability of 1 must take some value.

v_1	v_2	v_3
0	1	1
1	0	0
1	0	1
0	1	0
1	1	0
0	1	1
0	0	1
1	1	0

Fig. 6.7 Given observed values

This also holds for the initial state probability π_k which will be clarified in a bit sequence model example to come:

$$\sum_l^M a_{k,l} = 1 \text{ for any } k, \sum_l^M \pi_k = 1. \tag{6.5}$$

With the initial state transition probability and the rest of the transition probabilities, the hidden state sequence may be modeled as

$$P(s_1, s_2, \ldots, s_T) = P(s_1)P(s_2|s_1)P(s_3|s_2)\ldots.P(s_T|s_{T-1}) = \pi_k a_{s_1,s_2} a_{s_2,s_3} \ldots a_{s_{T-1},s_T}. \tag{6.6}$$

Given the state s_t, the observation v_t is independent of other observations and states [35]. To better understand the HMM concept, an example will be considered. In Fig. 6.7, a random bit stream, of bit-length $= 3$, is modeled as eight possible 3 bit sequences, $\boldsymbol{v} = (v_1, v_2, v_3)$.

Given our set $\boldsymbol{v} = (v_1, \ldots, v_T)$, the hidden state set is defined as $\boldsymbol{s} = (s_1, \ldots, s_t)$ where $1 \leq t \leq 3$ with maximum value of $T = 3$. Any s_t can take up only two values 1 or 0: the initial probability of starting with a 1, $\pi_{k=1} = P(s_0 = 1|\boldsymbol{v}) = 0.5$, and the initial probability of starting with a 0, $\pi_{k=1} = P(s_0 = 0|\boldsymbol{v}) = 0.5$. This initial relationship is depicted in Fig. 6.8.

Let us say that we want to probabilistically model the sequence $v = (v_1 = 1, v_2 = 1, v_3 = 0)$ so for $v_1 = 1$ we condition on $S_1 = 1$ by $P(s_1 = 1)$ so that $P(s_2) = P(s_2|s_1)$ can be calculated using data shown in Fig. 6.9.

Using Fig. 6.9, from this conditional distribution, we can extract the transition probability $a_{k,l}$ which states that

$$P(s_2|s_1 = 1) = 1 \tag{6.7}$$

where after using tables in Fig. 6.9, we have

$$P(s_2 = 1|s_1 = 1) = 0.5 \text{ and } P(s_2 = 0|s_1 = 1) = 0.5. \tag{6.8}$$

Graphically this can be represented as shown in Fig. 6.10.

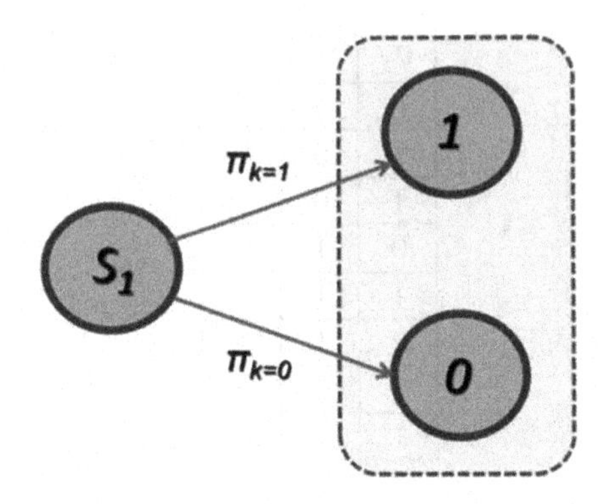

Fig. 6.8 Initial state of a hidden Markov chain

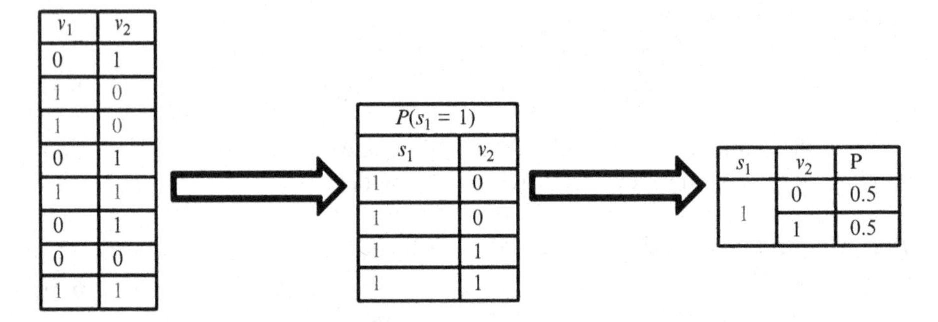

v_1	v_2
0	1
1	0
1	0
0	1
1	1
0	1
0	0
1	1

$P(s_1 = 1)$	
s_1	v_2
1	0
1	0
1	1
1	1

s_1	v_2	P
1	0	0.5
	1	0.5

Fig. 6.9 Summary of observed values

So far we have modeled

$$P(s_1, s_2) = P(s_1)P(s_2|s_1). \tag{6.9}$$

Upon value substitution we have

$$P(s_1 = 1, s_2 = 1) \quad = P(s_1 = 1)P(s_2 = 1|s_1 = 1) = (0.5)(0.5) = 0.25. \tag{6.10}$$

Finally what is left to model is the probability of $s_3 = 0$ that is the event $P(s_3 = 0|\mathbf{v})$. The tabular distribution of this event, or sequence, finally looks as shown in Fig. 6.11.

Using Fig. 6.11, from this conditional distribution, we can extract the transition probability $a_{k,l}$ which states that

$$P(s_3|s_2 = 1) = 1 \text{ where } P(s_3 = 0|s_2 = 1) = 1 \text{ and } P(s_3 = 1|s_2 = 1) = 0. \tag{6.11}$$

The finalized graph of state evolution is shown in Fig. 6.12.

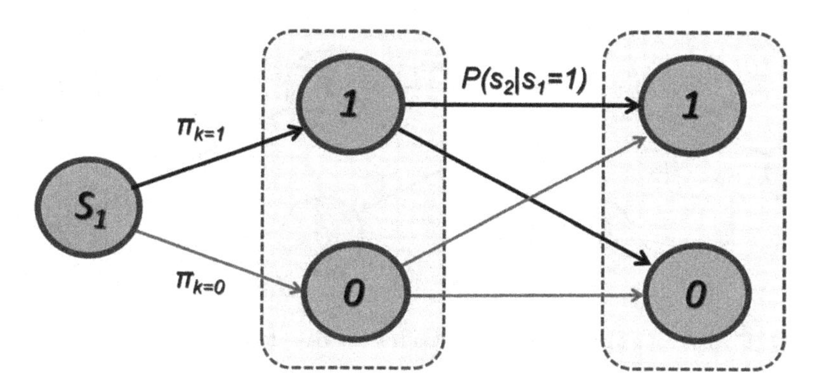

Fig. 6.10 Example of a Markov chain transition from s_1 to s_2

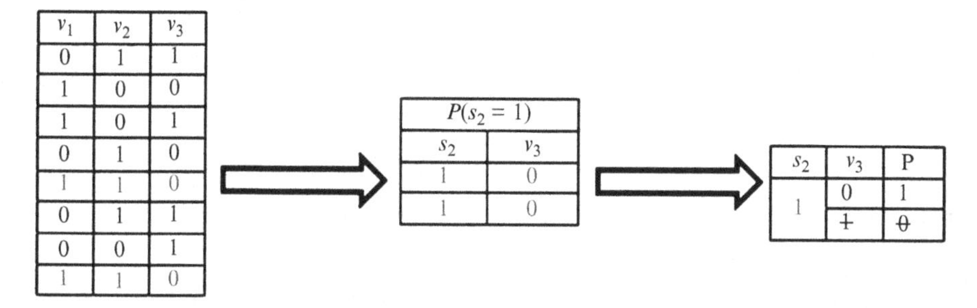

v_1	v_2	v_3
0	1	1
1	0	0
1	0	1
0	1	0
1	1	0
0	1	1
0	0	1
1	1	0

$P(s_2 = 1)$	
s_2	v_3
1	0
1	0

s_2	v_3	P
1	0	1
	~~1~~	~~0~~

Fig. 6.11 Summary of observed values

So far modeling of the joint state sequence probability,

$$P(s_1, s_2, s_3) = P(s_1)P(s_2|s_1)P(s_3|s_2). \tag{6.12}$$

With substituted values, we obtain

$$\begin{aligned} P(s_1 = 1, s_2 = 1, s_3 = 0) &= P(s_1 = 1)P(s_2 = 1|s_1 = 1)P(s_3 = 0|s_2 = 1) \\ &= (0.5)(0.5)(1) = 0.25. \end{aligned} \tag{6.13}$$

To prove this, we take a glimpse into the joint probability of the data set v shown in Fig. 6.13.

In an HMM, we do not know the state sequence that the model passes through, but we can know some probabilistic function of it. HMMs are most similar to a distribution mixture model. Mixture models are probabilistic models representing the presence of one or more minor subpopulations within a grander population. The key feature of HMMs and mixture models is that a probabilistic mixture model does not require the observed data set to directly identify the subpopulation to which an observation belongs [34]. In HMMs, the subpopulation is considered a hidden, not

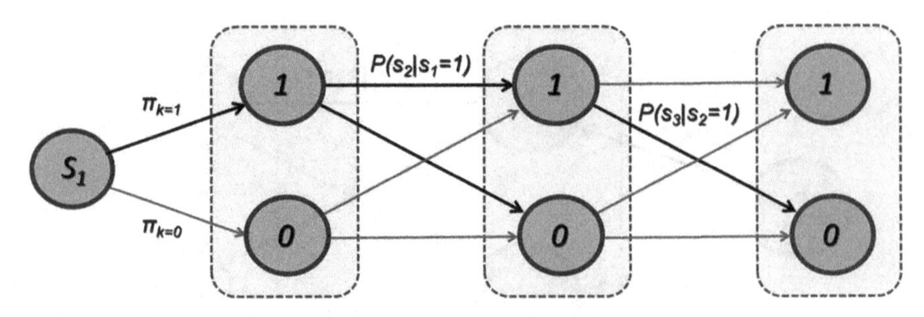

Fig. 6.12 Example of a Markov chain transition from s_1 via s_2 to s_3

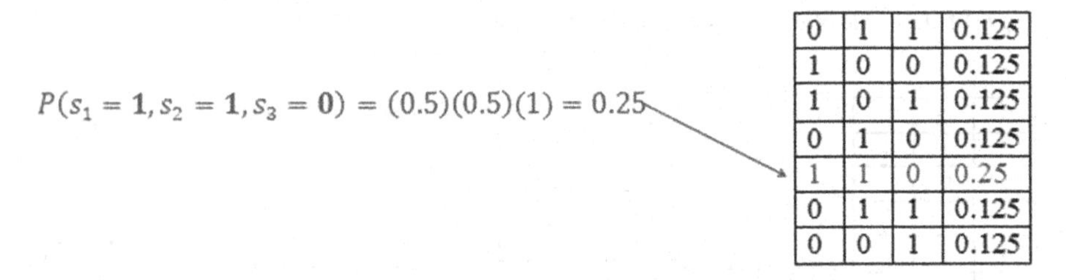

0	1	1	0.125
1	0	0	0.125
1	0	1	0.125
0	1	0	0.125
1	1	0	0.25
0	1	1	0.125
0	0	1	0.125

Fig. 6.13 Example of a Markov chain transition probability from s_1 via s_2 to s_3

observed state, while the observed data set should enable one to infer the state sequence from a sequence of observed data using the probability model. This fairly simple model has been found to be very useful in a variety of sequential modeling problems, most notably in speech recognition [36, 37] and computational biology [38]. A key practical feature of the model is the fact that inference of the hidden state sequence given observed data can be performed in linear fashion along the length of the sequence. Furthermore, this lays the foundation for efficient estimation algorithms that can determine the parameters of the HMM from training data.

Scheet and Stephens [39] present a statistical model for patterns of genetic variation in samples of unrelated individuals from natural populations. Their model is based on the idea that, over short regions, haplotypes in a population tend to cluster into groups of similar haplotypes. To capture the fact that clustering tends to be local in nature, their model allows cluster memberships to change continuously along the chromosome according to a hidden Markov model.

6.4.4 *Forward-Backward Algorithm*

The *forward-backward algorithm* or FB algorithm is a dynamic programming algorithm similar to the Viterbi algorithm, used for decoding, i.e., evaluating HMMs. The evaluation is performed by computing the posterior and anterior probabilities, i.e., posterior marginal probabilities of all hidden state variables given a sequence of observations, i.e., a sequence of observed states. Concisely, given a sequence of observations $v_{1:t} := v_1, \ldots, v_t$, the FB algorithm computes for all hidden state variables $s_k \in \{s_1, \ldots, s_t\}$ the distribution $P(s_k|v_{1:t})$.

The algorithm itself involves three steps:

1. Determining forward probabilities
2. Determining backward probabilities
3. Smoothing the determined values, i.e., evaluation

Definition. Forward probabilities are partial probabilities from the initial state to a current or desired state within the HMM. These probabilities are usually denoted by the letter α where $\alpha_t(j) = P(v_t|s_t = j)\ P(\mathit{all\ paths\ to\ state\ j\ at\ time\ t})$.

These probabilities are also known as *partial* probabilities and are calculated using a forward algorithm represented by α. In brevity the forward probability is the sum of all possible paths from the initial state through an HMM to a given state at a given time and hence is the probability of observing a particular sequence given an HMM. The number of paths needed to calculate α increases exponentially as the length of the observed sequence increases. However, αs at time $t - 1$ give the probability of reaching a particular observed state at time t in terms of the paths leading to time $t - 1$, i.e., the probability of the previously observed state at time $t - 1$ can be used to determine the probability of the observed state at current time t. By recursively calculating the probabilities of the sequence from $t - 1$ to t to $t + 1$ for a sequence of observations, the probability of a single complete or partial sequence can be determined without necessarily first determining the probabilities of all of the sequences individually, i.e., counting similar sequence repetitions. This method is much faster and much more tractable.

Definition. *Backward probabilities* or anterior probabilities are similar to the forward probabilities. The only difference is that the backward probabilities use the end or stop node as a start node used to calculate the future probabilities where the start node is a node, i.e., state being evaluated within the HMM.

Assume that we have some sequence of observations V of length t, $V = v_1, \ldots, v_t$, and some set of possible states, i.e., hidden states S. Each event or observation $v_j \in V$ is a result of some state $s_j \in S$. For any state sequence $s_1, \ldots, s_t$ where $s_j \in S$, a potential function is defined as such:

$$\psi(s_1, \ldots, s_t) = \prod_{j=1}^{t} \psi(s_{j-1}, s_j, j). \tag{6.14}$$

The potential function $\psi(s_{j-1}, s_j, j)$ or $\psi(s', s, j)$ returns the probability of transition from state s_{j-1} *to* s_j at position j. The potential function $\psi(s', s, j)$ might be defined as

$$\psi(s', s, j) = p_1(s|s')p_2(v_j|s); \tag{6.15}$$

then

$$\begin{aligned}\psi(s_1,\ldots,s_t) &= \prod\nolimits_{j=1}^{t} \psi(s_{j-1},s_j,j) \\ &= \prod_{j=1}^{t} p_1(s|s_{j-1})p_2(x_j|s_j) = p(v_1\ldots v_t, s_1\ldots s_t)\end{aligned} \tag{6.16}$$

where $p(v_1 \ldots v_t, s_1 \ldots s_t)$ is the probability mass function under HMM. The FB algorithm given sequence inputs of some length m, a set of states S, and the potential function $\psi(s_{j-1}, s_j, j)$ for s_{j-1}, $s_j \in S$, and $j \in \{1,..,t\}$, the following is computed:

1. *Forward probabilities*

   1.1 Initialization step
   For all states $s \in S$,

$$\alpha(1,s) = \psi(*,s,1). \tag{6.17}$$

   1.2 Recursion step
   For all $j \in \{2 \ldots t\}$, $s \in S$,

$$\alpha(j,s) = \sum\nolimits_{s' \in S} \alpha(j-1,,s') \times \psi(s',s,j). \tag{6.18}$$

2. *Backward probabilities*

   2.1 Initialization step
   For all $s \in S$,

$$\beta(t,s) = 1. \tag{6.19}$$

   2.2 Recursion step

$$\beta(j,s) = \sum\nolimits_{s' \in S} \beta(j+1,,s') \times \psi(s,s',j+1). \tag{6.20}$$

3. *Smoothing*

$$\begin{aligned}Z &= \sum\nolimits_{s_1\ldots s_t} \psi(s_1\ldots s_t) \\ &= \sum_{s \in S} \alpha(t,s).\end{aligned} \tag{6.21}$$

For all $j \in \{1 \ldots t\}$, $a \subset S$,

$$\mu(j,a) = \sum\nolimits_{s_1,\ldots,s_t:s_j=a} \psi(s_1,\ldots,s_t) = \alpha(j,a) \times \beta(j,a). \tag{6.22}$$

For all $j \in \{1 \ldots (t-1)\}$, a, $b \in S$,

$$\begin{aligned}\mu(j,a,b) &= \sum\nolimits_{s_1,\ldots,s_t:s_j=a,\, s_{j+1}=b} \psi(s_1,\ldots,s_t) \\ &= \alpha(j,a) \times \psi(a,b,J+1) \times \beta(j+1,b).\end{aligned} \tag{6.23}$$

In Eq. 6.17 since every HMM has a start point, or a starting initial state s_0, we label s_0 as *, which is a special case since * does not depend on any previous state. Functions 6.22 and 6.23 are pivotal to understanding the FB algorithm. In Eq. 6.22, the forward calculated probabilities are multiplied with the backward calculated probabilities. This probability includes the forward probabilities which cover all observations up to time j as well as the backward probabilities which include all of the future observations from time $j + 1$ to time t. The value delivered by the multiplication of the forward and backward probabilities is known as a *smooth* value, because these probabilities are combined to compute a final probability of being at position or time j and observing a particular observation v. Thus Eq. 6.22 delivers the probability of being in state a at time j and Eq. 6.23 delivers the transitional probability of being in present state a at time j and moving to state b at time $j + 1$, based on the sequences of observations.

In brevity, the FB algorithm is used to summarize the HMM and to utilize its structure fully. Using the FB algorithm, probabilities of events and states along a sequence can be easily determined.

6.5 Applied Imputation Algorithms

Two very popular imputation algorithms in use today are the fastPHASE algorithm and KNNimpute algorithm. The algorithms differ mostly in their approach to solving the missing value problem and their overall processing time. KNNimpute uses a K nearest neighbor approach which models a current row being worked on based on other rows within the data set with similar marker values. The fastPHASE algorithm first builds a probabilistic model of the data in the form of an HMM and then based on the HMM infers what is the best value to replace the blanks within the data set with.

6.5.1 *The KNNimpute algorithm*

KNNimpute is a popular iterative classification method for genotype imputation. It uses a nonparametric distribution free model, without any assumptions of drawn data probability distribution. Being based on the nonparametric statistical model, KNN inference is using nonparametric statistical tests [40]. All genotype imputation algorithms used previously described matrix setup of data elements for statistical analysis. An example of these matrices is shown in Fig. 6.14. Genotypes, the elements of the matrix, are encoded into three possible values based on their allelic definitions: *0, 1,* or *2*. Each numeric value corresponds to one of three unique genotypes of an SNP marker. For instance, *0* would encode the homozygous genotype X_aX_a, *1* would encode the heterozygous genotype X_aX_b or X_bX_a, and *2* would encode the homozygous genotype X_bX_b. There are four parameters

	A	B	C	D	E	F	G	H	I	J	K	L	M
1	GG	GG	AG	AA	GG	AA	TT	GG	TT	CC	TT	TT	AA
2	GG	GG	AA	AT	GG	AA	TT	GG	TG	CC	CT	CT	CA
3	GG	GG	AA	AA	GG	AA	TT	GG	GG	CC	TT	CT	AA
4	GG	GG	GG	TT	AG	AG	TT	GG	TT	CC	TT	CC	CC
5	GG	GG	AA	TT	GG	AA	TT	GG	TG	CC	TT	TT	AA
6	GG	GG	AA	TT	AA	GG	TT	CC	GG	CC	CC	CT	CA
7	GG	GG	AA	AA	GG	AA	TT	GG	TG	CC	TT	CT	CA
8	GG	GG	GG	TT	GG	AA	TT	GG	TT	CC	CT	CC	CA
9	GG	GG	AG	TT	AG	AG	TT	CC	TG	CC	TT	CT	CA
10	GG	GG	AG	AA	GG	?	TT	GG	TG	CC	TT	CT	CA
11	GG	GG	AA	AA	GG	AG	TT	GG	TT	CC	TT	CT	CA
12	GG	GG	AG	AT	AG	AG	TT	CG	GG	CC	TT	CT	CA
13	GG	GG	AG	TT	AG	AA	TT	GG	GG	CC	CT	CT	CA
14	GG	GG	AG	AA	GG	AA	TT	GG	TG	CC	TT	CC	CC
15	AA	GG	AA	AT	GG	AG	CC	CG	GG	TT	TT	CT	CA
16	GG	GG	AA	AT	GG	AA	TT	GG	TT	CT	TT	CT	CA
17	GG	GG	AG	AA	AG	AG	TT	CC	TG	CC	TT	CT	CA
18	GG	GG	AA	AT	GG	AA	TT	GG	GG	CT	TT	CT	AA

Fig. 6.14 An example of a map of genotypes of 18 individuals and 13 SNP markers. The *question mark* indicates a missing SNP genotype value. Its flanking markers and their matching flanking markers in other *rows* are highlighted. With the highlighted flanking markers in *row* 14, the value AA can be used to impute the missing genotype

followed by the algorithm. These parameters are Y, x, i, k. The letter Y corresponds to the SNP genotype being iteratively updated or inspected. This is usually the missing SNP genotype. The letter x resembles the set of flanking SNP markers of the SNP marker Y. The flanking markers correspond to most adjacent markers left and right of SNP marker Y. The letter i is the index of the ith individual in the matrix and k the number of nearest individuals to observe, both above and below of Y's row. The logic behind the algorithm is if the flanking SNP genotype markers of Y can predict Y accurately, then we can assume that the flanking markers and Y are in linkage disequilibrium, i.e., are inherited together. Subjects with similar flanking markers should have similar Y values. Therefore, patients with similar flanking markers can be used to predict the missing values of a marker, Y, under scrutiny (see Fig. 6.14). Y' is used to signify Y as an estimate or the prediction of Y. The original model for this algorithm is mathematically described as follows:

$$Y'(x) = \frac{1}{k}\sum\nolimits_{x_i \in N_k(x)} Y_i. \tag{6.24}$$

Here, $N_k(x)$ corresponds to the set of nearest neighbors of the missing SNP location. In this case rows of the data set represent individuals and "k" signifies how large this set is; k is the number of nearest neighbors to consider, specified by the given set of flanking markers x, based on Euclidean distance Eq. 6.24. The set of flanking markers of individual i whose marker Y_i is under scrutiny is represented by x_i. Hence, the statement $x_i \in N_k(x)$ simply is a condition that the flanking markers of the genotype Y_i belong to the nearest neighbor set $N_k(x)$. Euclidean distance in this case is a simple measure of distance in terms of matching flanking markers. The Euclidean distance between x as x_i increases with the number of mismatched flanking markers between them. The Eq. 6.24 simply indicates that given the set of matching flanking markers, sum of the column-encoded genotype values for all patients whose flanking markers are the same as that of the missing SNP genotype is divided by k. Division by k resembles an averaging, where the relevant column values, corresponding to the genotypes of nearest neighbors at position of column Y, are averaged by the size of the nearest neighbor set, k. The column used in averaging the encoded values is the same column that the missing value Y being estimated belongs to.

The x_i flanking markers and corresponding x flanking markers tend to have the same number of matching flanking genotypes.

The Y_i values of the corresponding members of the nearest neighbor set, defined by x_i, as mentioned are taken into account and the most frequent value of Y_i, determined by average, represents an estimate for the missing value.

KNN algorithm may be improved by weighing the contribution of Y_i to Y'. An improved version of the KNN impute algorithm known as the "Weighted KNN" or WKNN is defined by expression (6.24):

$$Y'(x) = \frac{\sum_{x_i \in N_k(x)} e^{-\|x_i - x\|} Y_i}{\sum_{x_i \in N_k(x)} e^{-\|x_i - x\|}}, \tag{6.25}$$

where $\|x_i - x\|$ represents the Euclidean distance between points x_i and x. This determines the number of mismatching SNP genotype markers between the flanking markers in x and those in x_i. If x and x_i have the exact matching flanking markers, then this difference is equal to *0* and the exponential value is equal to *1*, signifying a perfect match of flanking markers. This value is methodically the weight given to the Y_i value of x_i. In addition to the number of markers on each side (chosen to be *2*, *3*, *5*, or *10*), the number of nearest neighbors k is a tuning parameter. Previous studies [41] have considered $k = 3, 5, 10$, or *15*. Number of flanking markers chosen on each side of Y were *2*, *3*, *5*, or 10. All of these tuning parameters are considered in the average presented by the Weighted KNN approach.

6.5.1.1 Limitations

The KNN-based algorithms use the nearest neighbor rule to evaluate the best estimate value. This approach is designed to work on large data samples. Like with all approaches, it too has its advantages and disadvantages [42, 43]. The advantages of the KNN approach are:

1. Since KNNimpute essentially does not train at all, "the speed of training" is very fast.
2. The approach is simple and easy to learn.
3. By not assuming the underlined probability densities, and by avoiding any bias in favor of the chosen distribution, it is very robust to noise and irrelevant data

presence. Furthermore, when the data is log transformed, the Euclidean distance measure loses its sensitivity to drastic outliers that could bias the decision. The log-transformed data reduces the effect of outliers during genotype similarity determination.

The disadvantages of the KNN approach are the precision of the missing value estimation and certain hardware limitations [42]. Some important disadvantages of KNN are:

1. KNNs accuracy is limited by the parameter k, which is partly resolved by WKNN approach. KNN with strictly k neighbors tends to deliver sometimes biased results.
2. Large-size memory is required. As the data set grows, the algorithm encounters memory limitation problems and performance degradation with virtual memory management in the background using the hard disk more frequently.
3. The computational complexity of this approach is of the order $O\ (n\ d)$ where n is the number if flanking markers considered and d is the number of training samples used in forming the nearest neighbor set. If n is increased, $O\ (n\ d)$ implies that processing time will increase too [44]. With very large data sets, the computational complexity results in long execution times. In the other hand, the WKNN improved approach results in increased computational complexity yielding even longer execution times.
4. In addition to the limitations due to computational complexity, very slow learning algorithm adds to the run time algorithm performance, by further slowing it down.

When implementing the WKNN approach, weights are assigned to neighbors as per calculated Euclidean distance. WKNN overcomes some of the limitations of KNN by assessing equal weight to k neighbors, taking into account the k neighbors with more precisely matching flanking markers. To further improve the precision, WKNN also considers the entire training sample, that is, all of the flanking markers that match within their respective columns instead of just k instances of matching columns. This is the reason why with very large data sets, WKNN may perform poorly.

6.5.2 The fastPHASE Algorithm

The fastPHASE algorithm is a probabilistic model-based algorithm for finding the haplotype phase, which means determining haplotypes from genotypes, i.e., resolving the haplotype phase. The basic algorithm assumption is if the parental genotypes are known, the haplotype phase of the offspring can usually be determined. The task being given two parental haplotypes determine the possible child haplotype, a.k.a. “the phase.” Determining the phase of the offspring can contribute to imputing with high probability the value that might be missing in offspring genotypes. Without a strict experimental setting where the parents and the offspring are defined, phase determination, for the sake of imputation, is facilitated by assuming the *Hardy-Weinberg Equilibrium* (HWE). The HWE states that when all *agents of evolution*[1] are turned off, the proportions of alleles of genes and of markers, i.e., SNP markers, alike will remain constant from generation to generation. The HWE is assumed in GWAS data sets since such a data set is a fixed, unchanging snapshot of SNP genotype sequences recorded from several individuals, not involving or monitoring any changes in the SNP data. The data set can be treated as a population in HWE. The benefit of assuming HWE is that in the presence of constant allelic proportions, constant haplotype proportions are present as well. This facilitates the predictability of a haplotype occurring within a data set and thus the occurrence of specific SNP genotypes within them.

6.5.2.1 Preliminary Assumptions

To begin the discussion, let us assume that we are given a data matrix of SNP sequences of many individuals, considering that we are not changing the data set. Adding new individuals or modifying nonempty SNPs is not allowed. In other words we can assume that if the present data set was a model of the last generation, then the last generation would have had the same allelic proportions as this generation; hence the data set is in equilibrium or stationary state. This also assumes that the future generation would maintain the same proportions as the previous generations. Hence, given this closed set, HWE can be assumed. For fastPHASE, HWE is helpful in understanding SNP marker inheritance rates. Since we know the allelic proportions, we also know the haplotype proportions; we know that they are constant and unchanging. No new haplotypes will arise from the combination of any of the parents in the population, assuming that they are products of a HWE parent generation. Since we know the haplotype proportions, we know their frequencies and the probability estimates of them occurring within a population with relative certainty. Given individual haplotype probabilities, we can better guess or infer a whole or partially missing haplotype by finding the probability of the known sequence within the partially missing haplotype, of occurring in any of the known haplotypes.

6.5.2.2 Haplotype Clusters

The statistical model for patterns of genetic variation is based on the idea that over short regions of DNA, haplotypes in a population tend to cluster into groups of similar haplotypes; the shorter the haplotypes are, the more common they might be

[1] The agents of evolutionary change are mutation, genetic drift, sexual selection, natural selection, and gene flow.

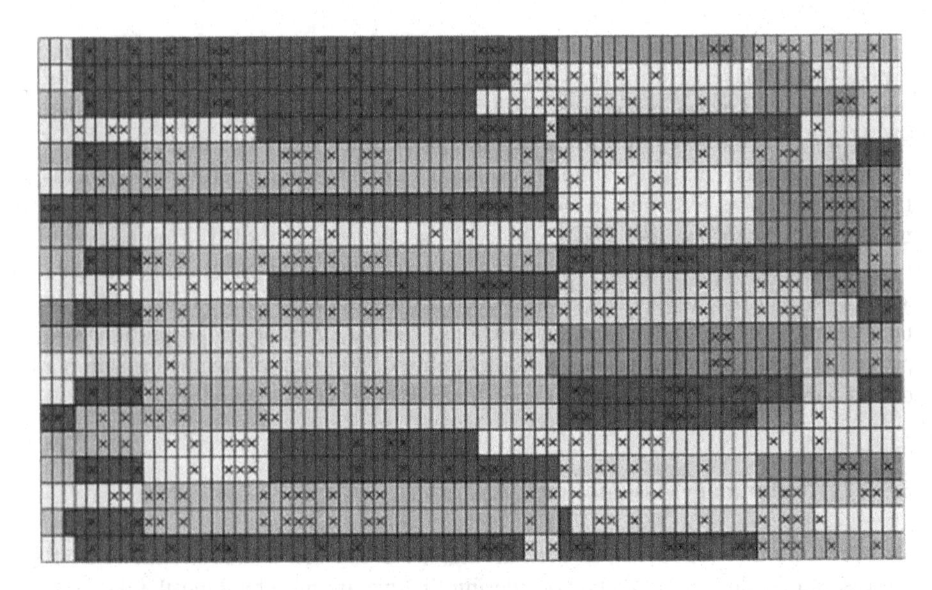

Fig. 6.15 Matrix data model allowing cluster membership to change continuously along a chromosome. Each *column* represents an SNP marker, with the two possible alleles indicated by *crossed squares* and *non-crossed or whole squares*. Successive pairs of rows represent the estimated pair of haplotypes for successive individuals. *Colors* represent the estimated cluster membership of each allele, which varies as the pointer would move along each haplotype. Locally, each cluster represents a common combination of alleles at tightly linked SNPs [39]

in a population. Many subjects share some short-sequence haplotypes. Scheet and Stephens in [39] observed that similar observable short haplotypes seem to cluster. The statistical model is thus motivated by the observation that, over short regions of let us say a few kilobases, haplotypes tend to cluster into groups.

As a result of recombination, those haplotypes that are closely related to one another and similar will vary as one moves along the chromosome, i.e., this clustering tends to be local in nature. The cluster's characteristics follow haplotype probabilities. Each cluster can be thought of as (locally) representing a common haplotype, or combination of SNP alleles, and the HMM assumption for cluster membership results in each observed haplotype being modeled as mosaic of a limited number of common haplotypes [39] (see Fig. 6.15).

Statistical models such as these are used to capture the complex patterns of correlation such as LD that exists among dense markers in samples from natural populations. The flexibility of changing cluster membership along a chromosome allows for both blocklike patterns of LD and gradual decline in LD with distance (the further markers are from each other, the less likely they are to be co-inherited). LD patterns are frequencies of several marker SNP alleles occurring together within a set of human subjects examined, with statistically close ties between SNP alleles. The purpose of the statistical models is to capture complex patterns of correlation between dense markers in samples. Such models employing markers correlation are tested on the genotype data to see if the pattern of variation has been accurately

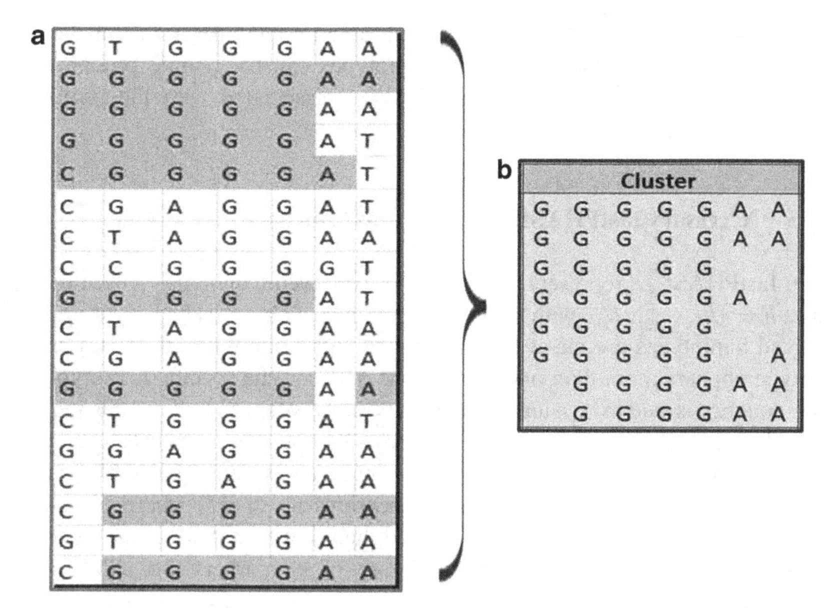

Fig. 6.16 (**a**) Similar haplotypes based on common SNP alleles shaded *gray*. (**b**) A summary of similar haplotypes, within the frames of a cluster. The cluster defined by the longest observed sequences of matching genotypes (e.g., the unique haplotype is GGGGGAA)

captured. It is tested against matrix genotype data with missing genotype values. The model is presented and its ability to accurately capture patterns of variation is assessed by applying the model to estimate missing genotypes and to infer haplotype phase from unphased genotype data. Since it is a phasing method, it looks at the SNP alleles present within the two haplotypes of a homologous chromosomal pair, i.e., it models the genotype of each individual SNP allele pair. There are many haplotypes on a chromosome. This sequence of haplotypes transcends to a sequence of clusters that may be modeled by an HMM. Each cluster represents a summarized haplotype for the group of similar haplotypes found within the whole data set (see Fig. 6.16.). Each cluster is viewed as a model haplotype, a summary of the similar haplotypes it contains.

When viewing haplotypes along a chromosome, each haplotype is a member of some defined haplotype cluster. The HMM assumption for haplotype marker sequences in a cluster of observed haplotypes results in each observed haplotype originating from a limited number of other similar observed haplotype sequences (cluster haplotype sequences). This model is modified to allow cluster membership to change along each haplotype and so to capture the fact that, although sampled haplotypes exhibit cluster-like patterns, they tend to be local in nature. The entire sequences of SNPs do not form haplotypes. Given the cluster of origin of each haplotype, the alleles at each marker of the observed haplotype are assumed to be independent draws based on the corresponding origin cluster's marker allele frequencies, depicted by θ as we shall see. The number of unique haplotypes is considerably smaller than the number of observed haplotypes in all of the rows of the data. Therefore, the observable haplotypes can have some probability of occurrence within the data. This probability is associated with the probability of their respective origin cluster.

6.5.2.3 Necessary fastPHASE Notation and Concepts

Within fastPHASE, h represents the set of all observed haplotypes within the data matrix, $h = (h_1, \ldots.., h_n)$, with n being the number of observed haplotypes present. Observed haplotypes are all of the haplotypes within the data set. If an observed haplotype appears more than once within the data set, the specific haplotype SNP allele sequence is said to be a unique haplotype and any appearance of this sequence within the data set is said to be an observed haplotype or an instance. Unique haplotypes may have more than one appearance or instance. Each haplotype is a sequence of SNP values at M marker locations $(1, \ldots, M)$ where h_{im} denotes the SNP allele value in the ith observed haplotype at SNP marker m, and $h_i = (h_{i1}, \ldots.., h_{im})$ is a haplotype consisting of m markers, i.e., SNP alleles. Clusters of haplotypes are defined by a unique haplotype. Each unique haplotype is the label of an individual cluster (see Fig. 6.16b – first row). Each cluster contains instances of the label of the cluster, i.e., labels correspond to the unique haplotype. This way we can pull out a probability measure of specific observable haplotype membership within a cluster, i.e., a haplotype frequency. In this case, the frequency of some unique or observable haplotype is equal to the number of members that its origin cluster contains divided by the total number of all observable haplotype members present in all clusters: all haplotypes. The markers found at locations h_{im} are "biallelic," meaning that they can take up one of two encoded values: *0* or *1*. Each sampled observable haplotype originates from one of K possible clusters. In the beginning of imputation, no information is available on clusters or on haplotypes. Clusters are defined on the fly using *expectation maximization* for parameter estimation. During the fastPHASE run, the unphased haplotypes that could not be assigned cluster membership determined their origin clusters by means of an HMM. The fastPHASE algorithm uses an SNP allele data set as opposed to a genotype data set, where each two consecutive rows of the data set resemble the homologous SNP allele sequences of a single individual, that is, each pair of vertically aligned alleles, one from each row, comprises of an SNP genotype (see Fig. 6.15). Each allele is treated independently, i.e., two alleles of a genotype are allowed to belong to two different haplotypes, thus may have different clusters of origin. In [39] z_i is a variable that denotes the cluster of origin for haplotype h_i. The set $\alpha = (\alpha_1,.., \alpha_k.., \alpha_K)$ is the set of cluster frequencies:

$$\alpha_k = \frac{|k|}{\sum_{i \in K} |i|}. \tag{6.26}$$

The relative frequency of cluster k is denoted by α_k in Eq. 6.26. It is the number of observed haplotypes in the cluster among all of the observed haplotypes in the entire data set. The probability that a cluster of origin is k corresponds to the size of the cluster (the number of observed haplotype members within it $p(z_i = k|\alpha) = \alpha_k$, where $\alpha = (\alpha_1, \ldots, \alpha_K)$. Considering that the SNPs of the haplotypes in fastPHASE are of a biallelic nature, a similar encoding scheme to KNNimpute is implemented, with a slight difference in semantics. Instead of coding genotypes, alleles are encoded, i.e., *0* is used for (*C-G*) and *1* for (*A-T*) or vice versa, as long as consistency is maintained. A whole table θ exists charting the frequencies of allele *1* at all the marker positions and for all of the clusters. Hence, the frequency of allele *1* in cluster k at marker position m is denoted by θ_{km}. For any given haplotype cluster, these frequencies are either very close to *0.01* or to *0.99*. This allows us to summarize or view the cluster as a haplotype; a unique haplotype as mentioned before is said to label the cluster. The label is a haplotype, denoting the cluster, denoting all of its members. Haplotype is a linear allelic definition. Given the cluster of origin of each haplotype, alleles observed at each marker are *independent* draws from cluster-specific allele frequencies, delivered by θ:

$$p(h_i|z_i = k, \theta) = \prod_{m=1}^{M} \theta_{km}^{h_{im}} (1 - \theta_{km})^{1-h_{im}}. \tag{6.27}$$

This conditional probabilistic distribution in Eq. 6.27 shows the probability of selecting h_i, given that we have cluster k as our origin cluster and θ_k from θ for our allelic frequencies. For any given h_i, inequality $0.01 \leq \theta_{km} \leq 0.99$ holds. This distribution is independent; therefore, the product of individual marker probabilities according to the cluster of origin z_i and θ is facilitated. It has no dependencies, but it is able to show associations between alleles at neighboring marker positions where $h_{im} : \{0,1\}$ and $\theta_{km}^{h_{im}} : \{1, \theta_{km}\}$. If $h_{im} = 1$, then $\theta_{km}(1) = \theta_{km}$, and if $h_{im} = 0$, then $(1)(1 - \theta_{km}) = (1 - \theta_{km})$. Abbreviated, when h_{im} assumes the allelic value *0*, the frequency of the value *0* at marker position m for origin cluster $z_i = k$ is equal to $1 - \theta_{km}$, where θ_{km} represents as previously defined the frequency of allelic value *1* at the marker position m for cluster k. With regard to expression (6.27), one problem exists. Namely, the origin cluster z_i for h_i is not known. Therefore, it must be determined. The only way we can determine whether z_i is truly the origin cluster of h_i is to test all possible clusters. Since the origin cluster of the haplotype is not known, we assume Eq. 6.28

$$p(h_i|\alpha, \theta) = \sum_{j=1}^{K} p(z_i = j|\alpha) p(h_i|z_i = j, \theta) = \sum_{j=1}^{K} \alpha_j \prod_{m=1}^{M} \theta_{jm}^{h_{im}} (1 - \theta_{jm})^{1-h_{im}}. \tag{6.28}$$

The haplotype frequency is tested in all clusters. We know then that the origin cluster z_i for some haplotype h_i is k when the product rule delivers an end result of one, given cluster k.

Table 6.2 Table of clusters of origin resembling the Markov chain. In the first row, z_{im}, $1 \leq m \leq M$, represents the origin cluster of allele m of haplotype i, i.e., z_{im} represents the origin cluster of marker h_{im}. For each z_{im} in the chain, there are K possible clusters to choose

Z_{i1}	*	*	Z_{im}	*	*	Z_{iM}
1	1	1	1	1	1	1
*	*	*	*	*	*	*
*	*	*	*	*	*	*
K	K	K	K	K	K	K

6.5.2.4 HMM for fastPHASE

The fastPHASE algorithm uses HMM to model [45] the fact that alleles at nearby markers are either likely to arise from the same cluster or not. The HMM is based on the classical Markov chain. Within this chain, the origin clusters z_i are hidden states, while the observed marker value is the observed state. Each allele at each marker may have its own origin cluster; therefore it is thus assumed that $z_i = (z_{i1}, \ldots, z_{iM})$ forms a Markov chain on $\{1, \ldots, K\}$ clusters. From the table in Table 6.2, we can see that for every marker, there is a cluster of origin which corresponds to one of the K possible clusters. Finding the origin cluster for M markers and K clusters has computation cost of the order of $O(nMK)$, where n is the number of observed haplotypes within the data set, M is the number of markers per haplotype, and K is the total number of clusters.

Specifically, if z_{im} denotes the cluster of origin for marker h_{im}, we assume that $z_i = (z_{i1}, \ldots, z_{iM})$ forms a Markov chain on $\{1, \ldots, K\}$ with initial state probabilities

$$p(z_{i1} = k) = \alpha_{k1} \tag{6.29}$$

which is the probability of starting with allele h_{i1} in cluster k with transition probabilities

$p_m(k \to k')$ given by

$$p_m\left(k \to k'\right) := p\left(z_{im} = k' \middle| z_{i(m-1)} = k, \alpha, r\right)$$
$$:= \begin{cases} e^{-r_m d_m} + \left(1 - e^{-r_m d_m}\right)\alpha_{k'm}, & k' = k \\ \left(1 - e^{-r_m d_m}\right)\alpha_{k'm}, & k' \neq k \end{cases}. \tag{6.30}$$

The initial state probability corresponds to the cluster frequency. The probability that some cluster k is the origin cluster of allele at marker position one, individual i, is the probability of cluster k which in turn is equal to the frequency of cluster k (6.29). The transition probabilities Eq. 6.30 follow the Markov rule. The origin clusters of each succeeding allele depend on the origin cluster of the present allele. If the origin cluster of the succeeding allele is not the same as that of the present allele, then the bottom probability measure is taken: $\left(1 - e^{-r_m d_m}\right)\alpha_{k'm}$, Eq. 6.30.

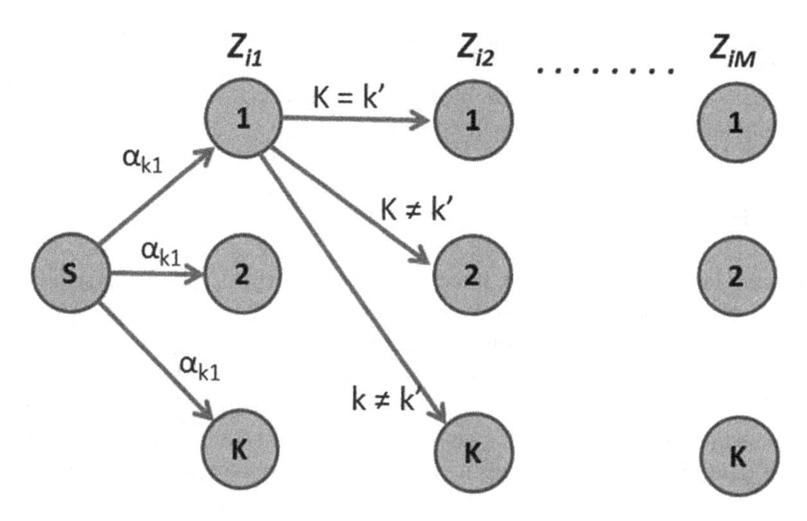

Fig. 6.17 The HMM model analogous to the Markov chain of Table 6.2. The probability of a change in origin cluster is represented by the edge-defined relationships between the nodes

If the origin cluster of the succeeding allele is the same as that of the present allele, then the top probability measure is taken: $e^{-r_m d_m} + \left(1 - e^{-r_m d_m}\right)\alpha_{k'm}$, Eq. 6.30. Variable d_m that specifies the physical distance between markers $m-1$ and m on the chromosome is assumed to be known, and $r = (r_1, \ldots, r_M)$ and α_{km} are parameters to be estimated. The Markov chain proposed is a discretized version of a continuous Markov jump process [135, 162], with jump rate r_m per base pair between markers $m-1$ and m with transition probabilities

$$P\left(z_{im} = k' \middle| z_{(im-1)} = k, jump\ occurs\right) = \alpha_{k'm}. \tag{6.31}$$

The r_m simply is an average rate at which m and $m + 1$ are not associated. It is informally thought of as being the recombination rate between m and $m + 1$. Nonetheless, the r_m values can be set to be all equal, constant, and if the distances between m and $m + 1$ are not known, parameter d_m can also be eliminated. So is the case in [39]. The transition in Fig. 6.17 is a graphical model of the HMM used in order to model allelic origin cluster changes during state transitions, performed during the move to the next observable marker. It corresponds to the tabular model presented in Table 6.2, but with specific relationships between the states and state values, expressed via the edges of the graph.

Figure 6.17 illustrates the probabilistic flow, as states change, from allele to allele within a haplotype. State changes can lead to cluster changes where the origin cluster of the current state k is not the origin cluster of the allele at the next state k', $k \neq k'$ modeled by the probability $k = \left(1 - e^{-r_m d_m}\right)\alpha_{k'm}$, Eq. 6.30. Some state changes do not yield new origin clusters when we have $k = k'$. These probabilities are modeled by $e^{-r_m d_m} + \left(1 - e^{-r_m d_m}\right)\alpha_{k'm}$, Eq. 6.30. Figure 6.17 depicts a cluster membership model for alleles. However, considering that we are dealing with genotypes and genotypes are tuples of alleles, this HMM must be expanded in order to accept two origin clusters for a genotype. In the following example, the genotypes will be modeled. This means that clusters of origin for two alleles from homologous DNA strands will be determined.

6.5.2.5 HMM for Two Alleles: The Genotype

Additional fastPHASE notation must be covered in order to understand how genotype cluster memberships are determined. First, $g = (g_1, \ldots..., g_n)$ denotes the set of genotype data on *n* diploid individuals where $g_i = (g_{i1}, \ldots, g_{iM})$ and g_{im} corresponds to the genotype at marker *m* for individual *i*. The values for any g_{im} can be *0, 1,* or 2. Since genotypes are a tuple of alleles that don't necessarily originate from the same cluster, origin clusters for both SNP marker alleles of a genotype must be determined. The two origin clusters for a genotype are represented by a tuple: $\dot{z} = (k_1, k_2)$ which is an unordered pair of clusters from which g_{im} originates. We can conclude from the table above that the running time for determining the origin clusters of haplotypes for SNP genotypes is of the order $O(K^2M)$. For genotype data $g = (g_1, \ldots..., g_n)$, $\dot{z}$ is the set of origin cluster tuples for every genotype at markers *1* through *M* for individual *i*. Thus $\dot{z} = \ \dot{z}_{i1}, \ldots, \dot{z}_{iM})$ forms Markov chain with initial state probabilities

$$p(\dot{z}_{i1} = \{k_1, k_2\}) = \begin{cases} (\alpha_{k_1 1})^2, & k_1 = k_2 \\ 2\alpha_{k_1 1}\alpha_{k_2 1}, & k_1 \neq k_2 \end{cases} \tag{6.32}$$

and transition probabilities

$$p_m(\{k_1, k_2\} \to \{k_1', k_2'\}) = \begin{cases} p_m(k_1 \to k_1')p_m(k_2 \to k_2') + p_m(k_1 \to k_2')p_m(k_2 \to k_1'), \\ \qquad k_1 \neq k_2 \text{ and } k_1' \neq k_2' \\ p_m(k_1 \to k_1')p_m(k_2 \to k_2'), \ otherwise. \end{cases} \tag{6.33}$$

In the initial state probability Eq. 6.32, since each of the two alleles of a genotype is independent draw from a cluster of origin, we treat the probabilities of the two drawing events as the product of the probabilities of drawing the two alleles.

The probability of drawing one allele from an origin cluster, *k*, corresponds to the origin cluster's frequency α_k, Eqs. 6.26 and 6.32. The probability of drawing two alleles of a genotype from the same cluster is equal to the squared product of the origin cluster's frequency $(\alpha_{k_1 1})^2$, where k_1 is one origin cluster Eq. 6.33. The probability of drawing two alleles from two separate clusters is equal to the product of the two cluster frequencies $2\alpha_{k_1 1}\alpha_{k_2 1}$, Eq. 6.33. In terms of transition probabilities, we have also two options. If the origin clusters initially are not the same and the transition clusters are also not the same, then refer to probability $p_m(k_1 \to k'_1)p_m(k_2 \to k'_2) + p_m(k_1 \to k'_2)p_m(k_2 \to k'_1)$. For all other cases, refer to probability $p_m(k_1 \to k'_1)p_m(k_2 \to k'_2)$, Eq. 6.33.

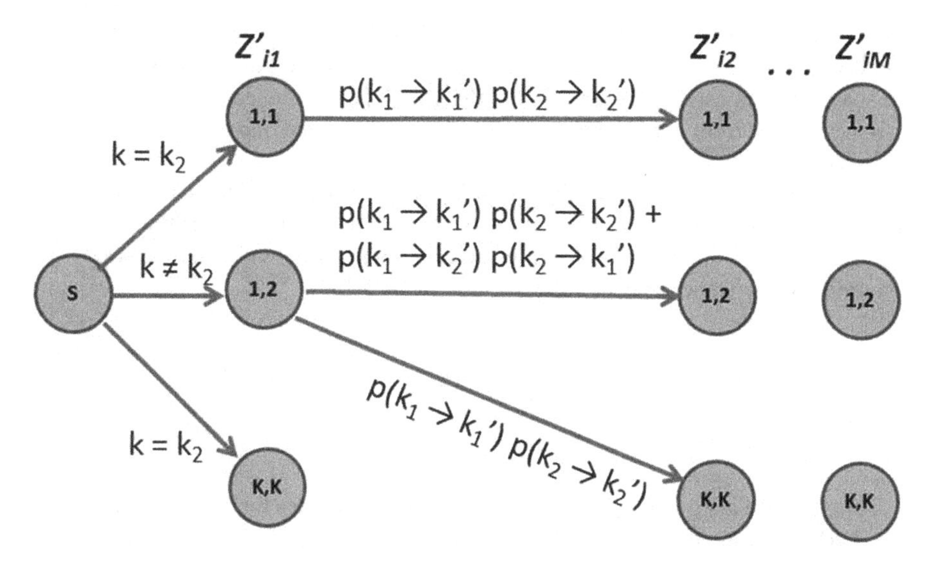

Fig. 6.18 The HMM that models genotype allele origin clusters. At each state there is a tuple $\{k_1,k_2\}$ defining the origin clusters for the two alleles of a genotype at each state z_i

Figure 6.18 shows an HMM and how allele tuple, i.e., genotype origin cluster memberships, traverses the probabilistic Markov model as we observe each allele tuple of each successive genotype along the chromosome. As the tree is traversed and a path is sketched, any straight horizontal path between two or more nodes, representing the states, means that the origin clusters of the successive genotypes are the same. The presence of zigzag paths between any two or more nodes means that the origin clusters for the genotypes are different. These models are of immense aid in resolving missing haplotype phase or determining the hereditary properties such as linkage and association between the SNP alleles comprising the individual haplotypes and the association and relationship between the homologous haplotypes themselves.

The missing genotype value problem poses great difficulties for haplotype reconstruction mechanisms such as these. Along the traversal, missing genotypes may be encountered. When missing values are encountered, the probabilistic model set forth for haplotype reconstruction must be modified. This modification is too time consuming, and taking into account the prevalence of missing data within such data sets, reducing data set sizes only decreases phasing accuracy of the model. Imputation of the missing genotype is the best solution. It is performed using a maximization step shown below, where the probability of a genotype $g_{im} = x$ given the set of genotypes g and parameters $v = (\theta,\alpha,r)$ is equal to the conditional probability Eq. 6.34 of g_{im} given the origin cluster(s) $\dot{z}_{im}$ of the genotype g_{im}. To recapitulate, here θ is the table of frequencies of the selected minor allele "1" for each recognized haplotype of SNPs, i.e., allele that appears least often at each marker. The α parameter is the set of cluster frequencies for each cluster, i.e., the number of haplotype instances each cluster has divided by the total number of haplotype instances for all of the clusters, and r is the jump rate of marker alleles; this parameter is usually silenced in fastPHASE. The mentioned set of parameters v is estimated from the given data using an *expectation maximization* or EM algorithm.

However, since the origin clusters of missing genotypes cannot be explicitly determined, all possible clusters are taken into account. The cluster that yields a probability $p(g_{im} = x) = 1$ is the cluster of origin which will later be used for maximizing the value of x:

$$p(g_{im} = x|g, v) = \sum_{k_1=1}^{K} \sum_{k_2=k_1}^{K} p\left(g_{im} = x|\dot{z}_{im} = \{k_1, k_2\}, v\right) \times p\left(\dot{z}_{im} = \{k_1, k_2\}|g_i, v\right). \quad (6.34)$$

Note that we are given a genotype sequence g_i. Given this fact, then by means of the HMM, we can determine whether for g_i, at position g_{im}, $\dot{z}_{im}$ origin clusters have changed or not from those of the genotype at g_{im-1}, given that g_{im-1} is observed. This speculation is made possible by the HMM imposed on the data set. Using the known origin clusters of genotypes in the sequence g_i, we can assume that the origin clusters are most likely the same for g_{im-1} and g_{im}. This assumption is put to the test by testing all possible clusters, for $\dot{z}_{im}$. Once the cluster has been found, an estimate of the genotype value for g_{im} is determined. Genotypes as we remember are encoded by the values {*0*, *1* or *2*}. The variable x in $p(g_{im} = x|g, v)$ can assume any one of these three values Eq. 6.35. Each of these values is individually introduced into the probabilistic function above Eq. 6.34. The x value that yields the largest or maximal probability is then selected. This process is repeated T times, where in [39], T was made constant at the value of $T = 20$. At each T, a different set of parameter estimates v is provided. By averaging $T = 20$ runs of this maximization procedure Eq. 6.35, a better estimate was achieved for g_{im}, the missing genotype:

$$\hat{g}_{im} = \text{argmax}_{x \in \{0,1,2\}} \frac{1}{T} \sum_{t=1}^{T} p(g_{im} = x|g, \hat{v}_t), \quad (6.35)$$

best estimate of genotype by maximization across T possible v parameter sets. T runs of an expectation maximization algorithm are used to estimate parameters, while Eq. 6.35 is used to maximize the expected value of missing genotype g_{im}.

6.5.2.6 Limitations

The fastPHASE algorithm doesn't have many functional limitations. Its major drawback concerns the amount of time it uses in order to process entire data sets. The clustering of similar SNP allelic sequences, which are necessary for estimating the cluster-relevant parameters prior to genotype imputation and phase determination of allele sequences, is resource demanding. The use of HMM and the T runs of the expectation maximization algorithm, used for estimating the "v" parameters, increase the processing time of fastPHASE as the data set size increases. The feasibility of applying the fastPHASE model to very large data sets was tested in [39]. A data set containing 3,150 individuals typed at 290,630 SNPs was created, amounting to 915.484.500 data elements within the entire data set. The fastPHASE software package required a total of 97 h (on a single 3-GHz Xeon processor with 8GB of RAM) to fit the model to the data. According to Scheet [39], fastPHASE, GERBIL, and HaploBlock are three very popular haplotype reconstruction methods using probabilistic models. The fastPHASE algorithm is among several imputation solutions widely used on SNP allele and genotype data sets today.

6.5.3 Comparison of fastPHASE and KNNimpute

In order to test genotype imputation algorithms, complete data sets with all known values are used for the creation of data sets with synthetically missing values. The original complete data set is known as the reference data set. The reference data set is used to create a new data set with missing values. The original data set can be put into a two-dimensional matrix, and using a random number generator, elements of the matrix can be selected at random and their values deleted, corresponding to missing values. This new set can be used to test the accuracy of the underlying imputation algorithms, by comparing the imputation results of these algorithms to the reference data set.

Many metrics for measuring the genotype imputation error rates exist. In order to measure error rates, reference values, i.e., real genotype values, must be known for synthetically missing genotypes in order to determine the accuracy of the imputed values. Two common metrics for quantifying imputation errors are known as the *error rate per locus* and the *error rate per allele* [22]. These measures measure the susceptibility of a particular marker, i.e., locus to imputation error. Such errors are not analyzed within this study. The main focus is on overall imputation accuracy, performed by testing fastPHASE and KNNimpute on independent missing value data sets with varying numbers of missing values. The overall error rate of the imputation algorithms will be taken into account, $e = x/m \cdot 10$, where the ratio "e" of falsely imputed values "x" to all imputed values "m" is taken and multiplied by 100 to give us an error percentage.

The error rates on seven data sets with independent distribution of missing values have been evaluated in Fig. 6.19. The size of the experimental data set contains 4.536 elements where 5 % of missing values amount to 266 elements and 10 % of missing values amount to 453 elements. It is apparent that the accuracy of the fastPHASE algorithm is better than the KNNimpute algorithm, especially as the number of missing values increases.

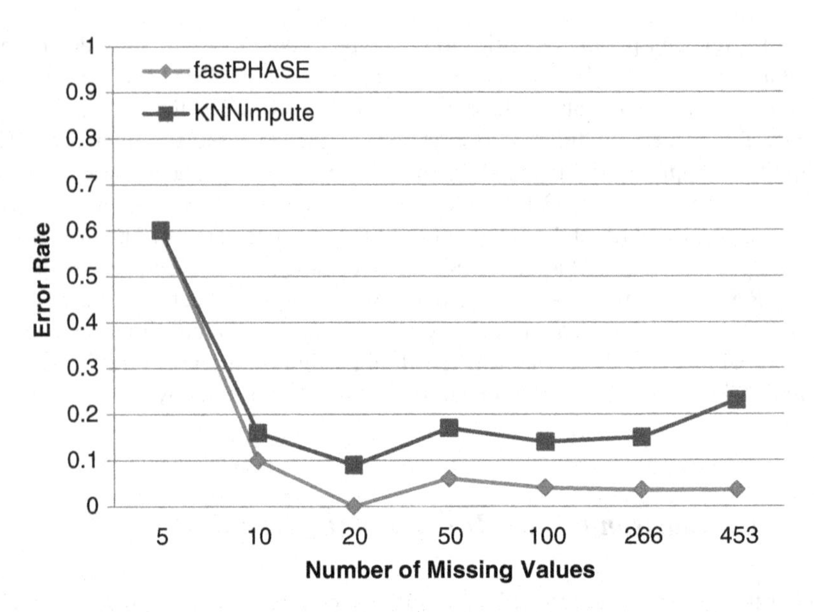

Fig. 6.19 Error rate comparison of KNNimpute and fastPHASE algorithms

In terms of speed, however, the average processing time of a data set of size 4.536 using fastPHASE amounts to a rough 3 min, while the KNNimpute on the same data set size completes imputation in roughly a second. One can imagine for data sets of extreme size, fastPHASE might even take days to complete. The main reason for such slow processing time is the preprocessing step the algorithm performs while creating the probabilistic model, i.e., HMM for the data set. This step is computationally and spatially intensive. When it comes to accuracy, the fastPHASE algorithm is preferred. However, when speed is the most crucial part of the analysis, then the KNNimpute algorithm is preferred as an alternative. The experiment itself was performed on a Intel Celeron 1,8 GHz processor platform using 2GB of RAM memory.

6.6 Summary

The missing value problem is a problem where data elements of a data set appear as missing. This leads to incomplete data sets and makes data analysis for the researcher/analyst increasingly difficult. The larger the data set size is, the greater the effect missing values may have on analysis. When dealing with genetic data sets, in particular SNP genotype data sets, where millions of SNP genotypes are present for thousands of patients, missing genotype values are something that must be efficiently resolved and with the greatest accuracy possible. The most effective way to infer missing genotypes with 100 % accuracy is to repeat the genotype calling or typing process. However, this process is very expensive, especially when thousands of patients are involved in a study, i.e., data set. The optimal solution is to impute or infer missing genotype values using a class of machine learning algorithms known as imputation algorithms. The most accurate imputation algorithms available today rely on probabilistically modeling data sets and base their missing value inferences, i.e., substitutions on value estimates, from probabilistic information provided by the probabilistic model. Two such algorithms are the KNNimpute algorithm and the fastPHASE algorithm. KNNimpute is a nonparametric approach to inferring missing values which relies on the k nearest neighbor algorithm. The fastPHASE algorithm uses a parametric approach with three key parameters, modeling the data set according to an HMM. The accuracy of both algorithms varies with the number of missing values present. In genotype data sets, it has been noticed that anywhere between 5 % and 10 % of genotype data may be missing. For smaller percentages of missing data (around 5 %), the KNNimpute algorithm has shown to be very fast and accurate. However, as the percentage of missing data increases towards 10 %, the fastPHASE algorithm, a slower algorithm than KNNimpute, shows to be much more accurate.

Even though it will most likely never be possible to infer missing values with 100 % accuracy, the goal of imputation algorithms is to infer values as fast as possible and as accurately as possible, reaching towards the 100 % accuracy mark. In continuous research in genotype imputation and improvement of computing technology, we can expect the realization of such goals.

References

1. Lon R. Cardon and John I. Bell, (2001),*Association study designs for complex diseases*, Nature Reviews, Genetics Vol 2, February 2001, pp.91–99.
2. (2012)*Variations in genes making them faulty (mutating)*, Fact sheet produced by the Centre for Genetics Education, March 2012, http://www.genetics.edu.au 555
3. Rubin D. B, (1976), *Inference and Missing Data*, Biometrika, Vol63, Issue 3, December 1975, pp.581–592
4. James Y. Dai, Ingo Ruczinski, Michael LeBlanc, Charles Kooperberg,(2006),*Comparison of Haplotype-based and Tree-based SNP Imputation in Association Studies*, 2006, Genet Epidemiol, 30(8): pp.690–702.
5. Kalla S., (2012), *Statistical Data Sets*, 27.07.2012, http://www.experiment-resources.com/statistical-data-sets.html
6. (2012) *What Is the Human Genome?*, Understanding cancer series - Lesson 1, National Cancer Institute. http://www.cancer.gov/cancertopics/understandingcancer/cancergenomics/AllPages
7. Sawyer SA, Parsch J, Zhang Z, Hartl DL., (2007), *Prevalence of positive selection among nearly neutral amino acid replacements in Drosophila*. Proc. Natl. Acad. Sci. U.S.A. 104 (16): pp.6504–6510.
8. Cooper, D.N., Smith, B.A., Cooke, H.J., Niemann, S., and Schmidtke, J., (1985), An estimate of unique DNA sequence heterozygosity in the human genome. Hum. Genet. 69, 1985: pp.201–205.
9. Collins, F.S., Guyer, M.S., and Charkravarti, A., (1997), *Variations on a theme: cataloging human DNA sequence variation*. Science. 278, 1997: pp.1580–1581.

10. Sachidanandam, Ravi; Weissman, David; Schmidt, Steven C.; Kakol, Jerzy M.; Stein, Lincoln D.; Marth, Gabor; Sherry, Steve; Mullikin, James C. et al., (2001), *A map of human genome sequence variation containing 1.42 million single nucleotide polymorphisms.* Nature 409 (6822), pp.928–33.
11. (2012), *Single Nucleotide Polymorphism*, Chinese medical and biological information (CMBI) site,http://cmbi.bjmu.edu.cn/cmbidata/snp/index00.htm http://cmbi.bjmu.edu.cn/
12. (2008) *SNP fact sheet*, Human genome project information, Last modified: Friday, September 19, 2008, http://www.ornl.gov/sci/techresources/Human_Genome/faq/snps.shtml
13. Kerchner C.F., (2005), *Haplotype vs. Haplogroup*, 29 Sep 2005 http://www.kerchner.com/haplotypevshaplogroup.htm
14. Jonathan Marchini, Bryan Howie, Simon Myers, Gil McVean, Peter Donnelly, (2007),*A new multipoint method for genome-wide association studies by imputation of genotypes,* Vol 39, No 7, July 2007, Nature Genetics, pp.906–913.
15. Aroon D Hingorani, Tina Shah, MeenaKumari, ReechaSofat, Liam Smeeth, (2010), *Translating genomics into improved healthcare*, Clinical Review, Science, medicine, and the future, BMJ, November 2010, pp.341
16. (2012), *Genome Wide Association Study (GWAS)*,National Human Genome Research Institute, Stanford school of medicine, http://med.stanford.edu/advance/phase2/
17. Guttmacher, A. E., Manolio, T. A., (2010),*Genomewide association studies and assessment of the risk of disease*. July 2010. N. Engl. J. Med. 363 (2): pp.166–76.
18. Pearson T., ManolioT., (2008),*How to interpret a genome-wide association study*. March 2008. JAMA 299 (11)
19. Gibson G.(2010),*Hints of hidden heritability in GWAS*. 2010. Nature Genetics 42 (7): pp.558–560.
20. Barrett J.,(2010), *How to read a genome-wide association study, genomes unzipped, public personal genomics*,http://www.genomesunzipped.org/2010/07/how-to-read-a-genome-wide-association-study.php
21. Broman K. W., (1999), *Cleaning Genotype Data*, December, 1999, Genetic Analysis Workshop 11: Analysis of genetic and environmental factors in common diseases. Genetic Epidemiology
22. Pompanon F., Bonin A., et. al. (2005), *Genotyping Errors: Causes, Consequences and Solutions*, 2005, Nature Reviews: Genetics, Nature Publishing Group, pp. 2
23. Kirk, K. M. &Cardon, L. R., (2002), *The impact of genotyping error on haplotype reconstruction and frequency estimation*, European Journal of Human Genetics, 10, 616–622
24. Akey, J. M., Zhang, K., Xiong, M. M., Doris, P., Jin, L., (2001),*The effect that genotyping errors have on the robustness of common linkage-disequilibrium measures*, Am. J. Hum. Genet. 68, 1447–1456 (2001): A study that investigates the effects of genotyping error on estimates of linkage disequilibrium, and shows that the robustness of the estimates depends on allelic frequencies and assumed error models.
25. Hackett, C. A. &Broadfott, L. B., (2003), *Effects of genotyping errors, missing values and segregation distortion in molecular marker data on the construction of linkage maps*, Heredity 90, 33–38
26. Douglas J. A., Boehnke M. & Lange K., (2000), *A multipoint method for detecting genotyping errors and mutations in sibling-pair linkage data*, Am. J. Hum. Genet. 66, 1287–1297
27. Manolio T. A., (2007), *Update on Genome-Wide Association Studies: We Live in Interesting Times*, U.S. Department of Health and Human Services, National Institute of Health, National Human Genome Research Institute, September 19, 2007
28. Huisman, M. (2000).*Imputation of missing item responses: Some simple techniques*. Quality and Quantity 34 331–351.
29. Marwala T. (2009), *Computational Intelligence for Missing Data Imputation*, Estimation, and Management Knowledge Optimization Techniques." (2009) Information Science Reference
30. Koler Daphne, Friedman Nir, (2009), *Probabilistic graphical models – Principles and techniques*, The MIT Press, Cambridge and London.

31. Gross, Jonathan L., Yellen, Jay, (2004), *Handbook of graph theory*, CRC Press, 2004, p. 35.
32. Gross, J. &Yellen, J., (2007),*Graph Theory and Its Applications*. CRC Press.
33. Markov, A. A., (1913), *An example of statistical investigation in the text of "Eugene Onyegin" illustrating coupling of "tests" in chains*. Proc. Acad. Sci. St. Petersburg VI Ser. 7:153–162.
34. LiseGetoor, Ben Taskar, (2007), *Introduction to statistical relational learning, Bioinformatics*, Adaptive computation and machine learning, MIT Press, Cambridge and London, pp. 28–71, Ch.2. Graphical models in a nutshell.
35. Padhraic Smyth, David Heckerman, Michael I. Jordan, (1996), *Probabilistic Independence Networks for Hidden Markov Probability Models*, Microsoft technical report, May 1, 1996.
36. L.R. Rabiner and B.H. Juang. (1986), *An introduction to hidden markov models*. In IEEE, ASSP Magazine, pp. 4{16.
37. Rabiner, L. R., (1989),*A tutorial on hidden Markov models and selected applications in speech recognition*. Proceedings of the IEEE 77:257–285.
38. Krogh, A., M. Brown, I. S. Mian, K. Sjolander, and D. Haussler. (1994). *Hidden Markov models in computational biology: applications to protein modeling*. J. Mol. Bio. 235: pp.1501–1531.
39. Paul Scheet and Matthew Stephens, (2006),A *Fast and Flexible Statistical Model for Large-Scale Population Genotype Data: Applications to Inferring Missing Genotypes and Haplotypic Phase,* Am J. Hum Genet. 2006 April; 78(4): 629–644. Published online 2006 February 17.
40. Gibbons J., Dickinson J. &Subhabrata S., (2003), *Nonparametric Statistical Inference*, 4th Ed. 2003. CRC Press
41. Zhaoxia Yu, Daniel J. Schaid, (2007), *Methods to impute missing genotypes for population data,* Hum Genet. 122: pp.495–504.
42. Bhatia N., Vandana, (2010), *Survey of Nearest Neighbor Techniques*, IJCSIS Intr. Jour. of Comp. Sci. and Inf. Sec., Vol. 8, No.2
43. Troyanskaya, O., Cantor, M., Sherlock, G., Brown, P., Hastie, T., Tibshirani, R., Botstein, D.& Altman, R.B., (2001),*Missing value estimation methods for dna microarrays*, Bioinformatics 17(6), pp.520–525.
44. Tohka J., (2011), *8001652 Introduction to Pattern Recognition*. Lecture 8: k-Nearest neighbors classification, Institute of Signal Processing Tampere University of Technology, lecture notes 2010–2011.
45. Anton Bovier,(2012), *Markov Processes Lecture Notes*, Ch.3 Stochastic Models of Complex Processes and Their Applications – Lecture Notes, Summer 2012, Universitat Bonn, Bonn, July 10, 2012.

Chapter 7
Computer Modeling of Atherosclerosis

Nenad Filipovic, Milos Radovic, Velibor Isailovic, Zarko Milosevic, Dalibor Nikolic, Igor Saveljic, Tijana Djukic, Exarchos Themis, Dimitris Fotiadis, and Oberdan Parodi

7.1 Introduction

Atherosclerosis is a disease of the large arteries characterized by the blood vessel endothelial dysfunction and lipid, cholesterol, calcium, and cell elements accumulations inside blood vessel wall [1]. It is commonly referred as plaque formation, vascular remodeling, acute and chronic obstruction of blood vessel lumen, blood flow disorder, and lower oxygenation of relevant tissues. Many studies confirmed different risk factors which contribute development and spreading of the atherosclerosis; the most common are hyperlipidemia, higher blood pressure and sugar values, cigarette consumption, age, and sex. Great contribution to atherosclerosis development gives mechanical quantities such as low shear stress areas which causes endothelium dysfunctions and atherogenesis [2]. The main objective of this study is to examine influence of low shear stress and arterial mass transport by modeling the blood flow and solution transport processes in arterial lumen and the wall. Transport processes of the atherogenic species such as low-density lipoprotein (LDL) from the bulk blood flow to and across arterial wall contributes to lipid accumulation in the wall [2].

Several mathematical models have recently been set up for the transport of macromolecules, such as low-density lipoproteins, from the arterial lumen to the

---

N. Filipovic (✉)
University of Kragujevac, Kragujevac, Serbia

Bioengineering R&D Center, BioIRC, Kragujevac, Serbia
e-mail: fica@kg.ac.rs

M. Radovic • V. Isailovic • Z. Milosevic • D. Nikolic • I. Saveljic • T. Djukic
Bioengineering R&D Center, BioIRC, Kragujevac, Serbia

E. Themis • D. Fotiadis
University of Ioannina, Ioannina, Greece

O. Parodi
National Research Council, Pisa, Italy

G. Rakocevic et al. (eds.), *Computational Medicine in Data Mining and Modeling*,
DOI 10.1007/978-1-4614-8785-2_7,

arterial wall and inside the wall [3–5]. These models are usually classified in three categories according to the level of description of the arterial wall. The simplest model is called the wall-free model, since in this model the arterial wall is simply described by means of an appropriate boundary condition. Kaazempur-Mofrad and Ethier [6] simulated the mass transport in a realistic human right coronary artery and Wada et al. [7] used a wall-free model to study the concentration polarization phenomenon. The wall-free model does not provide any information on the transmural flow and solute dynamics in the arterial wall. The fluid-wall models that can be either single layer or multilayer account for the solute dynamics not only in the lumen but also in the arterial wall. Stangeby and Ethier [8] analyzed the wall as single-layer porous medium and solved the coupled luminal blood flow and transmural fluid flow using Brinkman's equations. Sun et al. [9] used single and Al and Vafai [10] used multilayer models which represent intima and media separately. Olgac et al. [11] used a three-pore model for LDL transport.

It is now well known that the early stage of the inflammatory disease is the result of interaction between plasma low-density lipoproteins that filtrate through endothelium into the intima, cellular components (monocytes/macrophages, endothelial cells, and smooth muscle cells), and the extracellular matrix of the arterial wall [1, 12].

In this chapter, we performed computational study for plaque composition and initial progression. The aim is to connect LDL transport with macrophages and oxidized LDL distribution as well as initial plaque grow model inside the intimal area. We firstly described mass transport of LDL through the wall and the simplified inflammatory process. The Navier–Stokes equations govern the blood motion in the lumen, the Darcy law is used for model blood filtration, and Kedem-Katchalsky equations [13, 14] are used for the solute and flux exchanges between the lumen and the intima. Then we described the system of three additional reaction–diffusion equations that models the inflammatory process and lesion growth model in the intima. This model relies on a matter incompressibility assumption. The next sections are devoted to numerical simulation examples in two- and three-dimensional domain and comparison with experimental results from literature and our own experimental data from animal and human. Computational results and comparison with animal experiments are presented. Finally, the main conclusions of the work performed are given, with connecting between modeling and experimental work.

7.2 Numerical Model of Plaque Formation and Growing in 3D Space

In this section, we present a continuum-based approach for plaque formation and development in three dimensions. The governing equations and numerical procedures are given. The blood flow is simulated by the three-dimensional Navier–Stokes equations, together with the continuity equation

$$-\mu\nabla^2 u_l + \rho(u_l \cdot \nabla)u_l + \nabla p_l = 0 \tag{7.1}$$

$$\nabla u_l = 0 \tag{7.2}$$

where $\boldsymbol{u_l}$ is blood velocity in the lumen, p_l is the pressure, μ is the dynamic viscosity of the blood, and ρ is the density of the blood.

Mass transfer in the blood lumen is coupled with the blood flow and modeled by the convection-diffusion equation as follows:

$$\nabla \cdot (-D_l \nabla c_l + c_l u_l) = 0 \tag{7.3}$$

in the fluid domain, where c_l is the solute concentration in the blood lumen, and D_l is the solute diffusivity in the lumen.

Mass transfer in the arterial wall is coupled with the transmural flow and modeled by the convection-diffusion–reaction equation as follows:

$$\nabla \cdot (-D_w \nabla c_w + k c_w u_w) = r_w c_w \tag{7.4}$$

in the wall domain, where c_w is the solute concentration in the arterial wall, D_w is the solute diffusivity in the arterial wall, K is the solute lag coefficient, and r_w is the consumption rate constant.

LDL transport in lumen of the vessel is coupled with Kedem-Katchalsky equations:

$$J_v = L_p(\Delta p - \sigma_d \Delta \pi) \tag{7.5}$$

$$J_s = P\Delta c + (1 - \sigma_f) J_v \overline{c} \tag{7.6}$$

where L_p is the hydraulic conductivity of the endothelium, Δc is the solute concentration difference across the endothelium, Δp is the pressure drop across the endothelium, $\Delta \pi$ is the oncotic pressure difference across the endothelium, σ_d is the osmotic reflection coefficient, σ_f is the solvent reflection coefficient, P is the solute endothelial permeability, and $\overline{c}$ is the mean endothelial concentration.

The inflammatory process was solved using three additional reaction–diffusion partial differential equations:

$$\begin{aligned} &\partial_t Ox = d_2 \Delta Ox - k_1 Ox \cdot M \\ &\partial_t M + div(v_w M) = d_1 \Delta M - k_1 Ox \cdot M + S/(1 - S) \\ &\partial_t S = d_3 \Delta S - \lambda S + k_1 Ox \cdot M + \gamma(Ox - Ox^{thr}) \end{aligned} \tag{7.7}$$

where Ox is the oxidized LDL or c_w – the solute concentration in the wall from (7.4); M and S are concentrations in the intima of macrophages and cytokines, respectively; d_1,d_2,d_3 are the corresponding diffusion coefficients; λ and γ are degradation and LDL oxidized detection coefficients; and v_w is the inflammatory velocity of plaque growth, which satisfies Darcy's law and continuity equation [13]

$$v_w - \nabla \cdot (p_w) = 0 \tag{7.8}$$

$$\nabla v_w = 0 \tag{7.9}$$

in the wall domain. Here, p_w is the pressure in the arterial wall.

The Basic Relations for Mass Transport in the Artery. The metabolism of the artery wall is critically dependent upon its nutrient supply governed by transport processes within the blood. Two different mass transport processes in large arteries are addressed. One of them is the oxygen transport and the other is LDL transport. Blood flow through the arteries is usually described as motion of a fluid-type continuum, with the wall surfaces treated as impermeable (hard) boundaries. However, transport of gases (e.g., O2, CO2) or macromolecules (albumin, globumin, LDL) represents a convection-diffusion physical process with permeable boundaries through which the diffusion occurs. In the analysis presented further, the assumption is that the concentration of the transported matter does not affect the blood flow (i.e., a diluted mixture is considered). The mass transport process is governed by convection-diffusion equation

$$\frac{\partial c}{\partial t} + v_x \frac{\partial c}{\partial x} + v_y \frac{\partial c}{\partial y} + v_z \frac{\partial c}{\partial z} = D\left(\frac{\partial^2 c}{\partial x^2} + \frac{\partial^2 c}{\partial y^2} + \frac{\partial^2 c}{\partial z^2}\right) \tag{7.10}$$

where c denotes the macromolecule or gas concentration; v_x, v_y, and v_z are the blood velocity components in the coordinate system x,y,z; and D is the diffusion coefficient, assumed constant, of the transported material.

Boundary Conditions for Transport of the LDL. A macromolecule directly responsible for the process of atherosclerosis is LDL which is well known as atherogenic molecule. It is also known that LDL can go through the endothelium at least by three different mechanisms, namely, receptor-mediated endocytosis, pinocytotic vesicular transport, and phagocytosis [14]. The permeability coefficient of an intact arterial wall to LDL has been reported to be of the order of 10^{-8} $[cm/s]$ [15]. The conversion of the mass among the LDL passing through a semipermeable wall, moving toward the vessel wall by a filtration flow and diffusing back to the mainstream at the vessel wall, is described by the relation

$$c_w v_w - D\frac{\partial c}{\partial n} = K c_w \tag{7.11}$$

where c_w is the surface concentration of LDL, v_w is the filtration velocity of LDL transport through the wall, n is coordinate normal to the wall, D is the diffusivity of LDL, and K is the overall mass transfer coefficient of LDL at the vessel wall. A uniform constant concentration C_0 of LDL is assumed at the artery tree inlet as classical inlet boundary condition for Eq. 7.14.

Finite Element Modeling of Diffusion-Transport Equations. In the case of blood flow with mass transport, we have domination of the convection terms due to the low diffusion coefficient [16]. Then it is necessary to employ special stabilizing techniques in order to obtain a stable numerical solution. The streamline upwind/

Petrov-Galerkin stabilizing technique (SUPG) [17] is implemented within a standard numerical integration scheme. The incremental-iterative form of finite element equations of balance are obtained by including the diffusion equations and transforming them into an incremental form. The final equations are

$$\begin{bmatrix} \frac{1}{\Delta t}\mathbf{M_v} + {}^{n+1}\mathbf{K}_{\mathbf{vv}}^{(i-1)} + {}^{n+1}\mathbf{K}_{\mathbf{\mu v}}^{(i-1)} + {}^{n+1}\mathbf{J}_{\mathbf{vv}}^{(i-1)} & {}^{n+1}\mathbf{K}_{\mathbf{vp}}^{(i-1)} & \mathbf{0} \\ \mathbf{K}_{\mathbf{vp}}^{\mathbf{T}} & \mathbf{0} & \mathbf{0} \\ {}^{n+1}\mathbf{K}_{\mathbf{cv}}^{(i-1)} & \mathbf{0} & \frac{1}{\Delta t}\mathbf{M_c} + {}^{n+1}\mathbf{K}_{\mathbf{cc}}^{(i-1)} + {}^{n+1}\mathbf{J}_{\mathbf{cc}}^{(i-1)} \end{bmatrix}$$

$$\times \begin{Bmatrix} \Delta \mathbf{V}^{(i)} \\ \Delta \mathbf{P}^{(i)} \\ \Delta \mathbf{C}^{(i)} \end{Bmatrix} = \begin{Bmatrix} {}^{n+1}\mathbf{F}_{\mathbf{v}}^{(i-1)} \\ {}^{n+1}\mathbf{F}_{\mathbf{p}}^{(i-1)} \\ {}^{n+1}\mathbf{F}_{c}^{(i-1)} \end{Bmatrix} \tag{7.12}$$

where the matrices are

$$\begin{aligned} &(\mathbf{M_v})_{jjKJ} = \int_V \rho N_K N_J dV, && (\mathbf{M_c})_{jjKJ} = \int_V N_K N_J dV \\ &\left({}^{n+1}\mathbf{K_{cc}}^{(i-1)}\right)_{jjKJ} = \int_V D N_{K,j} N_{J,j} dV && \left({}^{n+1}\mathbf{K}_{\mu v}{}^{(i-1)}\right)_{jjKJ} = \int_V \mu N_{K,j} N_{J,j} dV \\ &\left({}^{n+1}\mathbf{K}_{\mathbf{cv}}^{(i-1)}\right)_{jjKJ} = \int_V N_K\, {}^{n+1}c_{,j}^{(i-1)} N_J dV && \left({}^{n+1}\mathbf{K}_{vv}^{(i-1)}\right)_{jjKJ} = \int_V \rho N_K\, {}^{n+1}v_j^{(i-1)} N_{J,j} dV \\ &\left({}^{n+1}\mathbf{J}_{\mathbf{cc}}^{(i-1)}\right)_{jjKJ} = \int_V \rho N_K\, {}^{n+1}v_j^{(i-1)} N_{J,j} dV && \left({}^{n+1}\mathbf{K}_{\mathbf{vp}}^{(i-1)}\right)_{jjKJ} = \int_V \rho N_{K,j} \hat{N}_J dV \\ &\left({}^{n+1}\mathbf{J}_{\mathbf{vv}}^{(i-1)}\right)_{jkKJ} = \int_V \rho N_K\, {}^{n+1}v_{j,k_{j,k}} N_J dV \end{aligned} \tag{7.13}$$

and the vectors are

$$\begin{aligned} {}^{n+1}\mathbf{F}_c{}^{(i-1)} &= {}^{n+1}\mathbf{F}_q + {}^{n+1}\mathbf{F}_{sc}{}^{(i-1)} - \frac{1}{\Delta t}\mathbf{M}_c\left\{{}^{n+1}\mathbf{C}^{(i-1)} - {}^{n}\mathbf{C}\right\} \\ &- {}^{n+1}\mathbf{K}_{cv}{}^{(i-1)}\left\{{}^{n+1}\mathbf{V}^{(i-1)}\right\} - {}^{n+1}\mathbf{K}_{cc}{}^{(i-1)}\left\{{}^{n+1}\mathbf{C}^{(i-1)}\right\}\left({}^{n+1}\mathbf{F}_q\right)_K \\ &= \int_V N_K q^B dV \qquad {}^{n+1}\mathbf{F_{sc}}{}^{(i-1)} = \int_S D N_K \nabla^{n+1}\mathbf{c}^{(i-1)} \cdot \mathbf{n} dS \end{aligned} \tag{7.14}$$

Note that $\hat{N}_J$ are the interpolation functions for pressure (which are taken to be for one order of magnitude lower than interpolation functions N_I for velocities). The matrices $\mathbf{M}_{cc}$ and $\mathbf{K}_{cc}$ are the "mass" and convection matrices; $\mathbf{K}_{cv}$ and $\mathbf{J}_{cc}$ correspond to the convection terms of Eq. 7.10; and $\mathbf{F}_c$ is the force vector which follows from the convection-diffusion equation in (7.10) and linearization of the governing equations.

7.2.1 Mesh-Moving Algorithm

In this section, we described Arbitrary Lagrangian-Euler (ALE) formulation which is used for blood flow simulation and mesh-moving algorithm. In order to make plaque formation and development algorithm and to connect blood flow simulation with bioprocess modeling, the 3D mesh-moving algorithm and ALE formulation for fluid dynamics are applied [18]. The governing equations, which include the Navier–Stokes equations of balance of linear momentum and the continuity equation, can be written in the ALE formulation as [18]

$$\rho\left[v_i^* + \left(v_j - v_j^m\right)v_{i,j}\right] = -p_{,i} + \mu v_{i,jj} + f_i^B \tag{7.15}$$

$$v_{i,i} = 0 \tag{7.16}$$

where v_i and v_i^m are the velocity components of a generic fluid particle and of the point on the moving mesh occupied by the fluid particle, respectively; ρ is fluid density, p is fluid pressure, μ is dynamic viscosity, and f_i^B are the body force components. The symbol "*" denotes the mesh-referential time derivative, i.e., the time derivative at a considered point on the mesh,

$$()^* = \frac{\partial()}{\partial t}\Big|_{\xi_i=const} \tag{7.17}$$

and the symbol "$_{,i}$" denotes partial derivative, i.e.,

$$()_{,i} = \frac{\partial()}{\partial x_i} \tag{7.18}$$

We use x_i and ξ_i as Cartesian coordinates of a generic particle in space and of the corresponding point on the mesh, respectively. The repeated index means summation, from 1 to 3, i.e., j = 1, 2, 3 in Eq. 7.19 and i = 1, 2, 3 in Eq. 7.16. In deriving Eq. 7.11, we used the following expression for the material derivative (corresponding to a fixed material point) $D(\rho v_i)/Dt$:

$$\frac{D(\rho v_i)}{Dt} = \frac{\partial(\rho v_i)}{\partial t}|_{\xi} + \left(v_j - v_j^m\right)\frac{\partial(\rho v_j)}{\partial x_i} \tag{7.19}$$

The derivatives on the right-hand side correspond to a generic point on the mesh, with the mesh-referential derivative and the convection term.

We now apply the conventional Galerkin procedure for space discretization of the fluid domain. The finite element equations for a 3D domain that follow from Eqs. 7.15 and 7.16 are

$$\rho\int_V h_\alpha v_i^* dV + \rho\int_V h_\alpha\left(v_j - v_j^m\right)v_{i,j}dV = -\int_V h_\alpha p_{,i}dV + \int_V \mu h_\alpha v_{i,jj}dV + \int_V h_\alpha f_i^B dV \tag{7.20}$$

$$\int_V \overline{h}_\beta v_{i,i}dV = 0 \tag{7.21}$$

where $h_\alpha(r,s,t)$ and $\overline{h}_\beta(r,s,t)$ are the interpolation functions for the velocities and pressure, respectively, as polynomials of the isoparametric coordinates r,s,t [16]. The number of interpolation functions is governed by the number of nodes of the selected finite element. In general, number of interpolation functions h_α and $\overline{h}_\beta$ are different. The integration is performed over the volume V of a finite element. We will further use Eq. 7.20 transformed by applying the Gauss theorem

$$\rho\int_V h_\alpha v_i^* dV + \rho\int_V h_\alpha\left(v_j - v_j^m\right)v_{i,j}dV - \int_V h_{\alpha,i}pdV + \int_V \mu h_{\alpha,j}v_{i,j}dV = \int_S h_\alpha\left[-pn_i + \mu v_{i,j}n_j\right]dS \tag{7.22}$$

We integrate incrementally the system of Eq. 7.22 over time period using a time step Δt, which can be constant or it can vary in the time period. Hence we need an incremental form of equations corresponding to time step. The system of equations (1.22) is nonlinear with respect to the velocities, and the element volume changes, thus we perform a linearization with respect to time, using the known values at a given time t. Since the computational algorithm we want to establish is implicit, we seek to satisfy the system of Eq. 7.26 at the end of time step Δt, i.e., at time $t + \Delta t$, where t is time at start of the current time step. For a generic quantity F defined at a mesh point, we can write the following approximation [16]:

$$^{t+\Delta t}F\Big|_{^t\xi} = {}^tF\Big|_{^t\xi} + F^*\Delta t \tag{7.23}$$

By applying this relation to the left- (LHS) and right-hand side (RHS) of Eq. 7.22, we obtain

$$^{t}(LHS) + (LHS)^{*}\Delta t = {}^{t+\Delta t}(RHS) \tag{7.24}$$

In calculating the mesh-referential time derivatives we use the following relations:

$$\left(\frac{\partial F}{\partial x_i}\right)^{*} = \frac{\partial F^{*}}{\partial x_i} - \left(\frac{\partial v_k^m}{\partial x_i}\right)\frac{\partial F}{\partial x_k} \tag{7.25}$$

$$(dV)^{*} = \frac{\partial v_k^m}{\partial x_k} dV \tag{7.26}$$

Also, we express the fluid velocities and pressure in the incremental form

$$^{t+\Delta t}v_i = {}^{t}v_i + \Delta v_i \tag{7.27}$$

$$^{t+\Delta t}p = {}^{t}p + \Delta p \tag{7.28}$$

where Δv_i and Δp are the velocity and pressure increments in time step. Of course we adopt interpolations for the velocities and pressure and have

$$\Delta v_i = h_\alpha \Delta V_i^\alpha \tag{7.29}$$

$$\Delta p = \overline{h}_\alpha \Delta P^\alpha \tag{7.30}$$

where ΔV_i^α and ΔP^α are the increments of nodal values, and summation over index α is implied.

Using the linearization (7.24) and the expressions (7.23) and Eqs. 7.25–7.30, we obtain from (7.21) and (7.22) the system of ordinary differential equations in the form

$$^{t}\mathbf{M}_{(1)}\mathbf{V}^{*} + {}^{t}\mathbf{K}_{(1)_{vv}}\Delta\mathbf{V} + {}^{t}\mathbf{K}_{vp}\Delta\mathbf{P} = {}^{t+\Delta t}\mathbf{F}_{(1)} - {}^{t}\mathbf{F}_{(1)} \tag{7.31}$$

and

$$^{t}\mathbf{M}_{(2)}\mathbf{V}^{*} + {}^{t}\mathbf{K}_{(2)_{vv}}\Delta\mathbf{V} = {}^{t+\Delta t}\mathbf{F}_{(2)} - {}^{t}\mathbf{F}_{(2)} \tag{7.32}$$

The matrices and vectors follow from the volume and surface integrals given in Appendix. The integrals are evaluated over the known volumes and surfaces at start of time step. The element matrices are evaluated at time t, and the right-hand side vectors consist of the terms which correspond to start and end of time step. The vectors $^{t+\Delta t}\mathbf{F}_{(k)},\ \ k = 1, 2$ contain terms with the prescribed values at the end of the time step, as given pressures on the boundary, or velocities of the mesh. On the other hand, the vectors ${}^t\mathbf{F}_{(k)}$, $k = 1, 2$ are evaluated with all values at start of time step. Further we use

$$\mathbf{V}^* = \Delta\mathbf{V}/\Delta t \tag{7.33}$$

and

$$\mathbf{P}^* = \Delta\mathbf{P}/\Delta t \tag{7.34}$$

so that the systems of Eqs. 7.31 and 7.32 have the unknown increments $\Delta\mathbf{V}$ and $\Delta\mathbf{P}$ only. The solutions of Eqs. 7.31 and 7.32 give the first approximation, and we form an iterative scheme of the form [16, 18]

$$ {}^{t+\Delta t}\hat{\mathbf{K}}^{(i-1)} \Delta\mathbf{U}^{(i)} = {}^{t+\Delta t}\mathbf{F}^{(i-1)} \tag{7.35}$$

where ${}^{t+\Delta t}\hat{\mathbf{K}}^{(i-1)}$ is the system matrix, ${}^{t\,+\,\Delta t}\mathbf{F}^{(i\,-\,1)}$ is the unbalanced force vector, and $\Delta\mathbf{U}^{i)}$ is the vector of nodal variables for the equilibrium iteration "i":

$$\left\{\Delta\mathbf{U}^{(i)}\right\} = \left\{\begin{array}{c}\Delta\mathbf{V}^{(i)}\\ \Delta\mathbf{P}^{(i)}\end{array}\right\} \tag{7.36}$$

with

$$\Delta\mathbf{U} = \Delta\mathbf{U}^{(1)} + \Delta\mathbf{U}^{(2)} + \cdots. \tag{7.37}$$

The iteration stops when a convergence criteria are satisfied, e.g., when $\|\Delta\mathbf{U}^{i)}\| \le \varepsilon_D$, where ε_D is a selected numerical tolerance.

We further use the penalty formulation and express the pressure in terms of the velocity components

$$p = -\lambda v_{i,i} \tag{7.38}$$

where λ is a large (penalty) number. We use this expression for pressure in Eq. 7.15 which then contains the velocities only, Eq. 7.16 is eliminated, and the corresponding terms in the finite element equations change accordingly.

7.2.2 2D Axisymmetric Model

In this section, we analyzed typical benchmark examples of mass transport as well as plaque formation and development. The first example is albumin transport in a large artery. The second example is LDL transport through a straight artery with the filtration through the wall, and third example is 2D axisymmetric plaque formation and development.

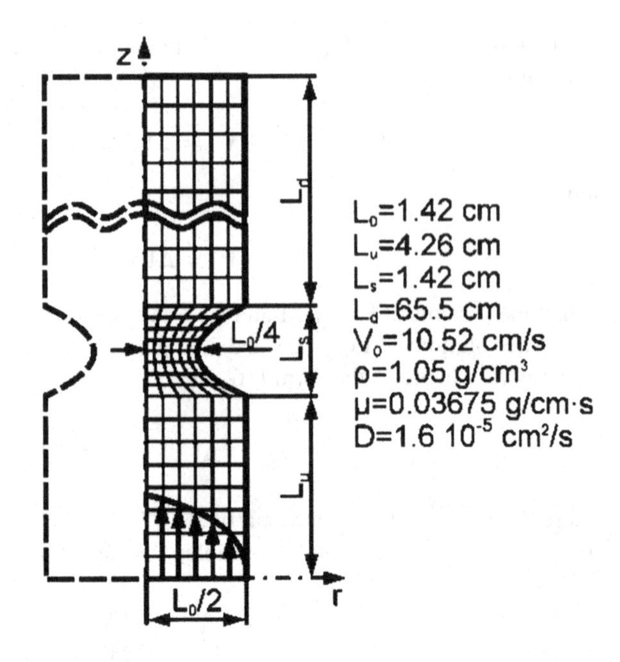

Fig. 7.1 Geometrical and material data for artery with stenosis

Example: Modeling Albumin Transport in a Large Artery The stenosed artery is shown in Fig. 7.1. Due to axial symmetry, only a half is modeled and the blood is considered to be a Newtonian fluid. Two-dimensional axisymmetric 2D elements are used. Data for blood density ρ, dynamic viscosity μ, diffusion coefficient for albumin transport D, the inflow mean velocity V_0, and geometry are given in the Fig. 7.1. It is assumed that the flow is steady.

Since the distance between the entrance and the stenosis position is large, it can be considered that the entering velocity profile is parabolic:

$$v(r) = 2V_0\left(1 - \left(\frac{2r}{L_0}\right)^2\right) \tag{7.39}$$

where r is the radial coordinate. At the wall, velocities are equal to zero, while at the axis of symmetry, only the radial velocity is equal to zero. At the outflow the zero traction stress is applied:

$$-p + \mu\frac{\partial v_z}{\partial z} = 0 \tag{7.40}$$

The boundary conditions for the concentration are (a) at the inlet, $c = c_0 = 2.58 \times 10^{-3}[mL/cm^3]$; and (b) $\partial c/\partial r = 0$ and $\partial c/\partial z = 0$ at the axis of symmetry. Note that to mean velocity $V_0 = 10.52$ $[cm]$ corresponds the Reynolds number $Re = 448$, and Pecklet number P_e is 934000 (defined as $P_e = L_0V_0/D$).

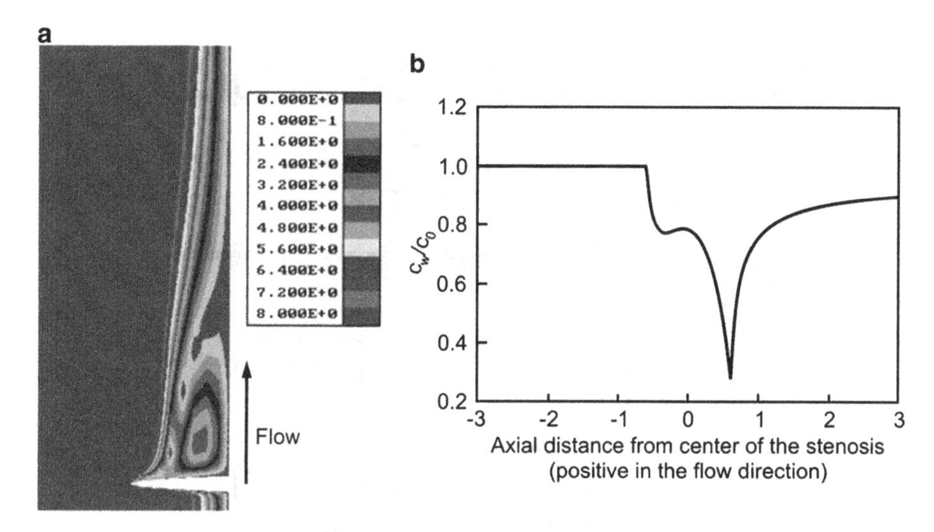

Fig. 7.2 Albumin transport in stenosed artery, the stenotic artery part: (**a**) velocity field; (**b**) normalized concentration at the wall c_w/c_0

Velocity field is shown in Fig. 7.2a, where the field of disturbed flow is noticeable after the stenosis. Albumin concentration at the wall c_w normalized with respect to the inlet concentration c_0 is given in Fig. 7.2b. A significant concentration increase in the domain of stenosis can be seen, with the peak at the distal region.

Example: Modeling the LDL Transport Through a Straight Artery with the Filtration Through the Wall The LDL transport through a straight artery is modeled in this example. The tube, representing the artery, has the diameter $d_0 = 0.6\ [cm]$ (Fig. 7.3a). The filtration velocity through the vessel wall is $v_w = 4 \times 10^{-6}\ [cm/s]$ and the overall mass transfer coefficient of lipoproteins at the arterial wall, $K = 2 \times 10^{-8}\ [cm/s]$. Blood was modeled as a Newtonian fluid with density $\rho = 1.0\ [g/cm^3]$ and viscosity $\mu = 0.0334\ [P]$. The steady-state conditions for fluid flow and mass transport are assumed. The entering blood velocity is defined by the Reynolds number Re (calculated using the mean blood velocity and the artery diameter).

The 2D axisymmetric elements are used. The boundary conditions include prescribed parabolic velocity profile (7.39) and concentration c_0 at the inlet, zero stress at the outlet (7.40), and filtration at the walls according to mass transfer equation.

The analytical solution for the axial and radial velocities, as well as for the concentration, is given in [19].

Figure 7.3b shows distribution of the surface concentration of LDL along the axis of the artery for three Reynolds numbers. It can be seen that the concentration of LDL at the wall boundary layer is increased with the axial distance from the

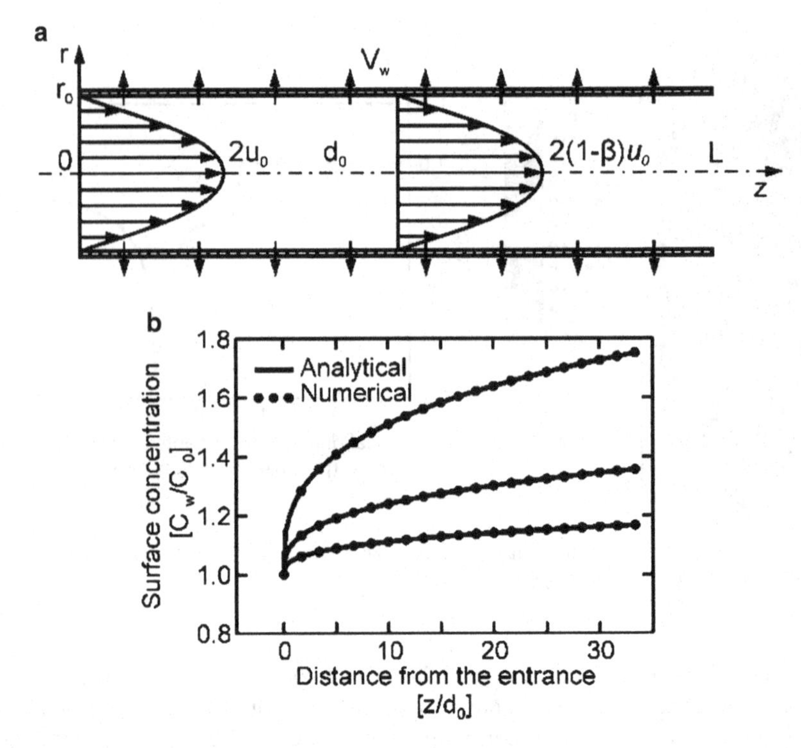

Fig. 7.3 Transport of the LDL through straight artery with semipermeable wall. (**a**) Schematic representation of velocity profiles (note that the profile changes in the flow direction since the wall is permeable); (**b**) normalized surface concentration of LDL, cw/co, in terms of the normalized distance from the entrance z/d0 (analytical and FE solutions)

entrance of the artery due the decrease of the velocity according to the expression shown in Fig. 7.3a, where $\beta = 2V_W z/V_0 r_0$.

Example: 2D Axisymmetric Plaque Formation and Development This is benchmark example for testing 2D axisymmetric plaque formation and development. The plaque formation and development is modeled through an initial straight artery with mild constriction of 30 %. The inlet artery diameter $d_0 = 0.4$ [*cm*]. Blood was modeled as a Newtonian fluid with density $\rho = 1.0$ [g/cm^3] and viscosity $\mu = 0.0334$ [*P*]. The steady-state conditions for fluid flow and mass transport are assumed. The entering blood velocity is defined by the Reynolds number Re (calculated using the mean blood velocity and the artery diameter). Velocity distribution for an initial mild stenosis 30 % constriction by area and for end stenosis process after 10^7 s is presented in Fig. 7.4. Similarly, pressure, shear stress, and LDL distribution inside the lumen domain for these two time stage of virtual stenosis are presented in the Figs. 7.5, 7.6, and 7.7, respectively. Also inside wall domain for oxidized LDL, intima wall pressure, macrophages, and cytokines distribution have been shown in Figs. 7.8, 7.9, 7.10, and 7.11 respectively.

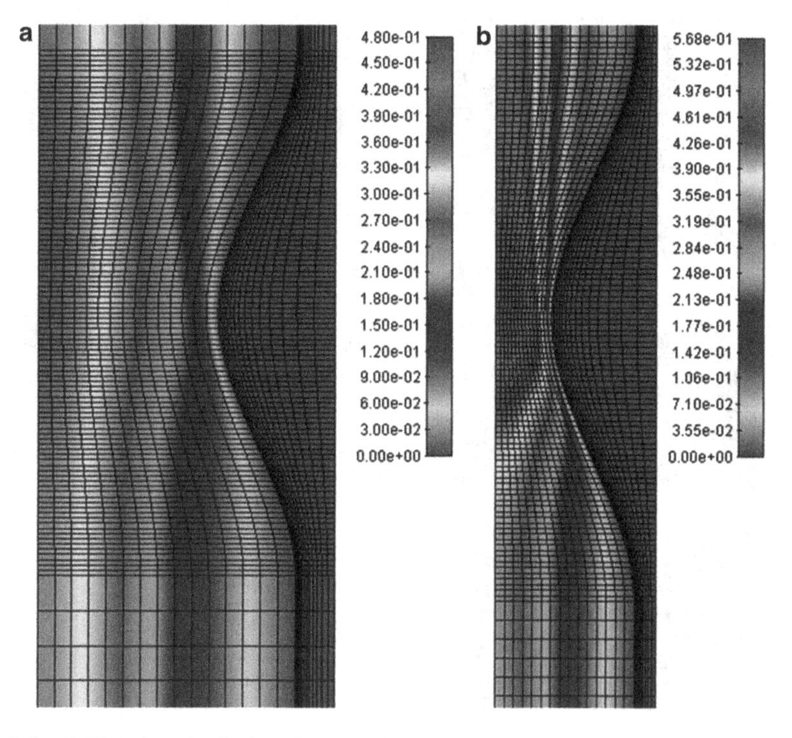

Fig. 7.4 (**a**) Velocity distribution for an initial mild stenosis 30 % constriction by area. (**b**) Velocity distribution for end stenosis process after 10^7 s [units m/s]

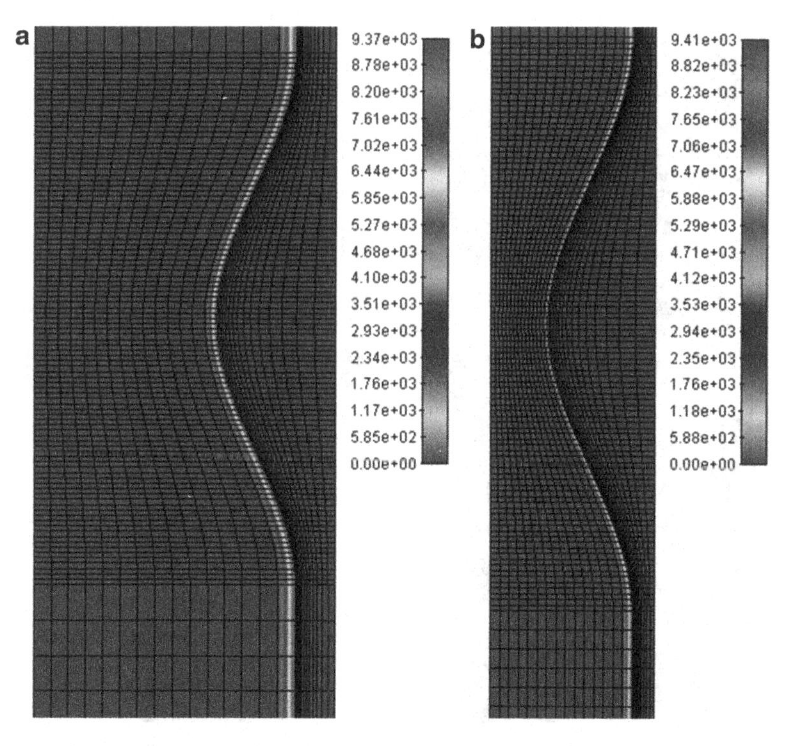

Fig. 7.5 (**a**) Pressure distribution for an initial mild stenosis 30 % constriction by area. (**b**) Pressure distribution for end stenosis process after 107 s [units Pa]

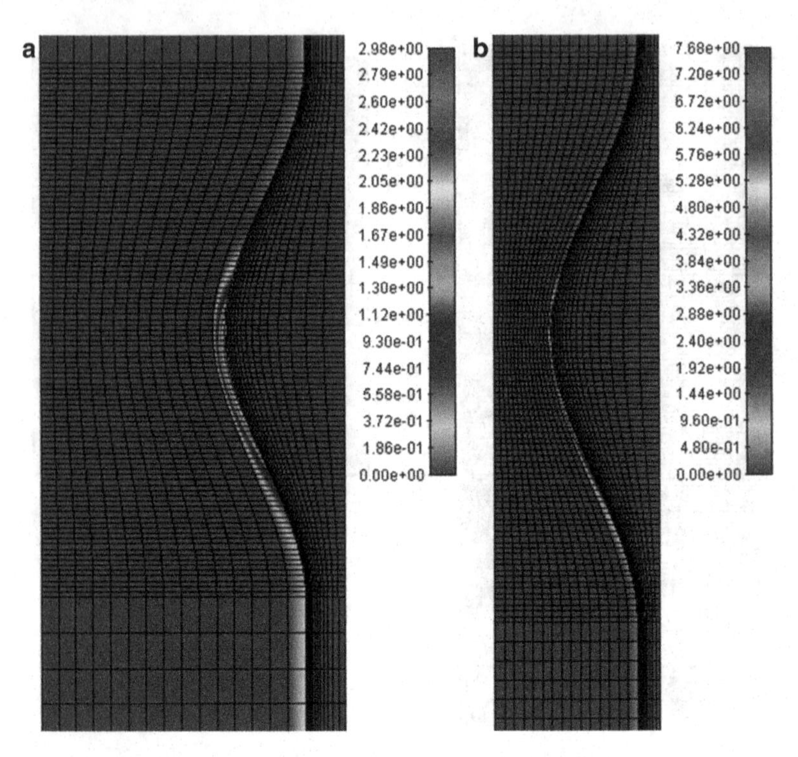

Fig. 7.6 (**a**) Shear stress distribution for an initial mild stenosis 30 % constriction by area. (**b**) Shear stress distribution for end stenosis process after 10^7 s [units dyn/cm^2]

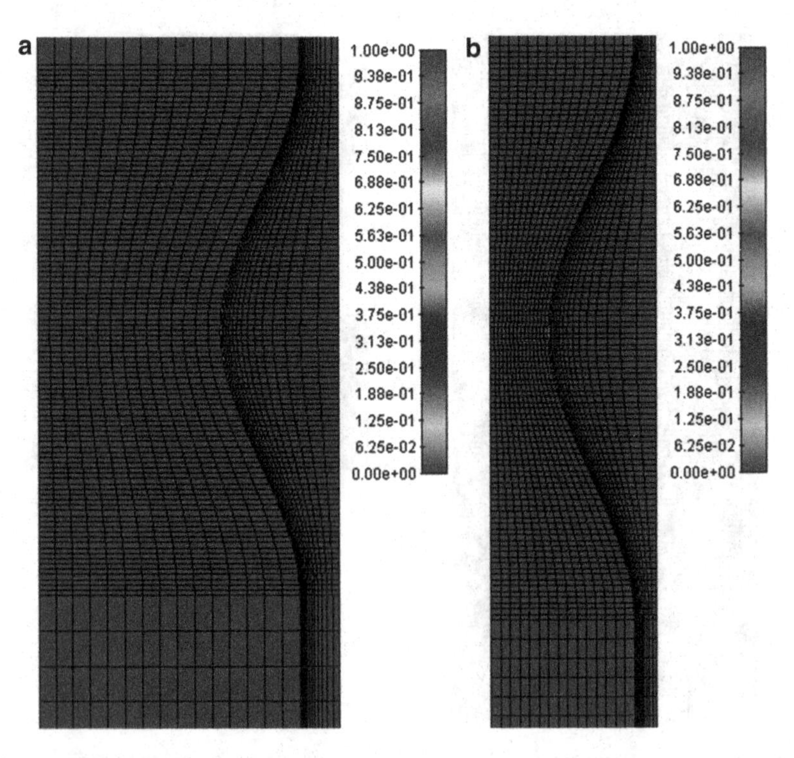

Fig. 7.7 (**a**) Lumen LDL distribution for an initial mild stenosis 30 % constriction by area. (**b**) Lumen LDL distribution for end stenosis process after 10^7 s [units mg/mL]

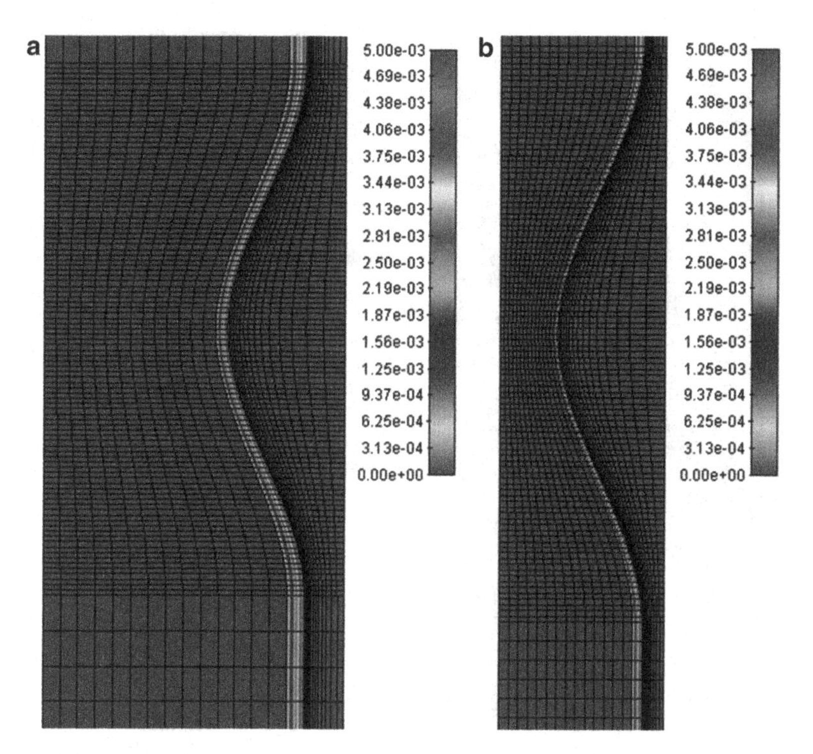

Fig. 7.8 (**a**) Oxidized LDL distribution in the intima for an initial mild stenosis 30 % constriction by area. (**b**) Oxidized LDL distribution in the intima for end stenosis process after 10^7 s [units mg/mL]

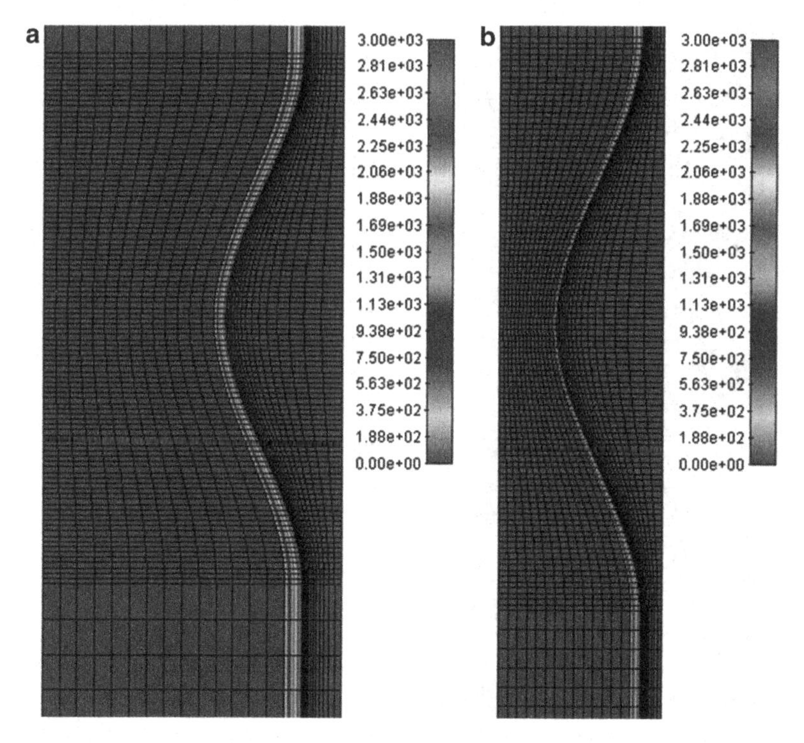

Fig. 7.9 (**a**) Intima wall pressure distribution for an initial mild stenosis 30 % constriction by area. (**b**) Intima wall pressure distribution for end stenosis process after 10^7 s [units Pa]

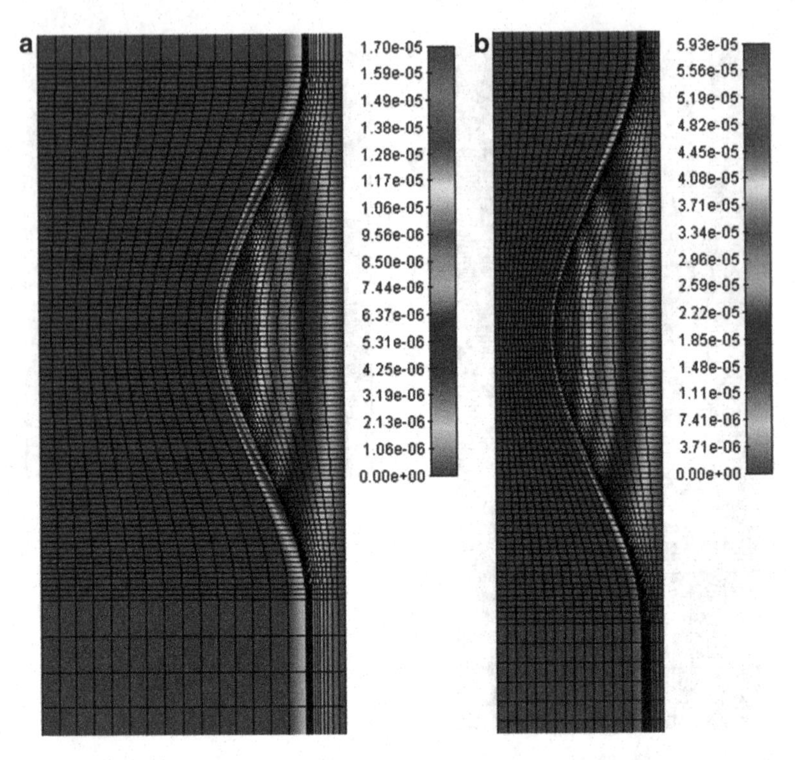

Fig. 7.10 (**a**) Macrophages distribution in the intima for an initial mild stenosis 30 % constriction by area. (**b**) Macrophages distribution in the intima for end stenosis process after 10^7 s [units mg/mL]

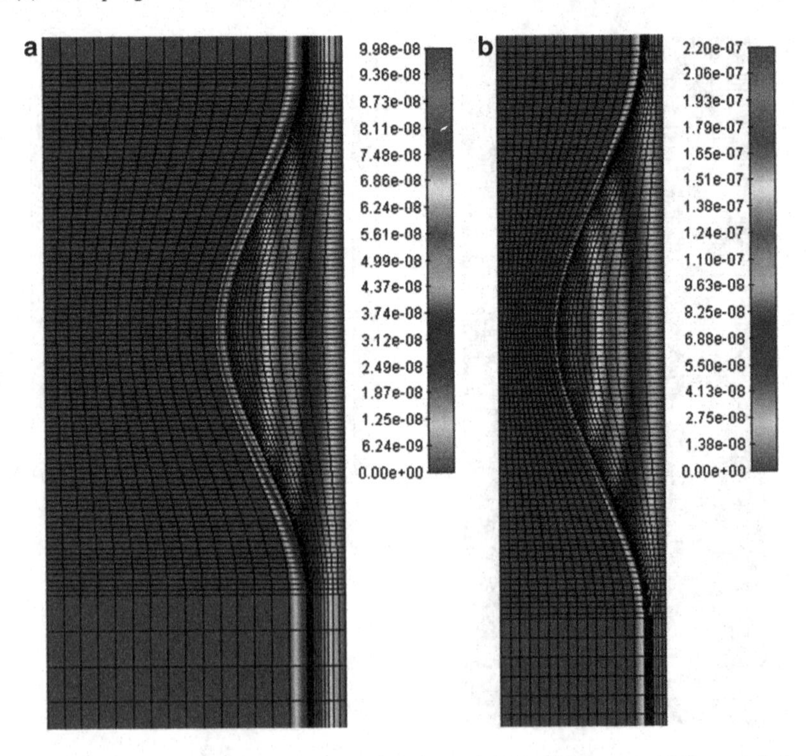

Fig. 7.11 (**a**) Cytokines distribution in the intima for an initial mild stenosis 30 % constriction by area. (**b**) Cytokines distribution in the intima for end stenosis process after 10^7 s [units mg/mL]

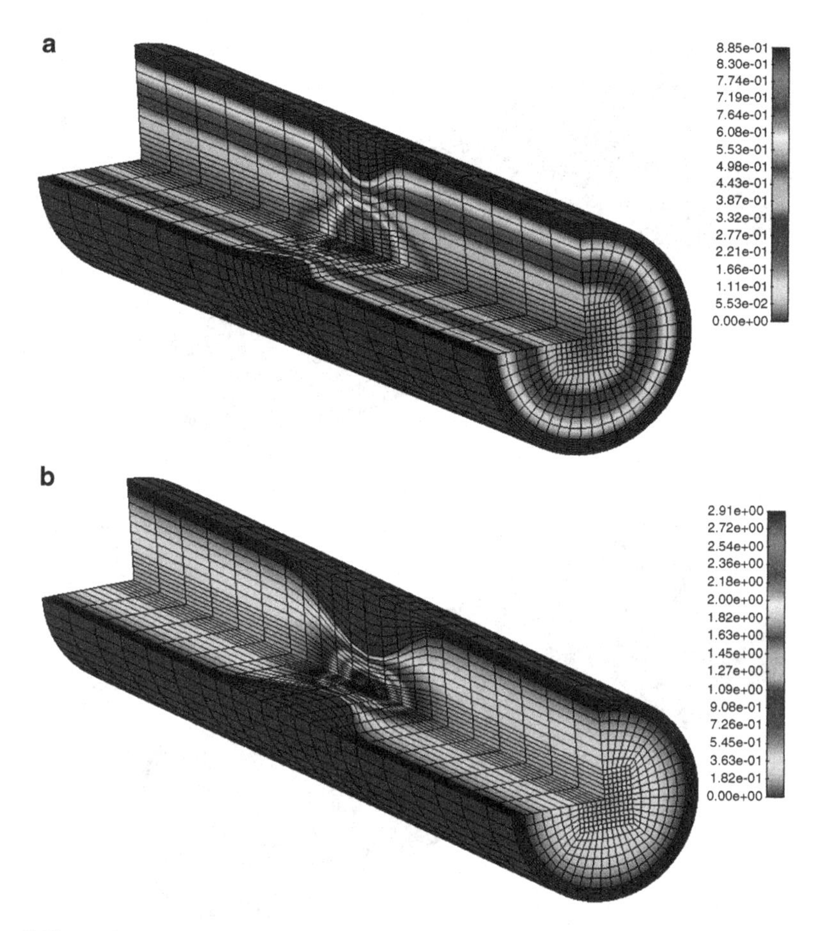

Fig. 7.12 (**a**) Velocity distribution for an initial mild stenosis 30 % constriction by area. (**b**) Velocity distribution for end stenosis process after 10^7 s [units m/s]

7.2.3 Three-Dimensional Tube Constriction Benchmark Model

In order to generate benchmark example for three-dimensional simulation, we tested simple middle stenosis with initial 30 % constriction for time period of $t = 10^7$ s (approximately 7 years) and compare results with the 2D axisymmetric model. The results for velocity distribution for initial and end stage of simulations are presented in Fig. 7.12a, b. The pressure and shear stress distributions for start and end time are given in Figs. 7.13 and 7.14. Concentration distribution of LDL inside the lumen domain and oxidized LDL inside the intima are presented in Figs. 7.15 and 7.16. The transmural wall pressure is presented in Fig. 7.17.

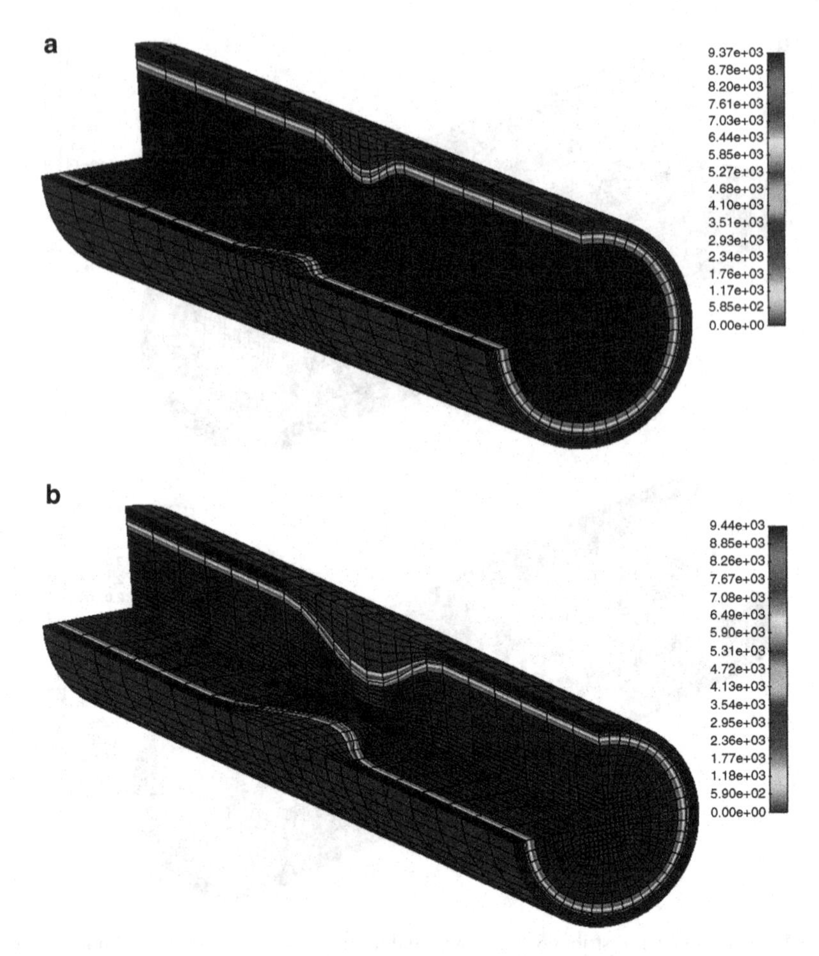

Fig. 7.13 (**a**) Pressure distribution for an initial mild stenosis 30 % constriction by area. (**b**) Pressure distribution for end stenosis process after 10^7 s [units Pa]

Macrophages and cytokines distributions are shown in Figs. 7.18 and 7.19. The diagram of three-dimensional plaque volume growing during time is given in Fig. 7.20. It can be seen that time period for developing of stenosis corresponds to data available in the literature [20].

From the above figures, it can be observed that during time period of the plaque progression, all variables as velocity distribution, shear stress, macrophages, and cytokines are increasing.

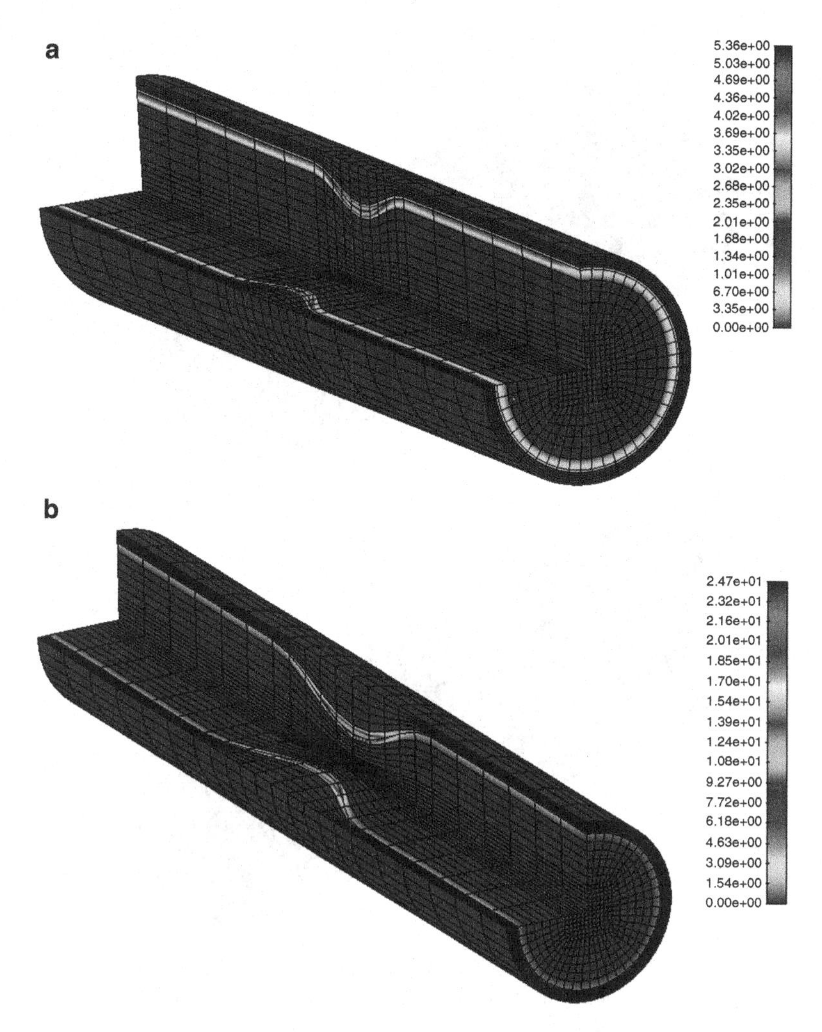

Fig. 7.14 (**a**) Shear stress distribution for an initial mild stenosis 30 % constriction by area. (**b**) Shear stress distribution for end stenosis process after 10^7 s [units dyn/cm^2]

7.3 Computational Modeling of Experiments

7.3.1 Cheng et al. 2006 Experiment in 3D

A validation of our ARTreat model of the inflammatory process is performed by comparing our solutions with the Cheng et al. 2006 experiment [21]. This experiment that was run on mice confirms that lesions occur in preferred locations such as bends and bifurcations and that biochemical composition of lesions depends on their location. In these experiments, the arterial geometry has

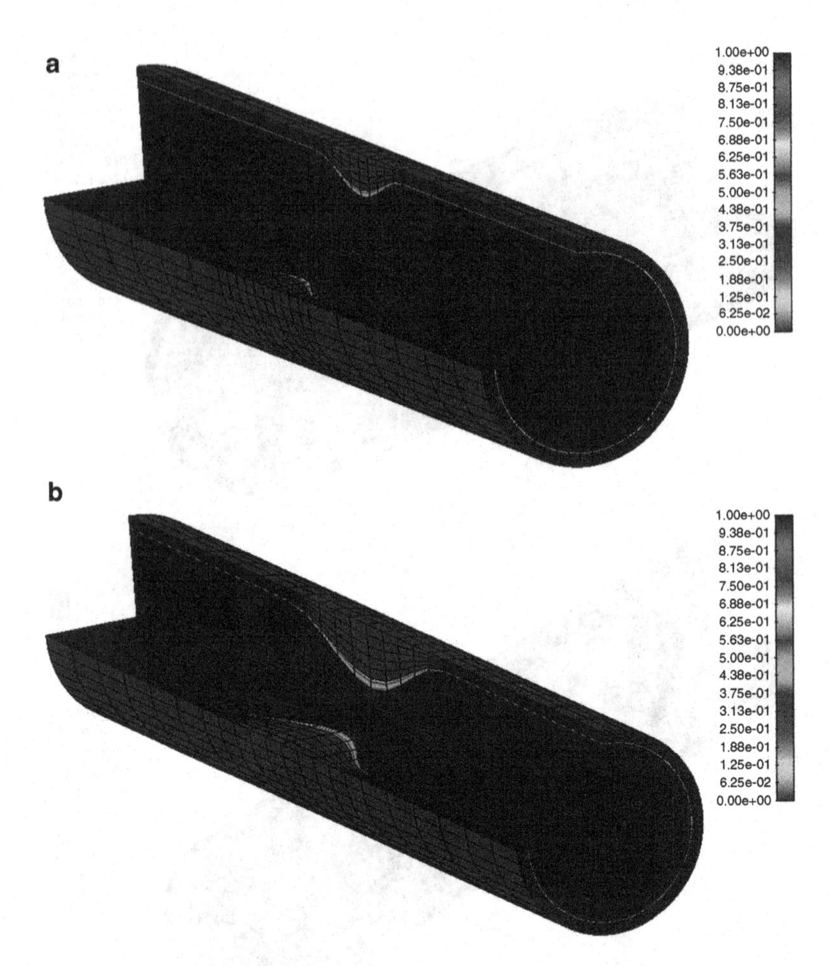

Fig. 7.15 (**a**) Lumen LDL distribution for an initial mild stenosis 30 % constriction by area. (**b**) Lumen LDL distribution for end stenosis process after 10^7 s [units mg/mL]

been modified by a perivascular cast that induces regions of lowered, increased, and lowered/oscillatory (i.e., with vortices) shear stresses (depicted in Fig. 7.21). Mice are fed by a rich cholesterol diet in order to assess plaque formation and composition. Our aim was to obtain numerical results that fit with the experiments. We here give the atherosclerotic plaque composition obtained by our numerical simulations.

A fully developed parabolic steady velocity profile was assumed at the lumen inlet boundary

$$u(r) = 2U_0\left(1 - (2r/D)^2\right) \tag{7.41}$$

where $u(r)$ is the velocity in the axial direction at radial position r, and $U_0 = 0.24$ m/s is the mean inlet velocity. The inlet artery diameter is $d_0 = 0.004$ m. Blood was modeled

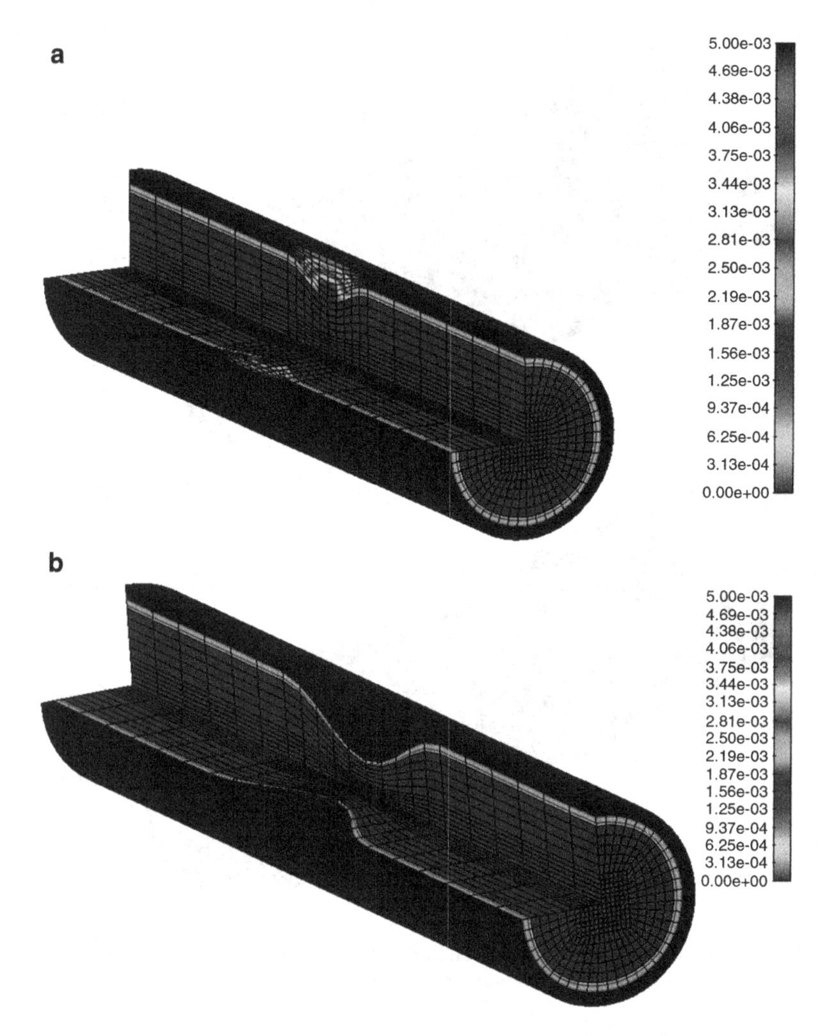

Fig. 7.16 (**a**) Oxidized LDL distribution in the intima for an initial mild stenosis 30 % constriction by area. (**b**) Oxidized LDL distribution in the intima for end stenosis process after 10^7 s [units mg/mL]

as a Newtonian fluid with density ρ =1,050 kg/m^3 and viscosity μ =0.0334*P*. The entering blood velocity is defined by the Reynolds number *Re* (calculated using the mean coronary blood velocity and the artery diameter). Basic values for this computer model are given in Table 7.1. It was taken that the permeability of the wall depends on residence time of solutes in the neighborhood of vascular endothelium and oscillatory shear index OSI [3, 22].

The results for oxidized LDL distribution are shown in Fig. 7.22. The concentration is presented in a dimensionless form, relative to the input concentration $\text{Co} = 2.86 \times 10^{-12}$ kg/m^3. It can be seen that for steady-state condition low wall

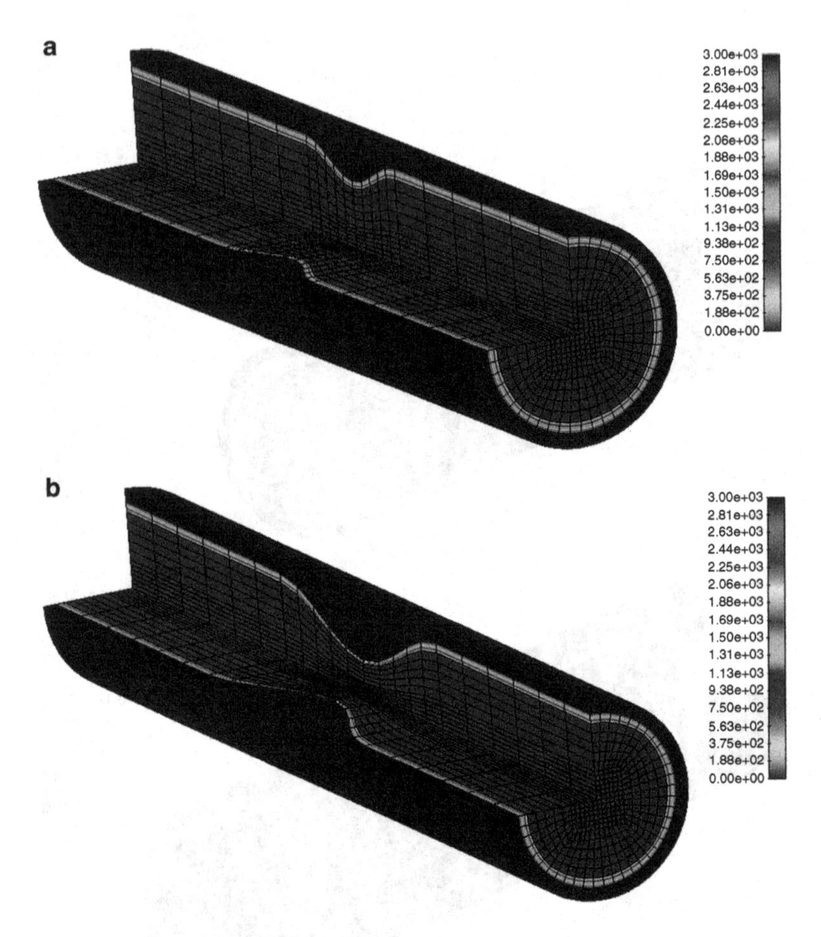

Fig. 7.17 (**a**) Intima wall pressure distribution for an initial mild stenosis 30 % constriction by area. (**b**) Intima wall pressure distribution for end stenosis process after 10^7 s [units Pa]

shear stress (WSS) appears after the cast which induces more LDL deposition in the recirculation zone. For unsteady calculation, we used residence time, and using our three-dimensional model, we found oxidized LDL distributions in both zones of low WSS and in the recirculation WSS. Increased oxidized LDL concentration was found in the zone of higher residence time, which corresponds to the plaque composition found in the Cheng measurement (Fig. 7.21).

The model of LDL transport includes convection-diffusion process within the blood lumen, followed by transport through tissue of the arterial where a mass consumption is also included. Further, a model of plaque initiation is presented which mathematically describes the LDL oxidation, coupled with formation of macrophages and foam cells that leads to plaque development and growth. The model parameters are used from literature as well as from our own investigation within the ARTreat consortium. These initial findings are in agreement with

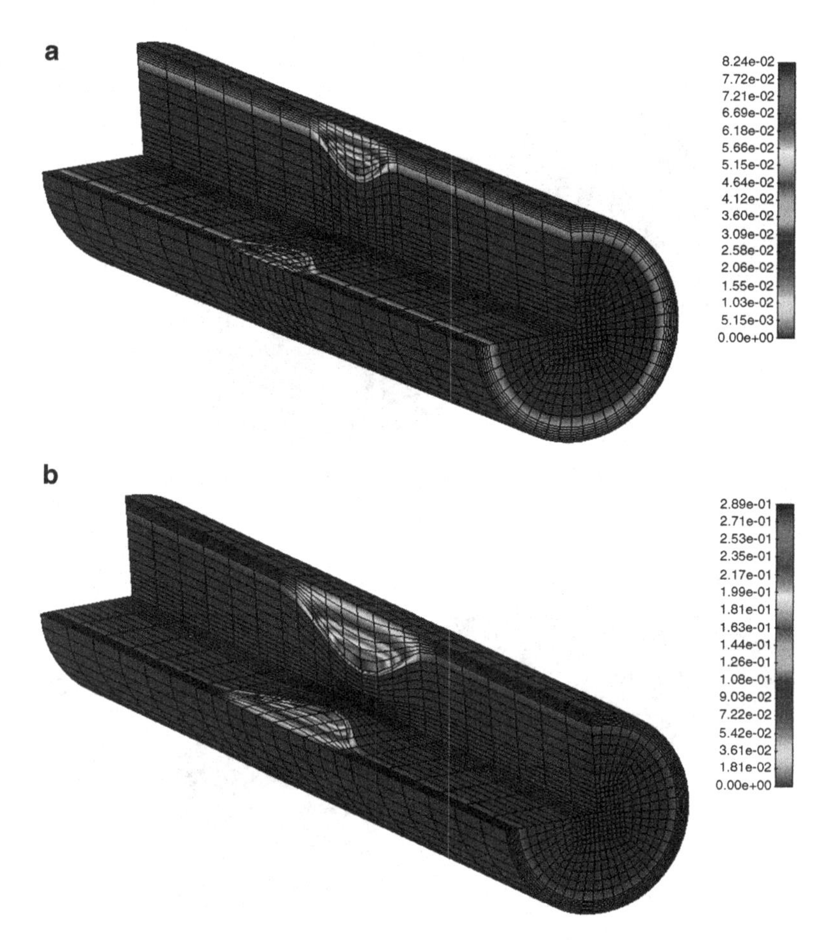

Fig. 7.18 (**a**) Macrophages distribution in the intima for an initial mild stenosis 30 % constriction by area. (**b**) Macrophages distribution in the intima for end stenosis process after 10^7 s [units mg/mL]

experimental observations, suggesting that mathematical models of very complex biochemical and biomechanical processes occurring in the plaque formation and growth could be of help in prevention and treatment of atherosclerosis.

7.3.2 Experimental and Computational LDL Transport Model from the University of Kragujevac

In this study, experimental model of LDL transport on the isolated blood vessel from rabbit on high-fat diet after 8 weeks is simulated numerically by using a specific model and histological data. The 3D blood flow is governed by the

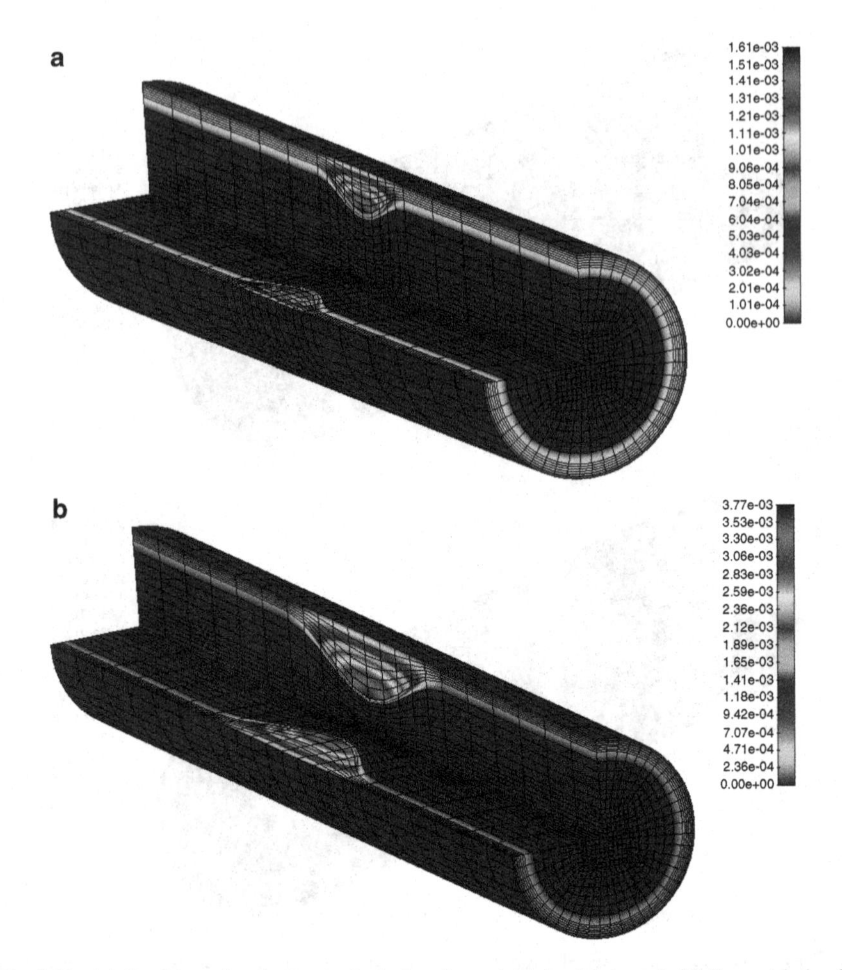

Fig. 7.19 (**a**) Cytokines distribution in the intima for an initial mild stenosis 30 % constriction by area. (**b**) Cytokines distribution in the intima for end stenosis process after 10^7 s [units mg/mL]

Navier–Stokes equations, together with the continuity equation. Mass transfer within the blood lumen and through the arterial wall is coupled with the blood flow by the convection-diffusion equation. LDL transport in lumen of the vessel is described by Kedem-Katchalsky equations. The inflammatory process is solved using three additional reaction–diffusion partial differential equations.

Matching of histological rabbit data is performed using 3D histological image reconstruction and 3D deformation of elastic body. Computed concentrations of labeled LDL of 15.7 % are in good agreement with experimental results. The understanding and the prediction of LDL transport through the arterial wall and evolution of atherosclerotic plaques are very important for the medical community.

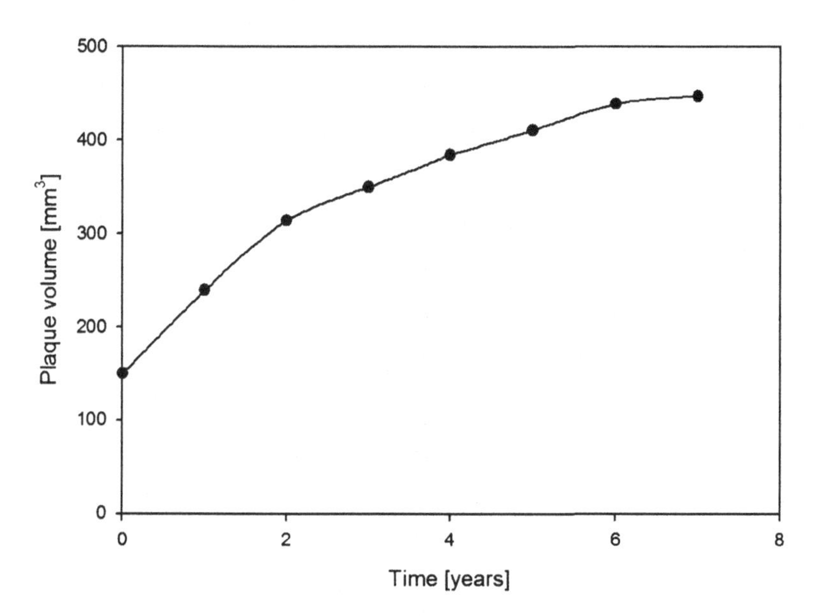

Fig. 7.20 Plaque progression during time (computer simulation)

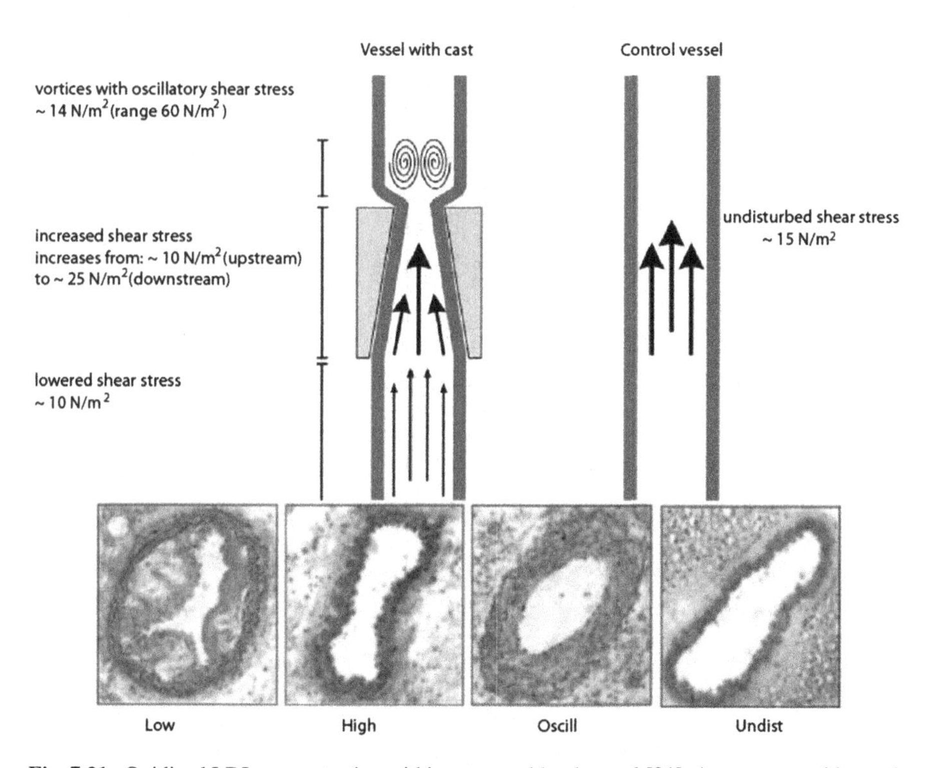

Fig. 7.21 Oxidized LDL concentration within a mouse blood vessel [21]. A mouse carotid vessel is partially obstructed with a cast. This modifies the blood flow and particularly the WSS patterns. The growth of atheromatous plaques is correlated with the reduction of WSS (right *before* and *after* the cast). Moreover, the composition of the plaques turn out to depend upon the WSS pattern: plaques associated with low WSS contain more oxidized LDL, whereas plaques located in zone of recirculating flow (*after* the cast) contain less oxidized LDL (According to [21])

Table 7.1 Values for Cheng experiment (Cheng et al. 2006 [21])

Lumen	$\rho = 1{,}050\ kg/m^3$	$\mu = 0.0334[P]$	$D_l = 2.8 \times 10^{-11}\ m^2/s$	$V_{max} = 0.24\ m/s$	Pout = 100 mmHg	$Co = 2.86 \times 10^{-12}\ kg/m^3$
Intima			$D_w = 1.3 \times 10^{-11}\ m^2/s$	$r_w = -1.4 \times 10^{-4}$	Pmed = 100 mmHg	
Inflammation	$d_1 = 10^{-7}\ m^2/s$	$d_2 = 10^{-9}\ m^2/s$	$d_3 = 10^{-7}\ m^2/s$	$k_1 = 10^{-5}\ m^3/kg\ s$	$\lambda = 1\ s^{-1}$	$\gamma = 1.e\text{-}5\ s^{-1}$

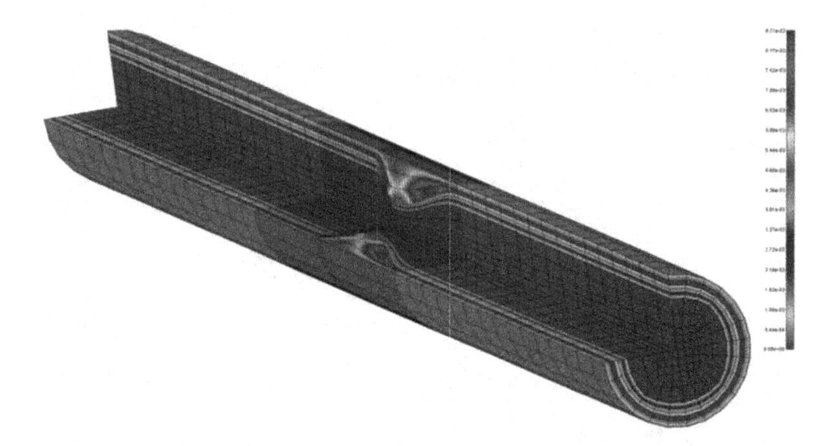

Fig. 7.22 Computed oxidized LDL distribution obtained by a 3D model of the Cheng experiment

7.3.2.1 Introduction

The position of the endothelium at the interface between blood and vessel wall with main role as a barrier to the transvascular convection and diffusion of blood-borne macromolecules is well known. The endothelial cells lining the blood vessels are flattened and elongated with nuclei that protrude into the lumen. They form a layer that prevents blood cell interaction with the vessel wall with a critical role in mechanics of blood flow, regulation of coagulation, leukocyte adhesion, and vascular smooth muscle cell growth. Damaged endothelium induces physiological and pathological changes [23, 24] such that decreased integrity of the endothelial barrier permits easier macromolecular transport into the intima [25].

Inflammatory process starts with penetration of low-density lipoproteins (LDL) in the intima. This penetration, if too high, is followed by leucocyte recruitment in the intima. This process may participate in formation of the fatty streak, the initial lesion of atherosclerosis, and then in formation of a plaque.

There are three major categories of LDL transport models. The simplest models are wall-free models, in which the arterial wall is substituted by a simplified boundary condition. Rappitsch [26] and Wada [27] applied these models for the analysis of the macromolecular transport in the arterial wall. A more realistic approach is lumen-wall models, where there is coupling of the transport within the lumen and the wall, [8, 28, 29]. Also there are multilayer models, which break the arterial wall down into several layers and model the transport within the wall, either at the microscopic or macroscopic levels. There are no so many numerical studies which rely on real experimental data for LDL transport.

In this study, we firstly described experimental setup for the LDL transport into the blood vessel wall in the isolated rabbit carotid artery under physiologically relevant constant pressure and perfusion flow on rabbit with 6 weeks high-fat diet. Mass transport of LDL through the wall and the simplified inflammatory process is

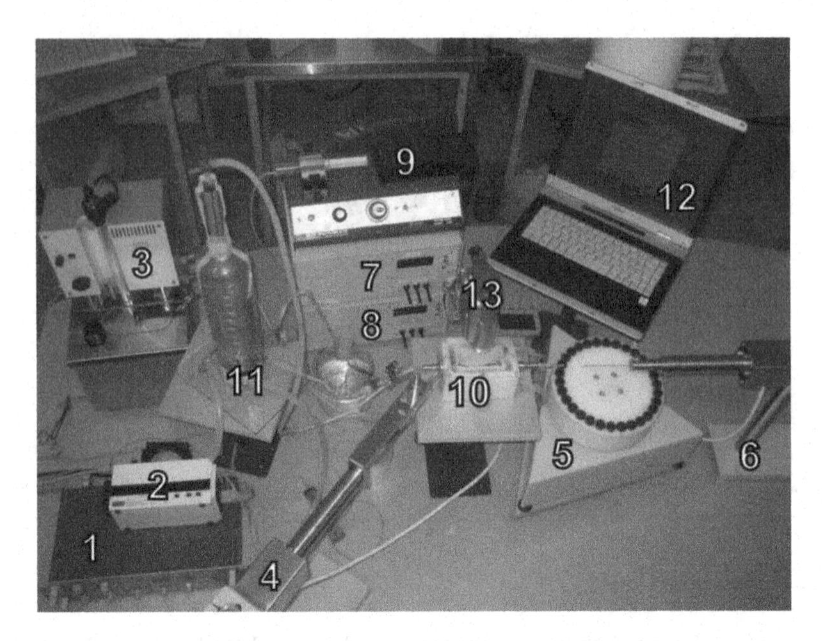

Fig. 7.23 Setting for ex vivo blood vessels experiments: *1*. Pressure and temperature A/D converter, *2*. peristaltic pump, *3*. heater thermostat, *4*. rapid infusion pump (RIP), *5*. automatic sampler, *6*. resistance changing device (RCD), *7*. control unit for RIP, *8*. control unit for RCD, *9*. syringe infusion pump, *10*. water bath, *11*. heating stabilizer, *12*. PC, *13*. digital camera

coupled with the Navier–Stokes equations, the Darcy equation for model blood filtration, and Kedem-Katchalsky equations [30, 31] for the solute and flux exchanges between the lumen and the intima. The next section presents numerical simulation and comparison with some initial experimental animal results of LDL transport and histological analysis.

7.3.2.2 Experimental Setup

Ex vivo blood vessels experiments of LDL transport were performed on the isolated rabbit a. carotis comm. All experiments were performed according to the Animals Scientific Procedures Act 1986 (UK) and local ethical guidelines. New Zealand White rabbits of both sex weighing 3.5–4 kg were anesthetized using Ketamine (Laboratorio Sanderson, Santiago, Chile), 4–6 mg per kg of body weight. Blood vessel was excised and placed in the water bath. Cannulas with equally matched tip diameters (2 mm) were mounted at proximal (cardial) and distal (cranial) ends of the blood vessel. The lumen was perfused with Krebs-Ringer physiological solution (KRS), using the peristaltic pump at 1 ml/min. The perfusate was continuously bubbled with a 95 % O_2 and 5 % CO_2 with the pH adjusted to 7.4 at 37 °C.

The distal cannula was connected to the resistance changing device. Perfusion pressure was measured with perfusion transducer (Fig. 7.23).

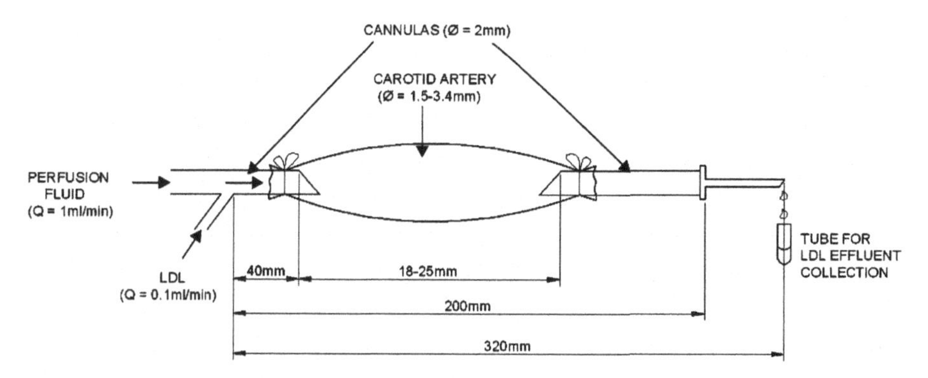

Fig. 7.24 Schematic presentation of the isolated blood vessel segment in the water bath

The blood vessel was stretched to its approximate in vivo length. The outer diameter of the blood vessel was measured using digital camera and originally developed software. The blood vessel wall thickness was measured at the end of each experiment using light microscope and microscopically graduated plate (see Fig. 7.24).

The blood vessel was considered to be viable if it contracted when 25 mMKCl was added to the bath, as well as if the presence of functional endothelium was verified by dilation with Ach (1 μM) at the end of experiment.

The isolated blood vessel was placed into the water bath with physiological buffer. After the equilibration period (20–30 min) at constant perfusion flow of 1 ml/min, 100 μl bolus was injected into the perfusion system containing ^{99m}Tc-Nanocis as an intravascular marker (referent tracer) or ^{125}I-LDL as a test molecule. The first 15 samples (three drops in each sample) and nine cumulative 3 min samples of perfusion effluent were sequentially collected. All samples were prepared for measurement of ^{125}I-LDL-specific activity by addition of physiological buffer until final volume of 3 ml/sample. Measurements of perfusion effluent samples containing ^{99m}Tc-Nanocis or ^{125}I-LDL were performed by means of the gamma counter (Wallac Wizard 1400).

The ^{125}I-LDL uptake is derived from the difference between the ^{99m}Tc-Nanocis value and that of ^{125}I-LDL recovery in each sample.

7.3.2.3 Histological Methods

Immunocytochemical staining was performed on 5 μm sections from formaldehyde-fixed paraffin-embedded blocks using a labeled streptavidin-biotin method with Thermo Scientific Detection System Anti-Mouse HRP (TM-060-HL). Sections were deparafinized and rehydrated. After microwave treatment of 21 min in citrate buffer pH 6.0, endogenous peroxidase activity was blocked with 3 % H_2O_2 for 15 min. The sections were first incubated with the primary antibody for 60 min (AbCam Mouse monoclonal (3G5) to LDL (MDA oxidized); ab63976;

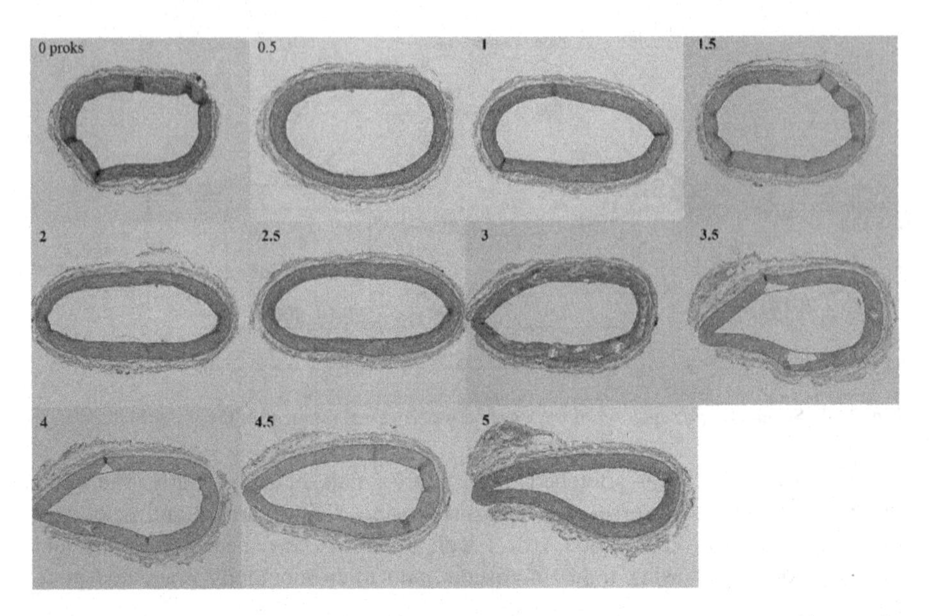

Fig. 7.25 Histological data (numbers on photos indicate distances from entry carotid artery in millimeters). *White* zones inside media denote labeled LDL localization. *Polylines* around media are segmentation lines produces by image processing software

dilution 1:50), then with biotinylated goat anti-mouse antibody (Thermo Scientific TM-060-BN) for 30 min at room temperature, after that with streptavidin peroxidase (Thermo Scientific TS-060-HRl) for 30 min at room temperature, and finally with chromogen 3-amino-9-ethylcarbazole (AEC) (Thermo Scientific TA-060-SA) for 10 min. Slides were counterstained with hematoxylin, washed in water, and mounted.

7.3.2.4 Results

The aim of our experiment was to determine distribution of accumulated ^{125}I-LDL radioactivity in the different segments of the isolated blood vessel. Specific software for 3D reconstruction of lumen domain and carotid wall artery was developed. Computer model of the artery is considered as a simple straight tube. The diameter of artery was $D = 0.0029$ m, the mean velocity $U_0 = 0.24$ m/s, dynamics viscosity $\mu = 0.0035$ Pa s, and density $\rho = 1{,}050$ kg/m^3. The transmural pressure under normal physiological condition was taken as 100 mmHg.

Histological images are shown in Fig. 7.25. The labeled LDL is localized in the white zones inside media which is probably due to destroyed radioactive LDL of tissue. Polylines around media are segmentation lines produces by in-house image processing software. Matching of histological data and computational simulation is presented in Fig. 7.26. The process of matching histological images was

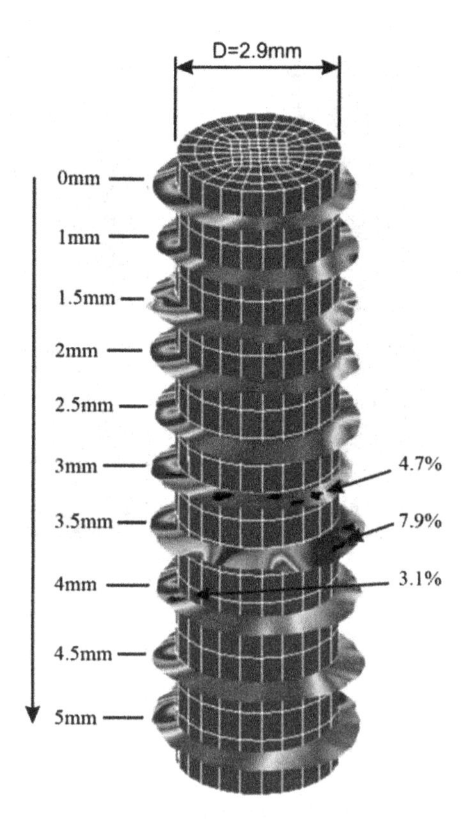

Fig. 7.26 Labeled LDL located in histology cross-section on each 0.5 mm for straight segment. Histology segments were obtained as deformable elastic rings opened from the current squeezed position to circle original tube. *Black* holes in these cross-sections show location of the labeled LDL. Percentages show labeled LDL area inside media and intima wall thickness

done by 2D deformation of each histological cross-section in order to keep the internal lumen approximately cylindrical shape. The maximum LDL was found at distal part of the carotid artery segment at 3.5 mm from entry segment. A full three-dimensional finite element analysis was performed using our in-house finite element code in order wall shear stress and function of permeability for the wall. Oxidized LDL, macrophages, and cytokines distribution is presented in Fig. 7.27. Diagrams of wall LDL, oxidized LDL, macrophages, and cytokines inside wall are shown in Fig. 7.28. Experimental LDL transport of 15.7 % was fitted with specific nonlinear least square analysis [32] in order to get numerical parameters. The fitted numerical parameters are given in Table 7.2.

Our results have shown a denudation of endothelial layer and a prominent accumulation of LDL in intima, especially in the layer right below basement membrane. In media we can notice a small quantity of LDL. In addition to extracellular lipid deposits, we also observed an immunoreactivity of certain VSMCs (Fig. 7.29). Using the TEM method, we can concur that these cells are in apoptosis (Fig. 7.30). A denudation of endothelial layer is confirmed with the method of transmission electron microscopy (Fig. 7.31).

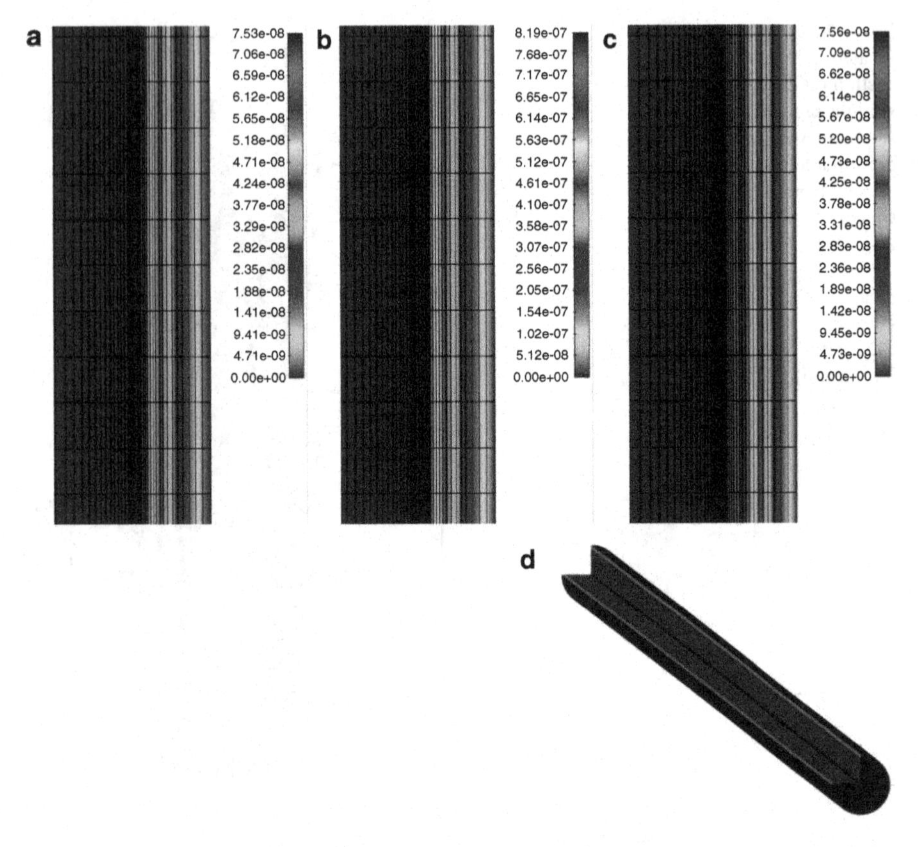

Fig. 7.27 (**a**) Oxidized LDL distribution 12.7 %; (**b**) macrophages distribution 9.2 % from media; (**c**) cytokines distribution 3.9 %; (**d**) three-dimensional representation of the model

7.3.2.5 Fluid–Solid Interaction Analysis

We used fluid–structure interaction analysis in order to match experimental results for UK experiments with rabbit artery. The aim of computational analysis was to determine material property of the wall in order to simulate maximal artery wall deformation. We assumed in the first approach linear elastic material wall property. The maximal displacements distribution for fluid pressure of 140 mmHg is presented in Fig. 7.32. Velocity vector distribution for input flow of 1.1 ml/min which included together perfusion flow and LDL flux has been shown in Fig. 7.33. Effective stress distribution for maximal artery deformation of 50 % which corresponding to the artery diameter of 3 mm is presented in Fig. 7.34. From fitting analysis, we got for the assumed linear elastic wall material Young's elasticity module $E = 150$ kPa. Further analysis from experimental data which suggested that during high-fat diet rabbits wall property are changed in order to make higher shear stress for the same input flow parameters. It can be concluded that increasing

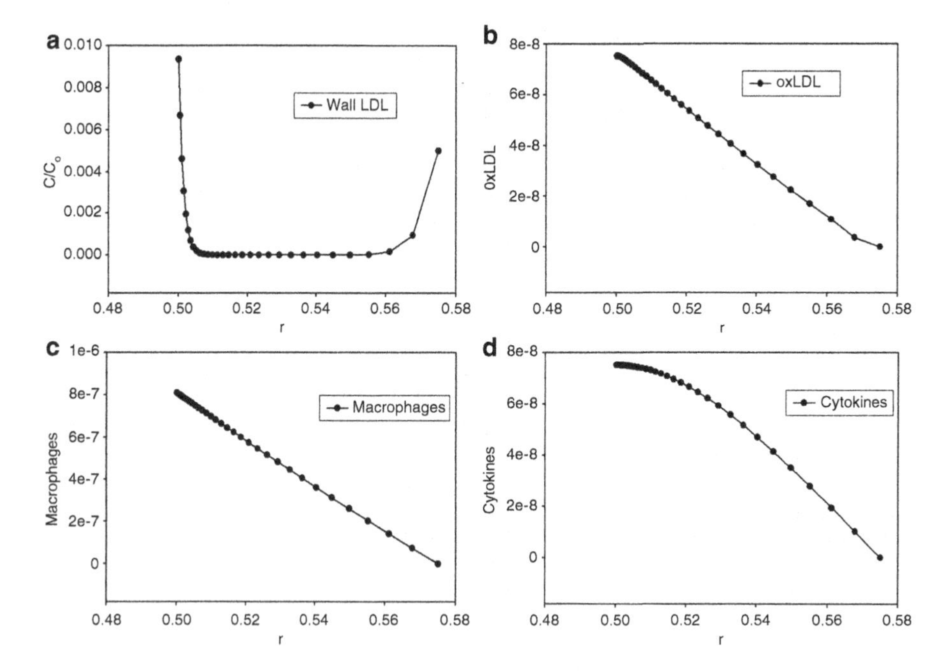

Fig. 7.28 (**a**) Dimensionless wall LDL concentration profile in the media; (**b**) oxidized LDL concentration profile in the media; (**c**) macrophages profile in the media; (**d**) cytokines profile in the media

of the shear stress leads directly to the smaller deformation and more rigid wall property, and we got that corresponding Young's elasticity modules is around 200 kPa for the rabbits with high-fat diet at week 12.

7.3.2.6 Discussion

In this chapter, we matched our experimental results of LDL transport through the rabbit carotid artery together with histological analysis and computer model. A full three-dimensional model of LDL transport as well as plaque initiation is coupled with the Navier–Stokes equations and continuity equation. We used Darcy law for model blood filtration and Kedem-Katchalsky equations for the solute and flux transfer. The arterial wall permeability was fitted with our experimental results of LDL transport. All parameters for computer model were fitted with nonlinear least square procedure. Matching of the labeled LDL location between experimental and computer model shows a potential benefit for future prediction of this complex process using computer modeling.

Table 7.2 Values for rabbit carotid artery experiment

Lumen	$\rho = 1{,}000\ kg/m^3$	$\mu = 0.035$ [P]	$D_l = 1.0e10^{-12}\ m^2/s$	$Umax = 0.4\ m/s$	$Pout = 120\ mmHg$	$Co = 3.0 \times 10^{-12}\ kg/m^3$
Intima			$D_w = 3.0e^{-12}\ m^2/s$	$r_w = -2.6 \times 10^{-4}$	$Pmed = 100\ mmHg$	
Inflammation	$d_1 = 10^{-7}\ m^2/s$	$d_2 = 10^{-7}\ m^2/s$	$d_3 = 10^{-7}\ m^2/s$	$k_1 = 1.9e^{-4}\ m^3/kg\ s$	$\lambda = 25\ s^{-1}$	$\gamma = 1\ s^{-1}$

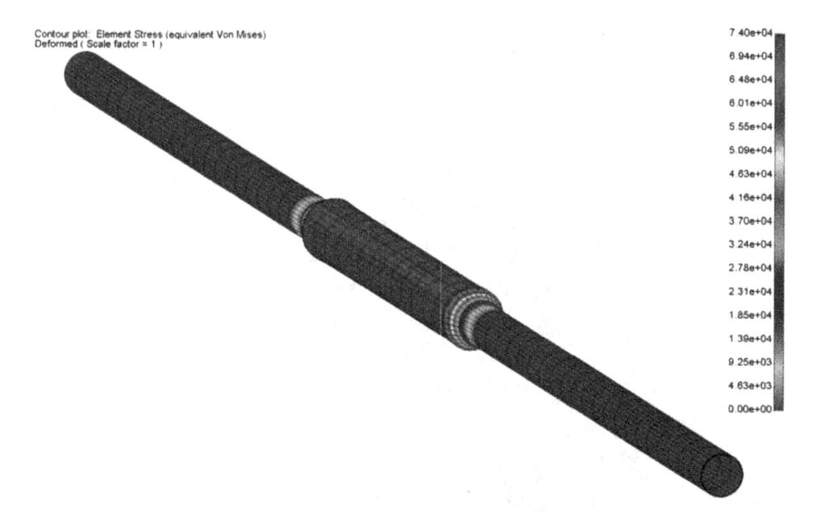

Fig. 7.29 Left common carotid artery of the rabbit after transport of ^{125}I-LDL. We can notice a denudation of endothelial layer and a prominent accumulation of LDL in intima, especially in the layer *right* below basement membrane. In media we can notice a small quantity of LDL. In addition to extracellular lipid deposits, we can also observe immunoreactivity of certain VSMCs of media. The immunoreactivity of wall structure decreases in the direction of intima adventitia (immunohistochemical staining of LDL, ×100)

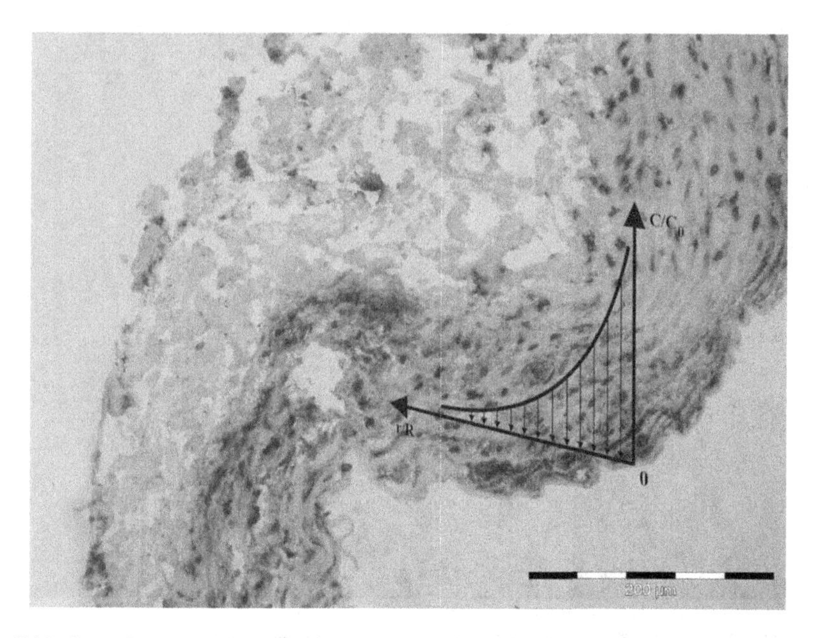

Fig. 7.30 Smooth muscle cell in media of left common carotid artery of the rabbit after transport of125I-LDL. We can notice intact plasma membrane and blebbing on the surface of the cell. Parts of the plasma membrane, i.e., blebs, are separated from the cell, taking a portion of cytoplasm with them, to become apoptotic bodies. In nucleus, we can notice a nuclear fragmentation and chromatin condensation (TEM)

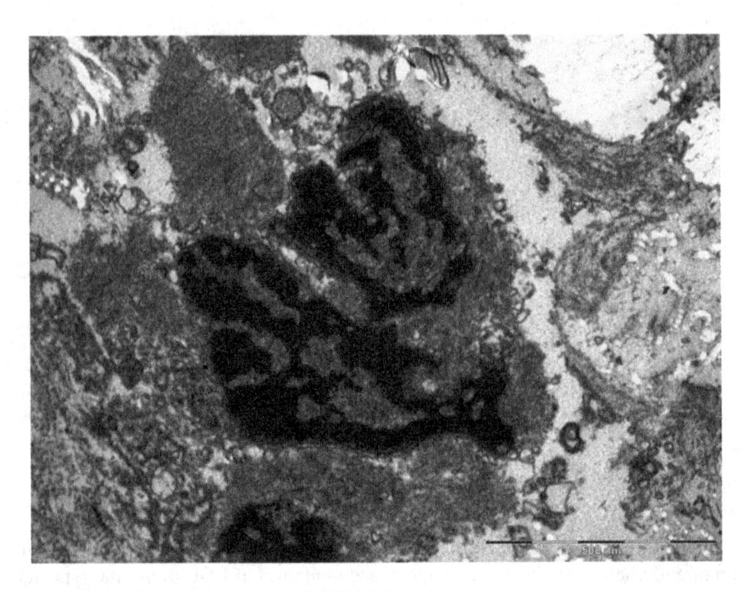

Fig. 7.31 Damage of the endothelial cell (TEM)

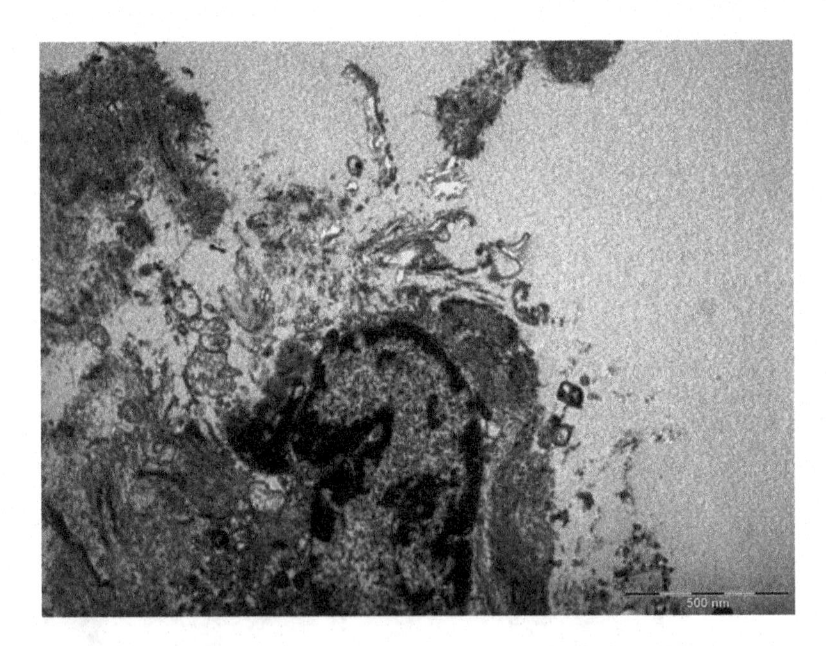

Fig. 7.32 Maximal displacements distribution for fluid pressure of 140 mmHg

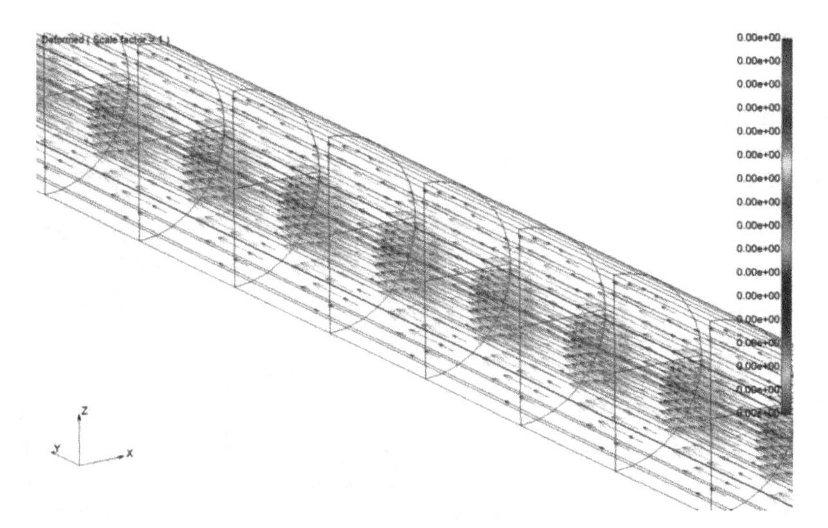

Fig. 7.33 Velocity vector distribution for input flow of 1.1 ml/min

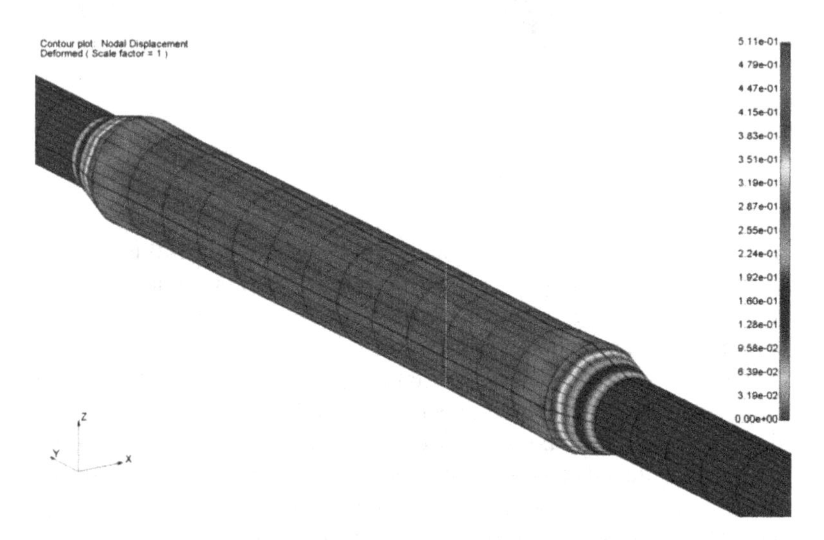

Fig. 7.34 Effective stress distribution for Young module E = 150 kPa in order to mach maximal artery diameter deformation of 50 % or 3 mm

7.4 Animal Experiments on the Pigs

In this chapter, a model of plaque formation on the pig left anterior descending coronary artery (LAD) is simulated numerically using a specific animal data obtained from IVUS and histological recordings. The 3D blood flow is described by the Navier–Stokes equations, together with the continuity equation.

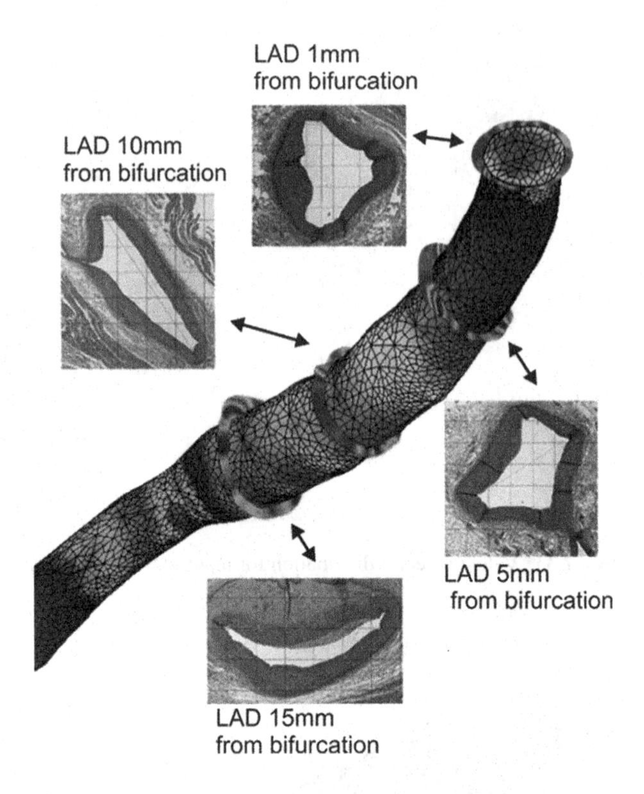

Fig. 7.35 Matching IVUS and histological cross-sectional geometry. Shear stress distribution is shown along the internal arterial wall

Mass transfer within the blood lumen and through the arterial wall is coupled with the blood flow and is modeled by a convection-diffusion equation. The LDL transports in lumen of the vessel and through the vessel tissue (which has a mass consumption term) are coupled by Kedem-Katchalsky equations [30, 31]. The inflammatory process is modeled using three additional reaction–diffusion partial differential equations. A full three-dimensional model was created which includes blood flow and LDL concentration, as well as plaque formation. Matching of IVUS and histological animal data is performed using a 3D histological image reconstruction and 3D deformation of elastic body. Computed concentration of macrophages indicates that there is a newly formed matter in the intima, especially in the LAD 15 mm region from bifurcation.

We used experimental data from pigs submitted to a high-cholesterol diet for 2 months. Specific software for 3D reconstruction of lumen domain and wall artery (coronary artery) was developed. Matching of histological data and IVUS slices is shown in Fig. 7.35. A 3D reconstruction was performed from standard IVUS and angiography images. After that, a full three-dimensional finite element analysis was performed using our in-house finite element code (www.artreat.kg.ac.rs) in order to find low and oscillatory WSS zones. The LAD was selected for this analysis. The process of matching with IVUS images was achieved by 2D modeling of tissue deformation for a number of cross-sections recorded by

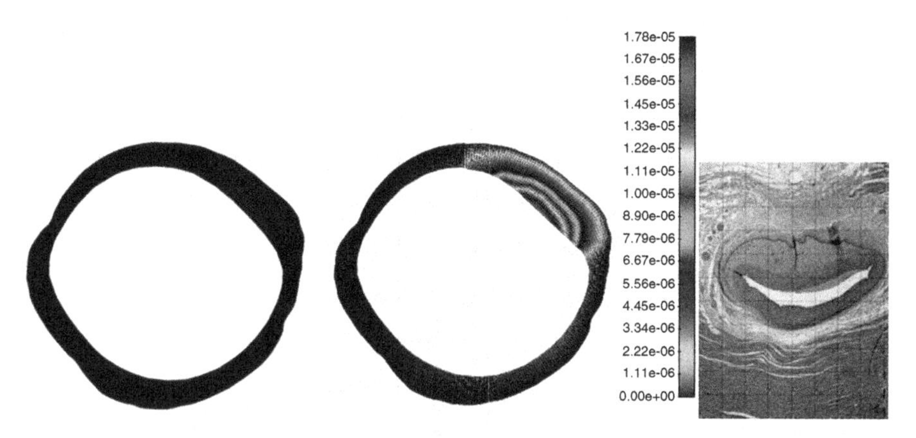

Fig. 7.36 Computer reconstruction of a cross-section of LAD at 15 mm after bifurcation (*left panel*), with computed concentration of macrophages [mg/ml] (*middle panel*); histological analysis (*right panel*) after 2 months of the high-fat diet

histological analysis (four cross-sections are shown in Fig. 7.35); those cross-sections are deformed until the internal lumen circumferential lengths in IVUS images are reached. Macrophages distribution shown in Fig. 7.36 corresponds to the low WSS zone at 15 mm below LAD bifurcation from left circumflex artery, where the largest plaque formation was found. Volume of the plaque obtained from histological analysis (after 2 months of high-fat diet for plaque formation) was fitted by employing a nonlinear least square analysis [32], in order to determine material parameters in equations of Sect. 7.2. The fitted numerical parameters are given in Table 7.3.

We examined experimental data obtained for the LAD artery of a pig after 2 months high-fat diet, in order to determine material parameters of the computer model. Matching computed plaque location and progression in time with experimental observations demonstrates a potential benefit for future prediction of this vascular disease by using computer simulation.

The results for shear stress distribution for pigs # 2, 3, 4, 5, 9 are shown in Fig. 7.37. It can be observed that there are low wall shear stress zones <5 dyn/cm^2 which are indicated in Fig. 7.37e in the proximal zones of the coronary arteries which is in good agreement with histological measurement from CNR (Table 7.4).

For a specific pig HF9, we tried to fit lesion area function with WSS obtained from simulation. The function has four parameters (a, b, c, d)

$$Lesion\ area = a \cdot \log\left(b + \frac{c}{wss + d}\right) \tag{7.42}$$

The fitted values for a, b, c, d are 100.95, 1.0, 0.00773, and −0.8849, respectively (Fig. 7.38).

Table 7.3 Values for animal experiment

Lumen	$\rho = 1{,}000\ kg/m^3$	$\mu = 0.035$ [P]	$D_l = 3.2 \times 10^{-11}\ m^2/s$	$Umax = 0.4\ m/s$	$Pout = 100\ mmHg$	$Co = 3.0 \times 10^{-12}\ kg/m^3$
Intima			$D_w = 1.3 \times 10^{-11}\ m^2/s$	$r_w = -2.6 \times 10^{-4}$	$Pmed = 100\ mmHg$	
Inflammation	$d_1 = 10^{-8}\ m^2/s$	$d_2 = 10^{-10}\ m^2/s$	$d_3 = 10^{-8}\ m^2/s$	$k1 = 20^{-6}\ m^3/kg\ s$	$\lambda = 25\ s^{-1}$	$\gamma = 1\ s^{-1}$

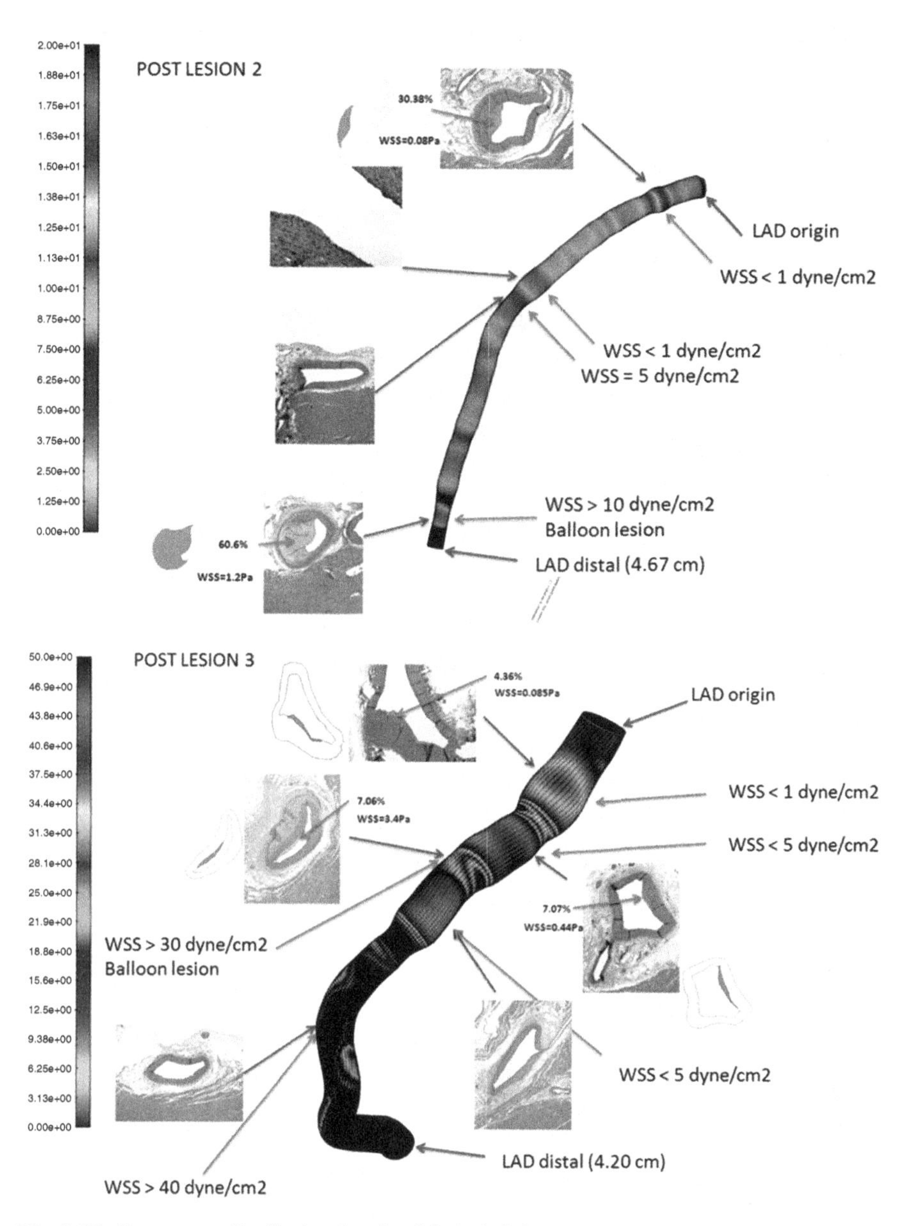

Fig. 7.37 Shear stress distribution for pigs # 2, 3, 4, 5, 9

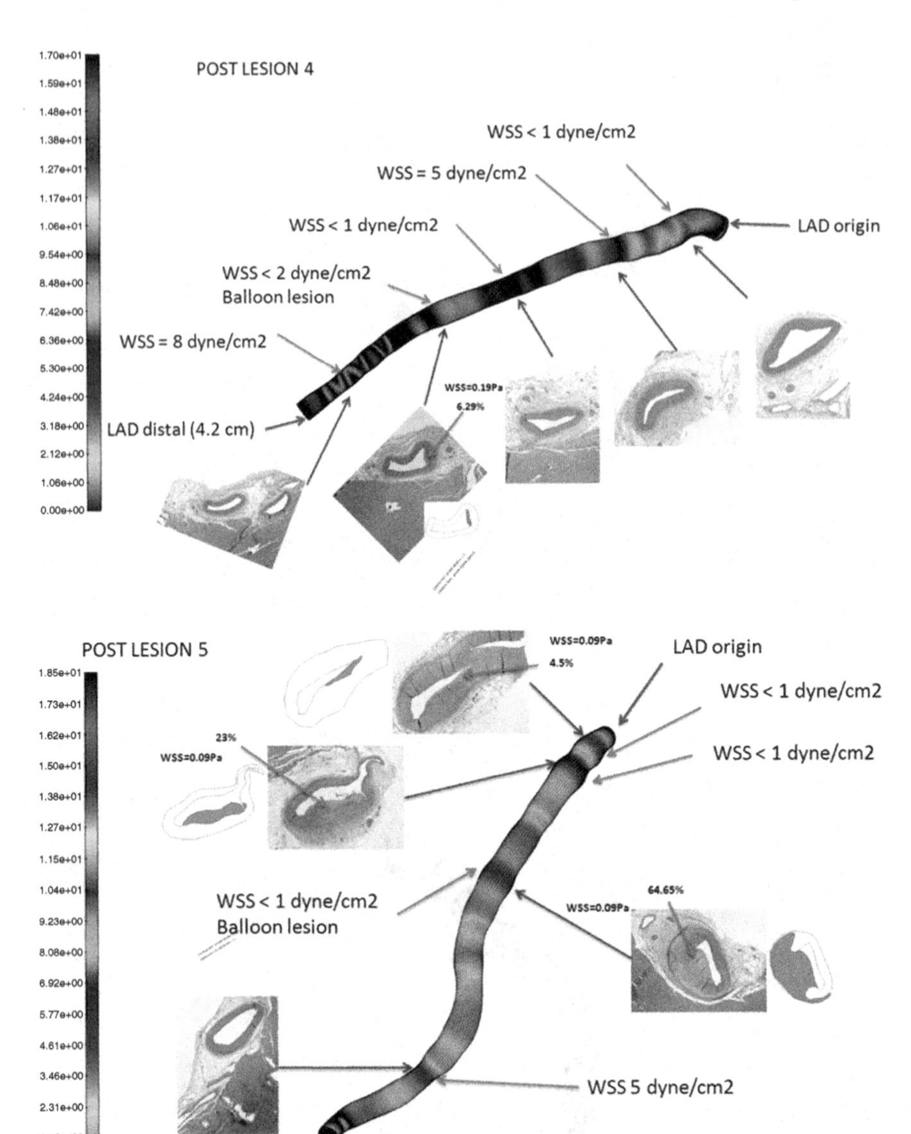

Fig. 7.37 (continued)

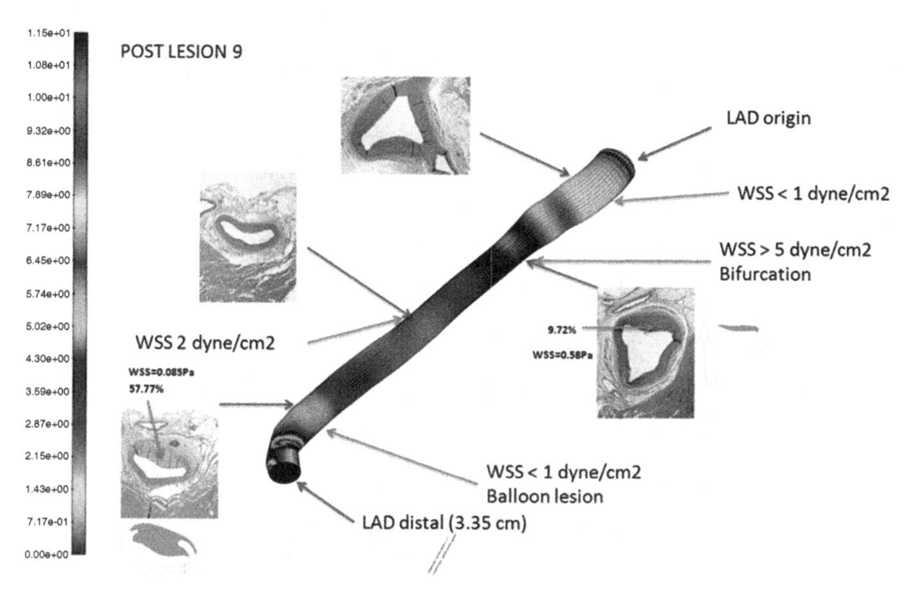

Fig. 7.37 (continued)

Comparison of numerical and experimental data for pig HF9 is presented in Figs. 7.39, 7.40, and 7.41. CXCR4 is considered to be initiator and plaque formation together with WSS distribution. The system of reaction–diffusion Eq. 7.7 is now

$$\begin{aligned} &\partial_t CXCR4 = d_1 \Delta CXCR4 - k_1 CXCR4 \cdot M \\ &\partial_t FC + div(v_w FC) = d_2 \Delta FC - k_1 CXCR4 \cdot FC \\ &\partial_t MIF = d_3 \Delta MIF - \lambda MIF + k_1 CXCR4 \cdot FC + \gamma\left(CXCR4 - CXCR4^{thr}\right) \end{aligned} \tag{7.43}$$

Boundary conditions for first equation are WSS function from Fig. 7.38 and MIF signal function for the second equation. For fluid domain, we used around 100,000 8-node finite elements. Wall domain was modeled with around 80,000 8-node finite elements. Boundary conditions as well as fitted parameters are given in Table 7.5. ALE formulation was applied for the moving mesh domains. Time domain of 2 months was achieved for each pig.

Numerical and experimental results for FC lipids percentage inside lesion area are presented in Fig. 7.39. It can be observed that good accuracy was achieved for lesion area, FC lipids as well as CXCR4.

Comparison of these experimental and numerical data is given in Table 7.6.

For pig HF11, comparison of numerical and experimental results for lesion area, FC lipids, as well as CXCR4 is presented in Figs. 7.42, 7.43, and 7.44. Boundary conditions and fitted model parameters have been shown in Table 7.7. Comparison of these data is given in Table 7.8.

Table 7.4 Experimental histological measurement for pigs HF6, HF9, HF11, and LIHF2

Case	CORO length (mm)	WSS (Pa)	SMCs estimate (aSMApositive area/ lesion area %)	FC lipids (area/ lesion area %)	CXCR4+ (area %)	MIF + (area %)	LAMP1 + (area %)	Nuclear area %
HF6	0.00	0.39	12.00	30.00	5.00	3.00	/	5.00
	5.00	0.97	18.00	29.00	10.00	5.50	/	/
HF9	0.00	2.37	18.00	27.00	7.00	5.00	14.00	10.00
	2.00	2.14	14.00	20.00	5.00	5.00	30.00	6.00
	3.00	2.20	13.00	20.00	4.00	6.00	25.00	5.00
	5.00	2.15	/	/	/	/	/	/
HF11	2.50	3.00	24.00	12.00	3.50	10.00	18.00	6.00
	5.00	1.85	16.00	20.00	2.00	12.00	14.00	7.00
	6.00	1.88	18.00	20.00	3.00	16.00	15.00	8.00
	10.00	1.77	26.00	15.00	5.00	8.00	4.00	10.00
LIHF2	5.00	2.69	35.00	5.00	1.00	2.00	1.00	5.00
	6.00	2.81	26.00	7.00	0.50	1.00	2.00	4.50
	10.00	1.38	20.00	10.00	0.70	1.00	1.00	6.00
	42.50	8.00	30.00	12.00	4.00	6.00	4.00	4.50
	43.50	15.00	22.00	15.00	/	/	/	/

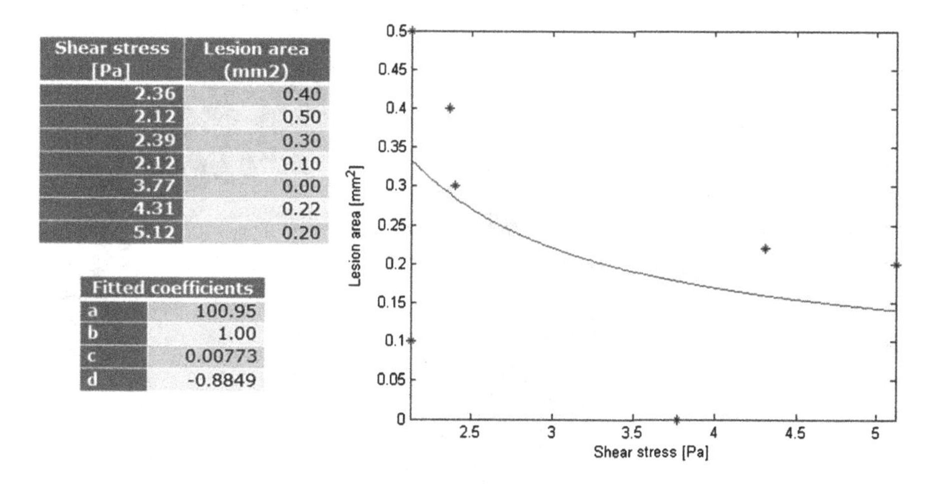

Shear stress [Pa]	Lesion area (mm2)
2.36	0.40
2.12	0.50
2.39	0.30
2.12	0.10
3.77	0.00
4.31	0.22
5.12	0.20

Fitted coefficients	
a	100.95
b	1.00
c	0.00773
d	-0.8849

Fig. 7.38 Fitted data for WSS function from lesion area $Lesion\ area = a \cdot \log\left(b + \frac{c}{wss+d}\right)$. Four parameters *a*, *b*, *c*, *d* were fitted

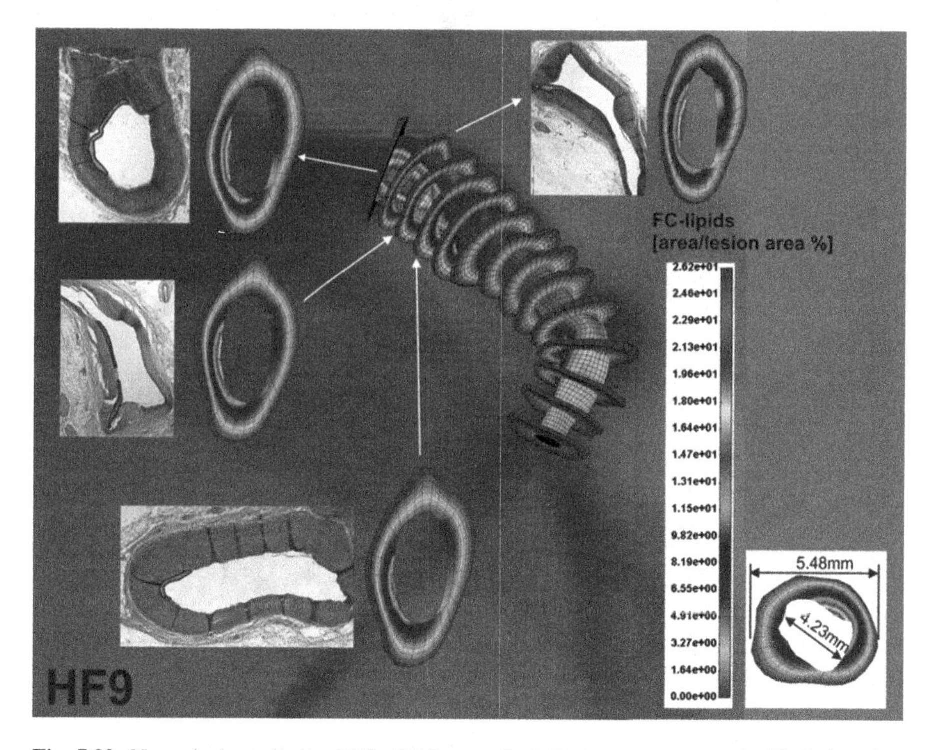

Fig. 7.39 Numerical results for *HP9*. *FC* foam cells lipids percentage area inside lesion area 26.2 %. Histological data per some characteristics cross-sections and comparison with numerical data. Numerical cross-sections dimension

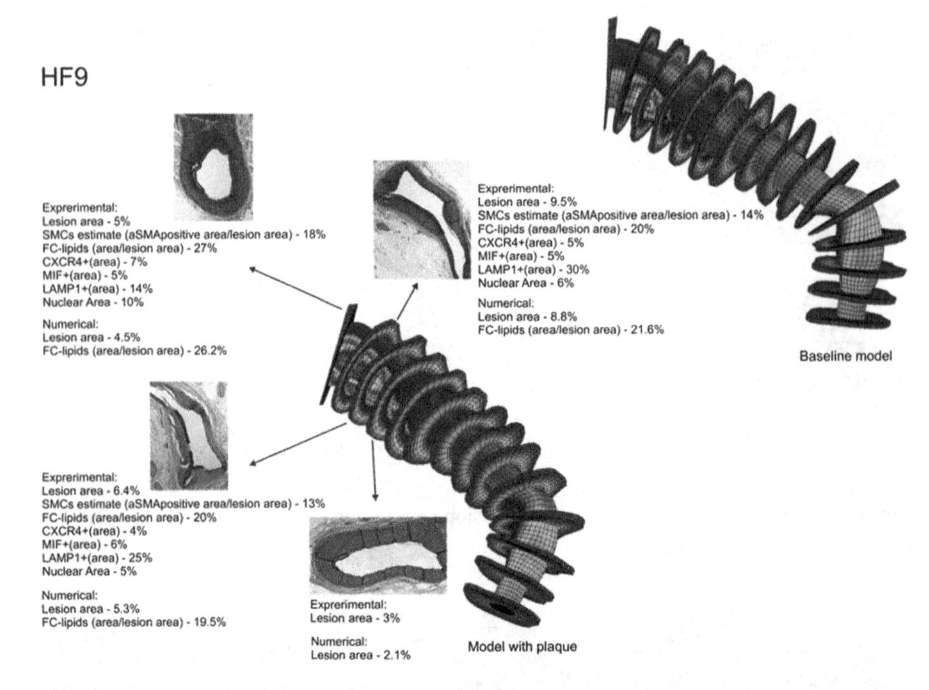

Fig. 7.40 Comparison between experimental and numerical data for pig *HF9*. Baseline and model with plaque. Cross-section presentation of histological and numerical data

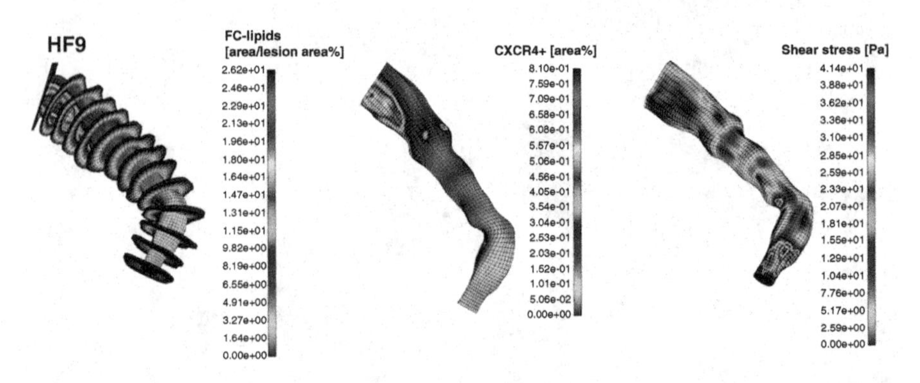

Fig. 7.41 Numerical results for pig *HF9*. Percentage of the FC lipids inside lesion area. Percentage of CXCR4 inside area. Shear stress distribution [Pa]

Table 7.5 Model parameter values for HF9 pig experiment

Blood domain	$\rho = 1{,}000\ kg/m^3$	$\mu = 0.035$ [P]	$D_l = 1.0 \times 10^{-11}\ m^2/s$	$Vsr = 11\ cm/s$	$Pout = 116\ mmHg$	$C_{LDL} = 629.2\ mg/dL$
Intima			$D_w = 1.0 \times 10^{-11}\ m^2/s$	$r_w = -2.6 \times 10^{-4}$	$Pmed = 100\ mmHg$	
Inflammation	$d_1 = 10^{-8}\ m^2/s$	$d_2 = 10^{-8}\ m^2/s$	$d_3 = 10^{-8}\ m^2/s$	$k_1 = 10^{-4}\ m^3/kg\ s$	$\lambda = 1.2\ s^{-1}$	$\gamma = 1.1e\text{-}5\ s^{-1}$
WSS function	$a = 100.95$	$b = 1.0$	$c = 0.0073$	$d = -0.8849$		

Table 7.6 Comparison of experimental and numerical data for pig HF9

HF9	Cross-section 0 mm, WSS = 2.37 Pa		Cross-section 2 mm, WSS = 2.14 Pa		Cross-section 3 mm, WSS = 2.20 Pa		Cross-section 5 mm, WSS = 2.15 Pa	
	Experimental	Numerical	Experimental	Numerical	Experimental	Numerical	Experimental	Numerical
Lesion area [%]	5	4.5	9.5	8.8	6.4	5.3	3	2.1
FC lipids (area/lesion area) [%]	27	26.2	20	21.6	20	19.5	–	11
CXCR4 (area) [%]	7	6.4	5	4.3	4	4.6	–	0.1

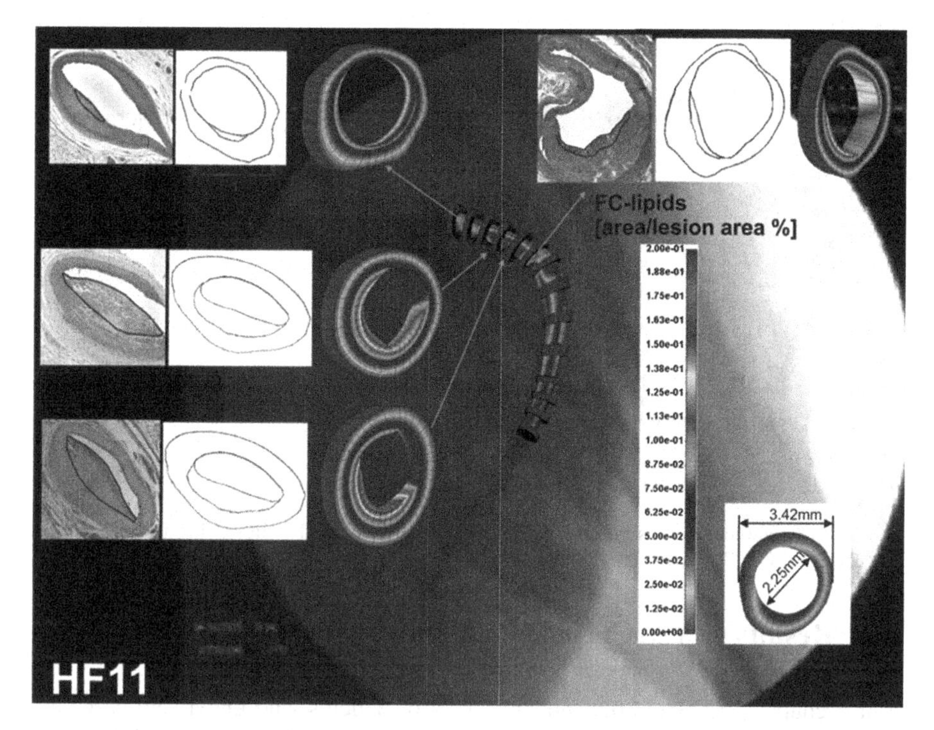

Fig. 7.42 Numerical results for *HP11*. FC lipids percentage area inside the lesion area. Histological data per some characteristics cross-sections and comparison with numerical data. Numerical cross-sections dimension

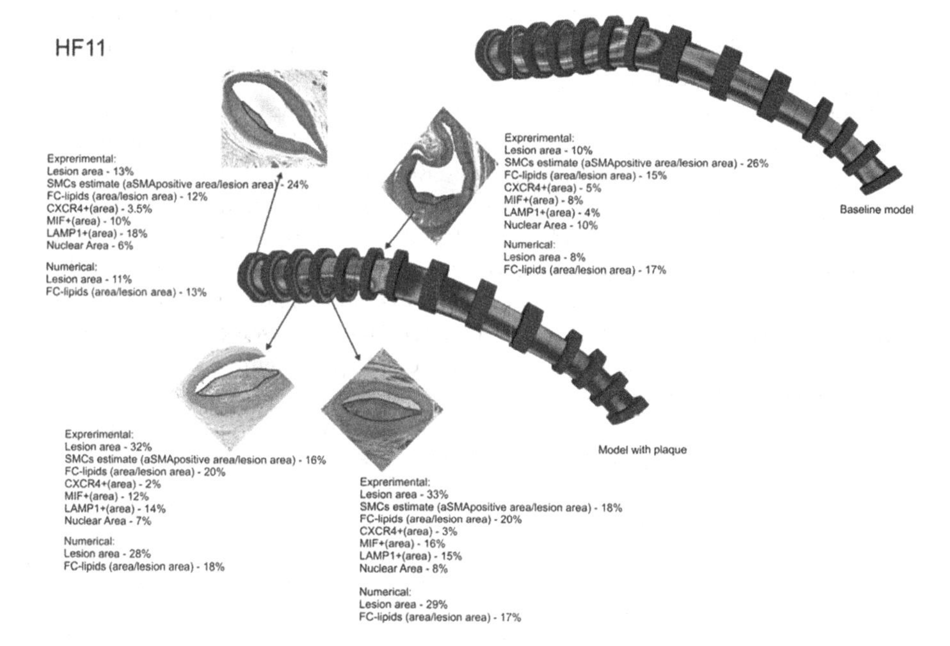

Fig. 7.43 Comparison between experimental and numerical data for pig *HF11*. Baseline and model with plaque. Cross-section presentation of histological and numerical data

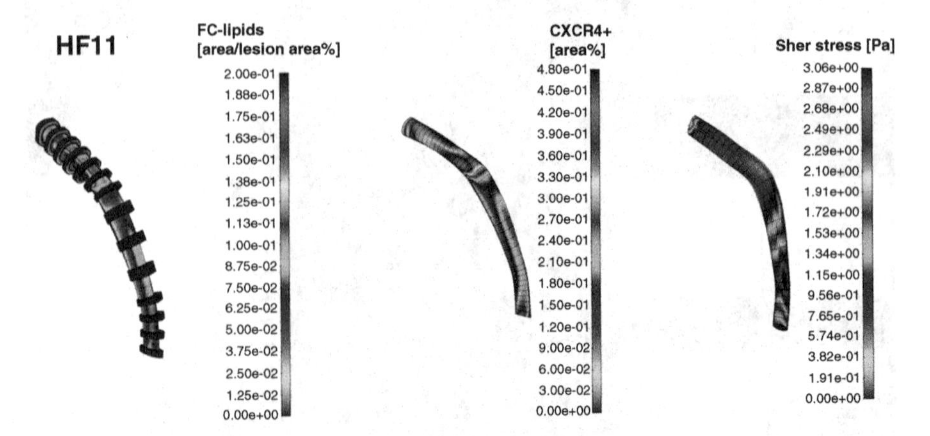

Fig. 7.44 Numerical results for pig *HF11*. Percentage of the FC lipids inside lesion area. Percentage of CXCR4 inside area. Shear stress distribution [Pa]

7.5 Results on the Coronary Patients

In this chapter, we tested our model in a set of patients who underwent coronary computed tomography angiography (CTA) for anginal symptoms [33]. The inflammatory process is modeled using three additional reaction–diffusion partial differential equations. The proof of concept of the model effectiveness was assessed by repetition of CTA, 6 months after the baseline evaluation. Beside the low values of local shear stress, plaque characteristics, risk profile, pattern of circulating adhesion molecules, and reduced coronary flow reserve at baseline appeared to affect plaque progression towards flow-limiting lesions at follow-up evaluation.

In the present study, a group of patients from CNR Pisa with coronary artery disease (CAD) and intermediate lesions was evaluated by CTA; clinical and imaging data of these patients have been described in detail in D4.4.1. An innovative approach to simulate the WSS-related low-density lipoprotein (LDL) transport across the endothelium and to identify LDL accumulation sites was used. The novelty of this work lies in the acquisition of systemic factors related to atherosclerosis evolution (risk profile, inflammation, circulating markers of endothelial activation), measurements of coronary microcirculatory vasodilating capability, and the systematic verification of prediction of plaque progression by repeated CTA, 6 months after the baseline evaluation.

For plaque volume progression, we need at least two different points in time from medical images. We use the following equations:

$$F_{t_{n(k+1)}}(i,j) = p_0(j) + p_1(j) * F_{t_{n(k)}}(i,j) + p_2(j) * \left.\frac{dF_{t_{n(k)}}(i,j)}{dt}\right|_{t_n} * \Delta t + p_3(j) * \left.\frac{d\tau^{wss}{}_{t_{n(k)}}(i,j)}{dt}\right|_{t_n} * \Delta t \tag{7.44}$$

Table 7.7 Model parameter values for HF11 pig experiment

Blood domain	$\rho = 1{,}000\ kg/m^3$	$\mu = 0.035$ [P]	$D_l = 1.0 \times 10^{-11}\ m^2/s$	$Vsr = 13\ cm/s$	$Pout = 118\ mmHg$	$C_{LDL} = 600\ mg/dL$
Intima			$D_w = 1.0 \times 10^{-11}\ m^2/s$	$r_w = -2.6 \times 10^{-4}$	$Pmed = 100\ mmHg$	
Inflammation	$d_1 = 10^{-8}\ m^2/s$	$d_2 = 10^{-8}\ m^2/s$	$d_3 = 10^{-8}\ m^2/s$	$k_1 = 2.1e^{-4}\ m^3/kg\ s$	$\lambda = 1.3\ s^{-1}$	$\gamma = 1.4e\text{-}5\ s^{-1}$
WSS function	$a = 100.95$	$b = 1.001$	$c = 0.0077$	$d = -0.884$		

Table 7.8 Comparison of experimental and numerical data for pig HF11

HF11	Cross-section 2.5 mm, WSS = 3.0 Pa		Cross-section 5 mm, WSS = 1.85 Pa		Cross-section 6 mm, WSS = 1.88 Pa		Cross-section 10 mm, WSS = 1.77 Pa	
	Experimental	Numerical	Experimental	Numerical	Experimental	Numerical	Experimental	Numerical
Lesion area [%]	13	11	32	28	33	29	10	8
FC lipids (area/lesion area) [%]	12	13	20	18	20	17	15	17
CXCR4 (area) [%]	3.5	3.0	2	1.5	3	2.3	3	2.5

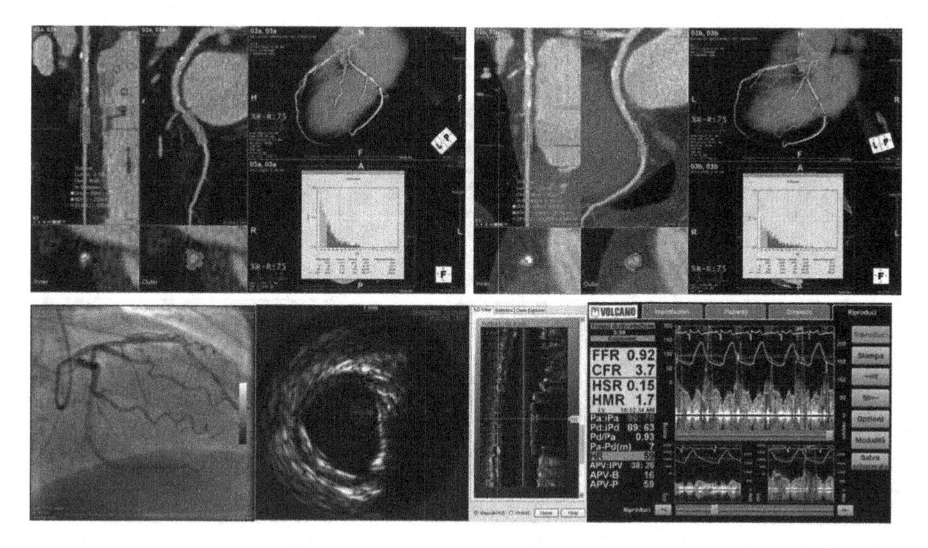

Fig. 7.45 Coronary CTA angiography at baseline and at 6 months (*upper*, *left*, and *right panels*, respectively), 2D coronary angiography (*lower panel*, *left*), IVUS study at the level of distal circumflex artery lesion (*lower panel*, *middle*), pressure and Doppler flow velocity of the same circumflex artery segments (*lower panel*, *right*) in patient 3 of D4.4.1. CTA detects a nonobstructive (35 % lumen diameter reduction) mixed plaque that progressed at 6 months (48 % lumen diameter reduction). CFR was markedly reduced at baseline evaluation (1.7), indicating an impaired microcirculatory vasodilating capability

where $F_{t_{n(k+1)}}(i,j)$ is function of the coordinate or wall thickness *I* for cross-section *j* at time t_n; k is iteration, $k = 0, 1, 2, 3\ldots$; *n* is the time point for image data; *p(0)*, *p(1)*,. . .are coefficients; τ^{wss} is the wall shear stress; and Δt is time step. A simple linear regression analysis with least square method is used for the estimation of the coefficient *p(0)*, *p(1)*. . . for each specific patient.

7.5.1 Clinical Validation by CTA Follow-up

CTA scans were acquired at follow-up paying attention to carefully reply the baseline acquisition procedure and similar patient's heart rate. The 3D artery reconstruction and plaque characterization were accomplished by using the baseline reconstructed images as mask, in order to obtain superimposable vessel pathway (Figs. 7.45 and 7.46). Quantitative image analysis and plaque characterization were undertaken by two independent expert cardio-radiologists, unaware of the clinical features. Discrepancies in measurements were solved by a third observer. IVUS analysis and multiple views coronary angiography were utilized as gold standard for plaque characterization and diameter lumen measurements, respectively.

Plaques were defined as noncalcified, calcified, or mixed. A lesion progression at 6 months follow-up was defined when the lumen diameter decreased of at least 30 % from baseline.

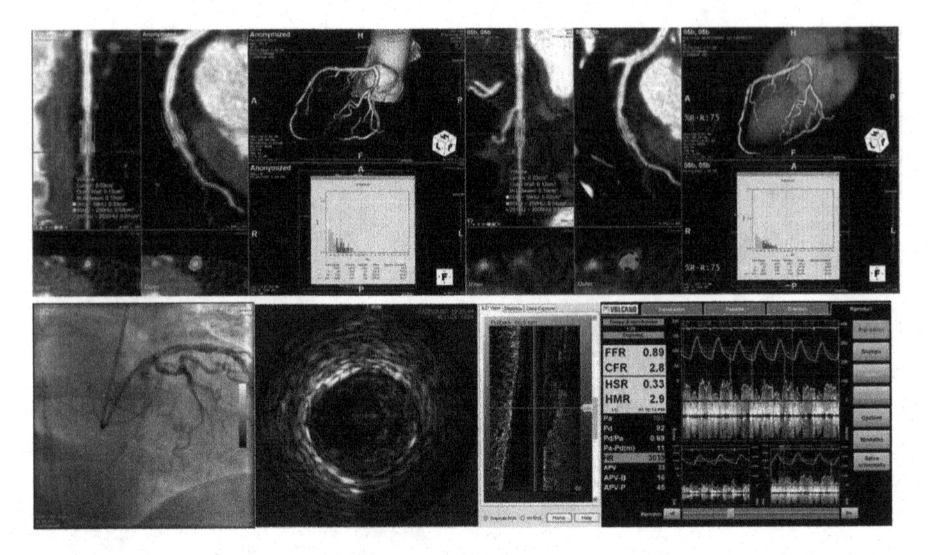

Fig. 7.46 Coronary CTA angiography at baseline and at 6 months (*upper*, *left* and *right panels*, respectively), 2D coronary angiography (*lower panel*, *left*), IVUS study at the level of distal circumflex artery lesion (*lower panel*, *middle*), pressure and Doppler flow velocity of the same circumflex artery segments (*lower panel*, *right*) in patient 5 of D4.4.1. CTA detects a nonobstructive (33 % lumen diameter reduction) soft plaque that progressed to an almost critical stenosis at 6 months (67 % lumen diameter reduction). CFR was almost normal, while FFR was reduced at baseline evaluation (0.84), indicating a slight hydraulic impact of the two adjacent lesions on the coronary flow

7.5.2 Patient Characteristics

Patient demographic and clinical characteristics, including risk factors and Framingham score, in the ten enrolled patients are reported in Table 7.9. Most of the patients had high Framingham risk score, but not significant coronary lesions at coronary angiography. Therefore, the target population was characterized by patients with risk factors and nonobstructive coronary plaques in CTA and IVUS studies.

7.5.3 Plaque Characterization and Detection of Lesion Progression

Four of the ten enrolled patients completed the 6 months follow-up by CTA, following the baseline evaluation (patients 1, 3, 4, 5). CTA allowed a complete evaluation of the three main coronary arteries, with high-quality visualization of vessel geometry and pathway, and definition of plaque characteristics. At baseline, IVUS and 2D coronary angiography permitted to confirm the presence of target lesions in all patients, with further details on plaque composition. Two of the four

Table 7.9 Patient demographic and clinical characteristics

Pt	Age	Sex	Familiarity	Hypertension	Hypercholesterolemia	Diabetes	Framingham risk %	Angina
01	73	M	N	Y	Y	Y	18	N
02	55	M	N	Y	Y	N	10	Y
03	70	F	N	N	Y	Y	25	Y
04	56	M	Y	Y	Y	Y	12	N
05	69	M	N	Y	Y	N	23	N
06	64	M	N	Y	Y	N	15	N
07	56	M	N	Y	Y	Y	8	N
08	70	M	Y	Y	Y	N	15	N
09	75	M	N	Y	Y	N	22	N
10	75	M	N	Y	Y	N	15	N

N no, *Y* yes

patients (patients 3 and 5) showed an obvious reduction of lumen diameter, indicative of plaque progression, both in the distal portion of the circumflex artery. Plaque was defined mixed in one patient and noncalcific in the other, according to CTA and IVUS (virtual histology) criteria. Plaque progression occurred in a bend region of the vessel and in its inner part in both patients. CFR was within the normal range (>3) in all investigated segments but one (CFR 1.7), corresponding to the circumflex artery with plaque progression (patient 3). FFR was within the normal values (>0.92) in all investigated segments but one (FFR 0.84), located in the circumflex artery with the most obvious plaque progression (patient 5). Thus, the two segments with plaque growth at 6 months follow-up in patients 3 and 5 were characterized by impaired CFR or reduced FFR, respectively. Angiographic and hemodynamic features at baseline and at 6 months in the four patients are depicted in Table 7.10. Plaque composition in the six segments evaluated by IVUS and virtual histology is reported in Table 7.11. Imaging and Doppler flow velocity data in the two patients with plaque progression are shown in Figs. 7.45 and 7.46.

7.5.4 Wall Shear Stress and Mass Transport Computation

Plaque progression for patient 3 at distal circumflex artery region was detected using CTA image analysis at baseline and after 6 months. Volume progression from 35 % to 48 % was observed with segmentation and registration of CT images. Downstream the bifurcation level with the second marginal branch predominantly low WSS values occur at baseline (Fig. 7.47a, arrow). A similar situation with a more obvious demonstration of the influence of WSS was observed on patient 5 in the distal portion of the circumflex artery (Fig. 7.48) that showed a marked progression of the baseline stenosis. Location of the lowest WSS (Fig. 7.48a) in the distal portion of the vessel corresponded to the site of plaque growth after 6 months (Fig. 7.48b). WSS distributions on the proximal left anterior descending artery at baseline and follow-up for patient 3 are shown in Fig. 7.49. It can be seen that intra-plaque WSS values were not reduced, as compared with the surrounding segments, and no significant change was observed with time, according to no change in plaque volume progression.

CFD data were used as input for a fitting procedure of volume plaque progression. Oxidized LDL distribution at baseline and follow-up study after 6 months in patient 5 is shown in Fig. 7.49. It can be observed that after 6 months, there is a significant increase in LDL distribution distal from the most narrowed part of the lumen domain. Due to complex lumen and wall domain, only LDL distribution for the joint boundaries is presented in Fig. 7.50. WSS in our model is used for plaque initiation and position at the wall for higher LDL penetration. There is a complex process of the macrophages transformation into the foam cells. Also foam cells directly created the intima volume increase. Fitting of the plaque volume increasing of 34 % from baseline to 6 months follow-up observed from medical image analysis was done with Eq. 7.44. The parameters for the numerical models of patient 5 are given in Table 7.12.

Table 7.10 Angiographic and hemodynamic features at baseline and at 6 months (CT)

	Basal CT			2D-Angio	6 months CT	IVUS	Doppler flow
Case	Stenosis %	Stenosis %	Stenosis %	Plaque	FFR	CFR	Resting AFV
Patient 01							
RCA middle	45	30	48	Mixed	0.96	3	11.34
Patient 03							
LAD proximal	40	30	40	Mixed	0.93	3.1	
CX distal	35	30	48	Mixed	1	1.7	15.24
Patient 04							
LAD middle	34	22	34	Mixed	0.94	3.1	13.45
Patient 05							
CX proximal	50	34	40	Mixed	0.84	3.3	14.02
CX distal	33	30	67	Non calcified	0.84	3.3	14.02

AFV average flow velocity (cm/s), *CFR* coronary flow reserve, *FFR* fractional flow reserve

Table 7.11 Plaque composition in the six segments evaluated by virtual histology (IVUS)

Case	NC	DC	FI	FF
Patient 01				
RCA middle	25.9	11.0	57.1	6.0
Patient 03				
LAD proximal	28.1	13.0	54.5	4.5
CX distal	16.6	6.1	53.5	23.8
Patient 04				
LAD middle	22.6	9.8	50.3	17.3
Patient 05				
CX proximal	9.4	1.2	56.2	33.2
CX distal	2.5	0.6	44.9	51.9

NC necrotic core, *DC* dense calcium, *FI* fibrotic, *FF* fibro fatty

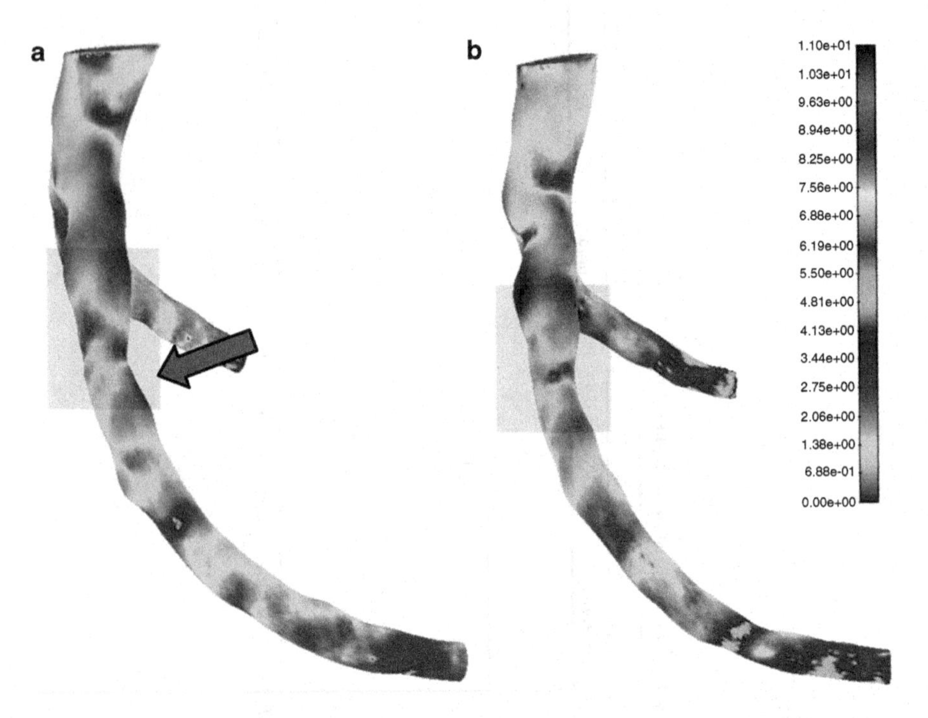

Fig. 7.47 Local shear stress distribution at baseline (**a**) and after 6 months follow-up (**b**) in patient 3. Area with plaque progression at follow-up showed at baseline the lowest shear stress value (*arrow*). Wall shear stress values are expressed in [Pa] units

7.5.5 Features Affecting Plaque Progression

The findings from ARTreat study point the attention on specific features that may affect plaque progression in a short midterm (6 months). Plaque progression occurred in the two patients with the highest Framingham risk score (25 and 23) of the enrolled population; both plaques developed at the inner side of curved

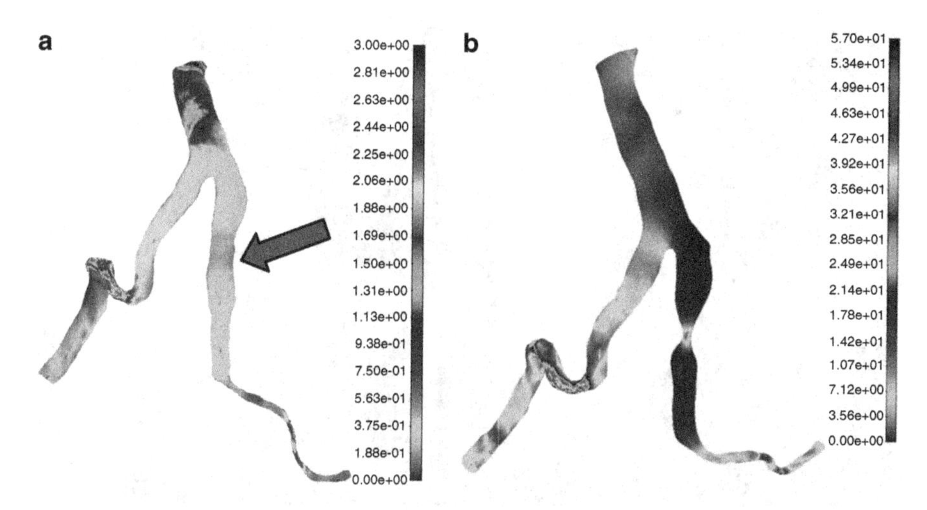

Fig. 7.48 Local shear stress distribution at baseline (**a**) and after 6 months follow-up (**b**) in patient 5. Area with plaque progression at follow-up towards critical stenosis (distal circumflex artery) showed at baseline the lowest shear stress value (*arrow*). Wall shear stress values are expressed in [Pa] units

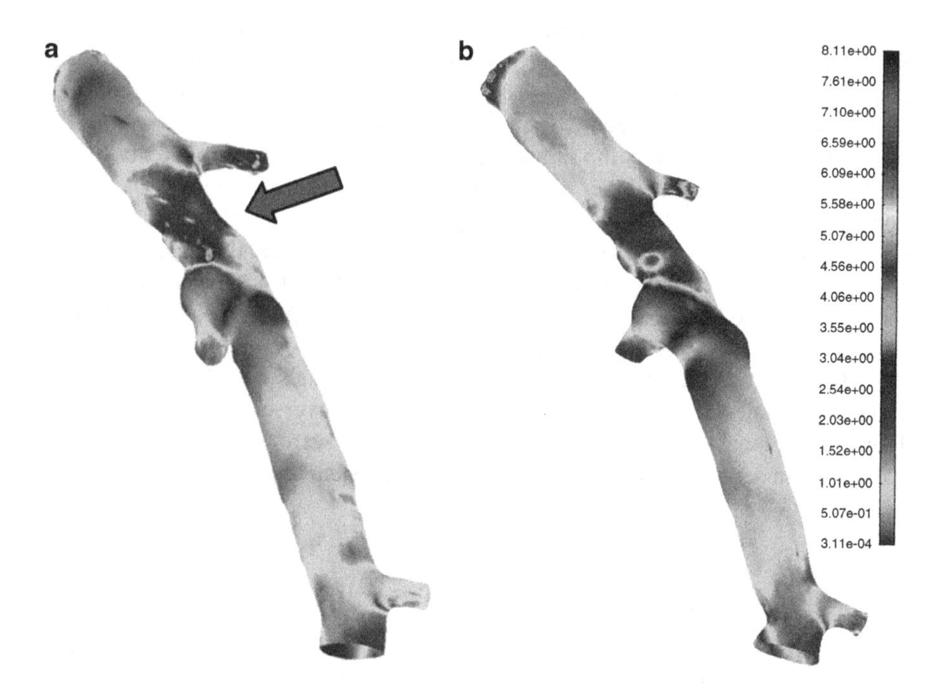

Fig. 7.49 Local shear stress distribution at baseline (**a**) and after 6 months follow-up (**b**) in a stable plaque of the *left* anterior descending artery in patient 3. The shear stress values in the stenotic region are higher than those in adjacent segments (*arrow*)

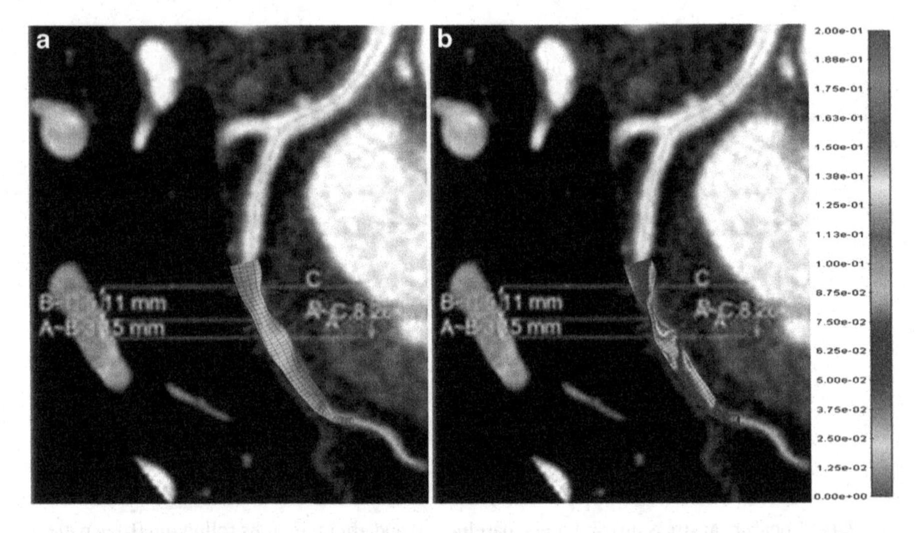

Fig. 7.50 LDL distribution at baseline and in the follow-up study in the distal circumflex artery with obvious lumen diameter reduction (patient 5). Angiography slices are depicted as background of the computer simulation results. Units for LDL concentration [mg/ml]

(myocardial side) distal circumflex artery segments. Interestingly, only these territories showed an impaired CFR or reduced FFR, indicative of an impaired coronary vasodilating capability. The greatest plaque growth occurred in a noncalcified, eccentric plaque (patient 5). At virtual histology (IVUS study), the fibro-fatty component was the prevalent content of this plaque, markedly higher than that observed in the other patients. Conversely, necrotic core was the scarce component. Patient 5 showed the highest increase in E-selectin plasma concentrations, at 6-month evaluation, largely above the normal range, among the four patients that completed the study, while only patient 4 showed an obvious increase in VCAM-1. No changes were observed in ICAM-1 levels in the four patients. Baseline WSS values in the progressive lesions (patients 3 and 5) were 3.36 dyn/cm^2 and 1.85 dyn/cm^2, respectively. These values were lower than those calculated immediately before and after the lesions. Conversely, WSS values in the four stable plaques averaged 4.2 ± 0.4 dyn/cm^2 and were not lower than values calculated in segments immediately adjacent to the stenoses.

7.5.6 Discussion

A multiscale model for the biological process of plaque formation and progression has been applied to CTA imaging in patients with nonobstructive coronary plaques. The model includes the 3D reconstructed arterial model, the blood flow, the WSS distribution, the molecular/cell model of the arterial wall/blood composition, and

Table 7.12 Values for case 5, distal circumflex artery

Lumen	Blood density $\rho = 1{,}000\ kg/m^3$	Blood viscosity $\mu = 0.0035$ [Pa s]	Lumen diffusivity $D_l = 2.8 \times 10^{-11}\ m^2/s$	Mean inflow Qmean = 0.45 ml/s	Outlet lumen pressure Pout = 82 mmHg	Inlet LDL concentration Co = $2.6 \times 10^{-12}\ kg/m^3$
Intima			Wall diffusivity $D_w = 2.6 \times 10^{-12}\ m^2/s$	Consumption rate $r_w = -2.6 \times 10^{-4}$	Media pressure Pmed = 80 mmHg	
Inflammation	OxLDL diffusivity $d_1 = 10^{-3}\ m^2/s$	Macrophages diffusivity $d_2 = 10^{-10}\ m^2/s$	Cytokines diffusivity $d_3 = 10^{-8}\ m^2/s$	Plaque growth coefficient $k_1 = 2 \times 10^{-6}\ m^3/kg\ s$	Degradation coefficient $\lambda = 30\ s^{-1}$	OxLDL detection coefficient $\gamma = 1\ s^{-1}$

the biological mechanism involved in the generation and growth of atherosclerotic plaque. Our model starts with passive penetration of LDL in particular areas of the intima. We assume that once in the intima, LDL is immediately oxidized. When the oxidized LDL exceeds a threshold, there is recruitment of monocytes. The incoming monocytes immediately differentiate into macrophages. Transformation of macrophages into foam cells contribute to the recruitment of new monocytes. This yields the secretion of a pro-inflammatory signal (cytokines), self-support inflammatory reaction. Newly formed foam cells are responsible for the local volume increase. Under a local incompressibility assumption, when foam cells are created, the intima volume is locally increasing. Volume change of the wall affects the fluid lumen domain which means that fully coupling is achieved. The specific numerical procedures using ALE were developed for this purpose. The regression analysis was employed for plaque volume development at two different measurement times for baseline and 6 months follow-up study. Nonlinear least square fitting procedure was used for a plaque composition.

The two time points obtained in our study population allowed to characterize the native shear stress in the two vessel site of plaque progression with time. In fact, the baseline sub-significant lesions in distal circumflex artery (lumen diameter reduction of 35 % and 33 %) subtended to low WSS values that markedly changed in patient (case 5) with plaque progression towards an almost critical stenosis. Most of the fitting parameters were obtained from patient-specific measurements at the time of catheterization procedure (usually 1–2 weeks following CTA evaluation), namely, intracoronary flow velocity and pressures. Furthermore, plaque composition was assessed by virtual histology in each patient, providing a degree of accuracy much higher than CTA that will allow better definition of models for plaque progression. This approach is also improving the reliability of WSS computation and inter-patient comparison. Further support to reliability of CTA image reconstruction and shear stress measurement will derive from investigations under way on comparison of WSS measurements by 3D segment reconstruction from IVUS and angiography. Information gained by measurement of CFR, as pointed out previously, would more properly address the issue on correlation between local WSS and plaque formation. The heterogeneous course of each plaque in the same patient underscores the role of vessel geometry and local hemodynamic forces in determining the natural history of plaque remodeling. Patient-specific local and systemic features, as assessed in this study, can provide insights on the nature of arterial remodeling and, potentially, on plaque vulnerability.

Determination of plaque location and composition and computer simulation of progression in time for a specific patient shows a potential benefit for prediction of disease progression. The proof of validity of 3D reconstructed coronary CTA scans in the evaluation of atherosclerotic plaque burden may shift the clinical information of coronary CTA from morphological assessment towards a functional tool [33].

7.6 Results on the Carotid Artery Patients

In this chapter, we choose two specific patients from MRI study with significant plaque progression. Plaque volume progression using three time points for baseline, 3, and 12 months follow-up is fitted. Our results for plaque localization correspond to low shear stress zone, and we fitted parameters from our model using nonlinear least square method. Determination of plaque location and composition and computer simulation of progression in time for a specific patient shows a potential benefit for prediction of disease progression. The proof of validity of three-dimensional computer modeling in the evaluation of atherosclerotic plaque burden may shift the clinical information of MRI from morphological assessment towards a functional tool. Understanding and prediction of the evolution of atherosclerotic plaques either into vulnerable or stable plaques are major tasks for the medical community.

Our working hypotheses are that the local arterial wall volume increases as the result of newly formed foam cells. Monocytes evolve in macrophages which phagocyte oxidized LDL and evolve in foam cells by massive ingestion of oxidized LDL. Transformation of macrophages into foam cells contribute to the recruitment of new monocytes. It yields the secretion of a pro-inflammatory signal (cytokines). It is self-support inflammatory reaction. Newly formed foam cells are responsible for the local volume increase. Under a local incompressibility assumption, when foam cells are created, the intima volume is locally increasing. We further assume that arterial wall permeability increases in the zone of low wall shear stress and elastic property of the arterial wall for fluid–structure interaction problem.

7.6.1 Patients' Data

Fifty patients with carotid atherosclerotic disease underwent high-resolution MRI of their carotid arteries in a 1.5 T MRI system (Signa HDx GE Healthcare, Waukesha, WI) with a four-channel phased array neck coil (PACC, Machnet BV, Elde, The Netherlands) [34]. The study protocol was reviewed and approved by the regional research ethics committee, and all patients gave written informed consent.

After an initial coronal localizer sequence, axial 2D time-of-flight (TOF) MR angiography was performed to identify the location of the carotid bifurcation and the region of maximum stenosis. The following sequences were used to depict the various contents within the plaque structure: T1 weighted (repetition time/echo time: 1 × RR/7.8 ms) with fat saturation; T2 weighted (repetition time/echo time: 2 × RR/100 ms) with fat saturation; proton density (PD) weighted (repetition time/echo time: 2*RR/7.8 ms) with fat saturation; and short tau inversion recovery (STIR) (repetition time/echo time/inversion time: 2 × RR/46/150 ms). The field of view was 10 cm × 10 cm and matrix size 256 × 256. The in-plane spatial resolution achieved was of the order of 0.39 mm × 0.39 mm × 3 mm. Plaque components, i.e., lipid and fibrous tissue were manually delineated by two

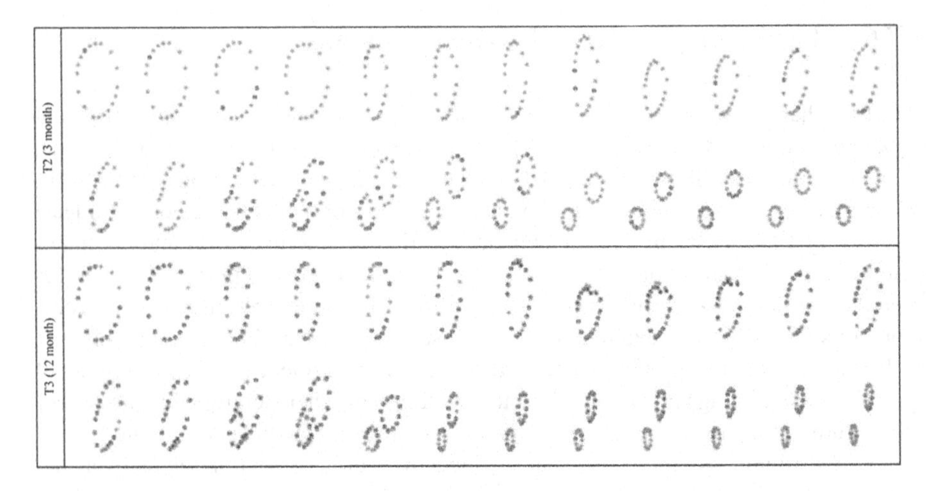

Fig. 7.51 Simulated contour plots compared with target contours at time steps T2 (3 months) and T3 (12 months). *Green*: simulated contours; *Red*: target contours

experienced MR readers using CMR Tools (London, UK) with previously published criteria [34]. Two specific patients with high plaque progression were chosen. We did manual segmentation of plaque components, such as fibrous cap and chronic hemorrhage tissue, calcium, and PH, using carotid MR images.

7.6.2 Fitting Procedure for Plaque Volume Growing Function

For plaque volume growing a fitting procedure for growth function which takes into account change of the coordinates, shear stress as well as effective wall stress data from fluid–structure interaction calculation is developed.

In this methodology, we used three known times T1, T2, and T3 [35] for estimation of plaque volume growth.

Starting from the plaque geometry at T1, we used three different growth functions to simulate plaque progression and tried to reach best agreement with plaque geometry obtained from image reconstruction at T2 and T3:

- GF1 – Growth function which uses nodal coordinates data only
- GF2 – Growth function which uses nodal coordinates and shear stress data
- GF3 – Growth function which uses nodal coordinates, shear stress, and solid stress data

These growth functions as well as the fitting procedure are described in detail in the appendix section.

Overlapping contour plots of the target and simulated results for time steps T2 and T3 are presented in Fig. 7.51.

7.6.3 Growth Functions and the Fitting Procedure

For simulation of the plaque growth, the following procedure was used:

Step 1: Start from the original in geometry at T1.

Step 2: Set

$$f_{t1_0}(i,j) = f_{T1}(i,j)$$

$$f_{t1_1}(i,j) = f_{T1}(i,j) + (f_{T2}(i,j) - f_{T1}(i,j))/m$$

$$\tau_{t1_0}(i,j) = \tau_{T1}(i,j)$$

$$\tau_{t1_1}(i,j) = \tau_{T1}(i,j) + (\tau_{T2}(i,j) - \tau_{T1}(i,j))/m$$

$$\sigma_{t1_0}(i,j) = \sigma_{T1}(i,j)$$

$$\sigma_{t1_1}(i,j) = \sigma_{T1}(i,j) + (\sigma_{T2}(i,j) - \sigma_{T1}(i,j))/m$$

Step 3: We use m time steps to go from T1 to T2 and n time steps to go from T2 to T3. This means that we use total $n + m$ time steps to go from T1 to T3. For $k = 1, \ldots n + m$ do the following:

- GF1:

$$f_{t1_{(k+1)}}(i,j) = a_0(j) + a_1(j) \cdot \left(w(j) \cdot f_{t1_k}(i,j) + (1 - w(j)) \cdot f_{t1_0}(i,j)\right) + a_2(j) \cdot \left.\frac{df}{dt}\right|_{T_{1_k}} (i,j) \cdot \Delta t_k \tag{7.45}$$

- GF2:

$$f_{t1_{(k+1)}}(i,j) = a_0(j) + a_1(j) \cdot \left(w(j) \cdot f_{t1_k}(i,j) + (1 - w(j)) \cdot f_{t1_0}(i,j)\right) + a_2(j) \cdot \left.\frac{df}{dt}\right|_{T_{1_k}} (i,j) \cdot \Delta t_k + a_3(j) \cdot \tau_{t1_k}(i,j) + a_4(j) \cdot \left.\frac{d\tau}{dt}\right|_{T_{1_k}} (i,j) \cdot \Delta t_k \tag{7.46}$$

- GF3:

$$f_{t1_{(k+1)}}(i,j) = a_0(j) + a_1(j) \cdot \left(w(j) \cdot f_{t1_k}(i,j) + (1 - w(j)) \cdot f_{t1_0}(i,j)\right) + a_2(j) \cdot \left.\frac{df}{dt}\right|_{T_{1_k}} (i,j) \cdot \Delta t_k + a_3(j) \cdot \tau_{t1_k}(i,j) + a_4(j) \cdot \left.\frac{d\tau}{dt}\right|_{T_{1_k}} (i,j) \cdot \Delta t_k + a_5(j) \cdot \sigma_{t1_k}(i,j) + a_6(j) \cdot \left.\frac{d\sigma}{dt}\right|_{T_{1_k}} (i,j) \cdot \Delta t_k \tag{7.47}$$

where $\frac{df}{dt}\Big|_{T1_k}(i,j) = \frac{f_{t1_k}(i,j) - f_{t1_{(k-1)}}(i,j)}{t_k - t_{k-1}}$, $\frac{d\tau}{dt}\Big|_{T1_k}(i,j) = \frac{\tau_{t1_k}(i,j) - \tau_{t1_{(k-1)}}(i,j)}{t_k - t_{k-1}}$, and $\frac{d\sigma}{dt}\Big|_{T1_k}(i,j) = \frac{\sigma_{t1_k}(i,j) - \sigma_{t1_{(k-1)}}(i,j)}{t_k - t_{k-1}}$ are derivatives of displacement, shear stress, and solid stress, respectively; $\Delta t_k = t_{k+1} - t_k = \frac{T3 - T1}{m+n}$ is a time step; f are x and y coordinates; τ are wall shear stress values; σ are solid stress values of nodal points; $j = 1, 2, ...24$ is the slice number; and i is the index for the points on each slice. $a_0(j)$, $a_1(j)$, $a_2(j)$, $a_3(j)$, $a_4(j)$, $a_5(j)$, $a_6(j)$ and $w(j)$ are coefficients of growth functions GF1, GF2, and GF3 to be determined in such a way to obtain the best match of calculated geometries and experimental geometries at times T2 and T3. Since we use m time steps to go from T1 to T2 and n time steps to go from T2 to T3, we compared f_{t1_m} with experimental geometry at time T2 and $f_{t1\_(m+n)}$ with experimental geometry at time T3. The previous formulas of growth functions are very similar with formulas that Yang used in his paper [35].

Coefficients of the plaque volume growth functions (GF1, GF2, and GF3) $a_0(j)$, $a_1(j)$, $a_2(j)$, $a_3(j)$, $a_4(j)$, $a_5(j)$, $a_6(j)$ and w are calculated, independently, for all 24 slices by using simplex optimization method, method which does not involve derivative calculations, developed by John Nelder and Roger Mead [36]. We minimized sum of the squared errors between calculated and real geometry at times T2 and T3 for each of 24 slices.

$$ESS(j) = \sum_{i=1}^{N_j}\left(\left(x_{T2,i}(j) - \overline{x}_{T2,i}(j)\right)^2 + \left(y_{T2,i}(j) - \overline{y}_{T2,i}(j)\right)^2\right) + \sum_{i=1}^{N_j}\left(\left(x_{T3,i}(j) - \overline{x}_{T3,i}(j)\right)^2 + \left(y_{T3,i}(j) - \overline{y}_{T3,i}(j)\right)^2\right) \tag{7.48}$$

where N_j is the number of nodes for slice; $x_{T2,i}(j)$, $y_{T2,i}(j)$, $x_{T3,i}(j)$, and $y_{T3,i}(j)$ are real x and y coordinates at time steps T2 and T3 for slice j; and $\overline{x}_{T2,i}(j)$, $\overline{y}_{T2,i}(j)$, $\overline{x}_{T3,i}(j)$, and $\overline{y}_{T3,i}(j)$ are calculated x and y coordinates at time steps T2 and T3 for slice j.

The best results were obtained by using growth function GF3 which takes into account both wall shear and solid stress. Total squared error is calculated as:

$$ESS = \sum_{j=1}^{24} ESS(j) \tag{7.49}$$

Total squared errors for all growth functions are GF1 = 36.02, GF2 = 29.98, and GF3 = 26.31.

Total squared error does not give a picture of how our model is really accurate; it only serves to compare results obtained with different growth function. Because of that we calculated mean relative percent error:

$$RE(j) = \frac{1}{2N_j}\left(\sum_{i=1}^{N_j}\frac{\Delta P_{T2,i}(j)}{r_{T2,i}(j)} + \sum_{i=1}^{N_j}\frac{\Delta P_{T3,i}(j)}{r_{T3,i}(j)}\right) \cdot 100 \tag{7.50}$$

$$RE = \frac{1}{24}\sum_{j=1}^{24} RE(j) \tag{7.51}$$

where $\Delta P_{T2,i}(j)$ and $\Delta P_{T3,i}(j)$ are distances between real and predicted position of i-th point of j-th slice at times T2 and T3. $r_{T2,i}(j)$ and $r_{T3,i}(j)$ are distances between real position of i-th point and center of gravity for j-th slice at times T2 and T3.

Mean relative percent errors for all growth functions are GF1 = 2.7 %, GF2 = 2.62 %, and GF3 = 2.51 %. It can be observed that model which uses growth function GF3 is the most accurate. Mean relative percent errors for GF3 is 2.51 %, which means that distance between predicted position of point and real position of point is in average only 2.51 % of distance between real position of point and slice center of gravity. This seems to be very good result.

7.6.4 Results

We compared changes in the cross-section areas for different patients with carotid artery progression. From 50 patients, we choose two with significant evidence of MR plaque progression in order to estimate parameter for our model of plaque formation and development. From MR slices, we segmented the inner and outer wall at nine cross-sections for baseline, 3, and 12 months. Segmented data for patient #1 are presented in Fig. 7.52. Trends for increasing or decreasing cross-section areas versus time for patient #1 have been shown in Fig. 7.53. It can be seen that almost all cross-section areas are increasing during follow-up time. For the same patient, the correlation with wall shear stress zones is shown in Fig. 7.54. We used three categories as colors for the light: red color denotes large decreasing in the cross-section area changes and middle wall shear stress, yellow color denotes small decreasing in the cross-section area changes and middle wall shear stress, while increasing in the cross-section area changes and low wall shear stress is denoted by green color. Obviously from Fig. 7.54, it can be concluded that there is a significant correlation with large increasing of the cross-section areas and low wall shear stress for patient #1. Similar analysis was done with patient #2. Cross-section areas vs. time (0, 3, and 12 months) for patient #2 has been shown in Fig. 7.55. Green color light which denotes large increasing in the cross-section area changes and low wall shear stress also is mostly dominant for patient #2 (see Fig. 7.56).

The different material properties for patient #1 have been shown in Figs. 7.57 and 7.58. There are three different parts in the carotid arterial wall and Young's elasticity modules: artery tissue E = $3.0e^6$ Pa, fibrous cap E = $6.0e^6$ Pa, and subacute hemorrage E = $4.5e^6$ Pa.

Effective wall stress for three diferent times (0, 3, and 12 months) at maximum peak systole is presented in Fig. 7.59.

Volume of the plaque progression obtained from MRI system was fitted by employing a nonlinear least square analysis [32], in order to determine

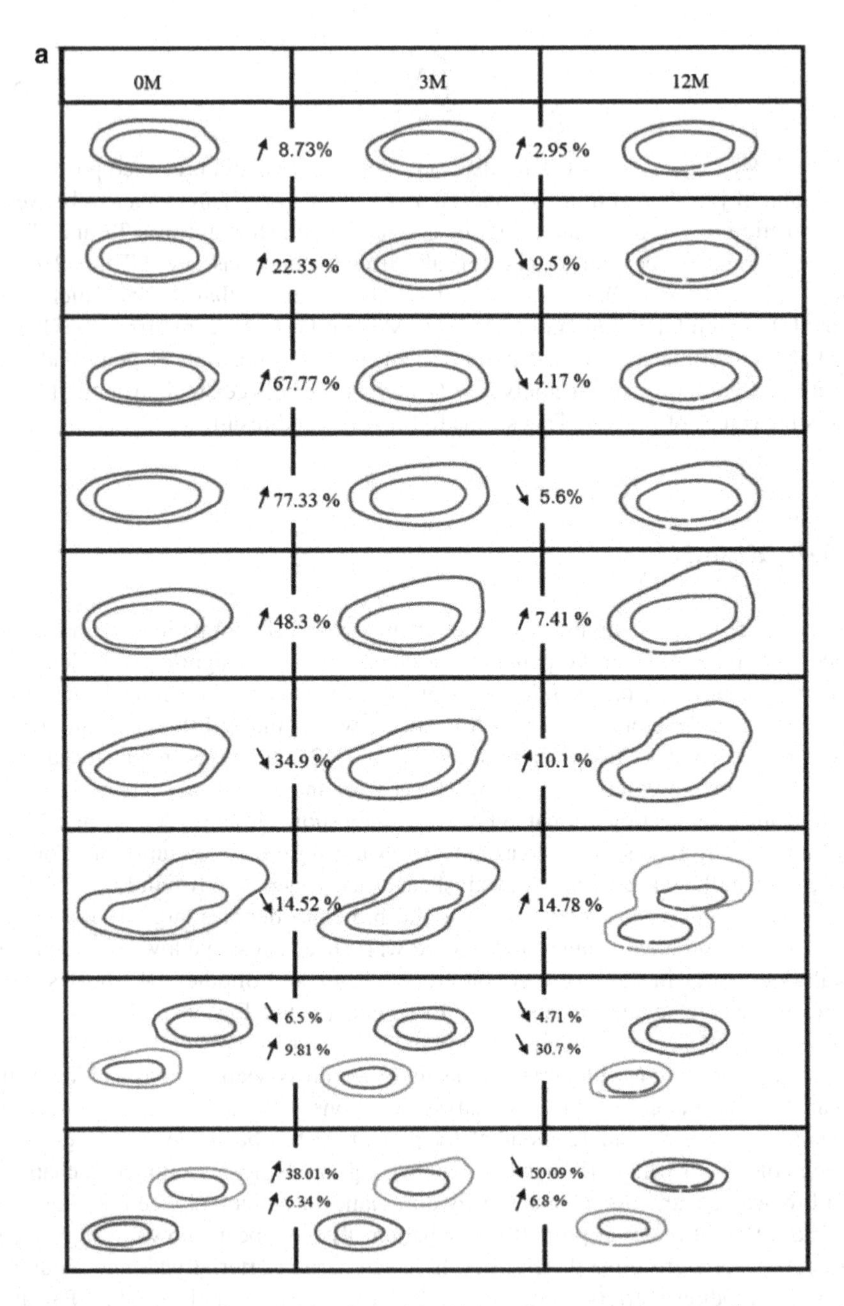

Fig. 7.52 Cross-section areas changes for patient #1 (area between the *inner* and the *outer* wall)

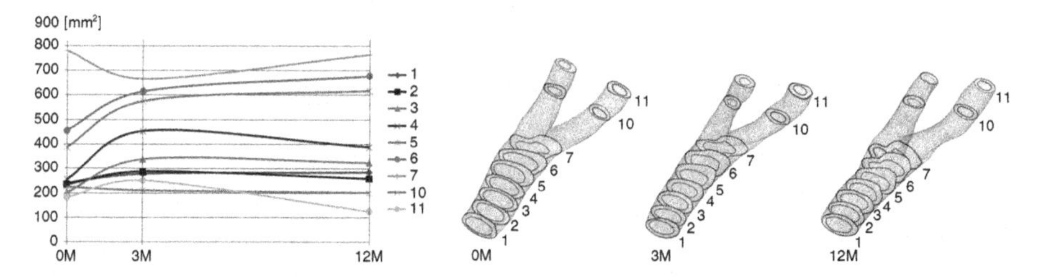

Fig. 7.53 Cross-section areas versus time (0, 3 and 12 months) for patient #1

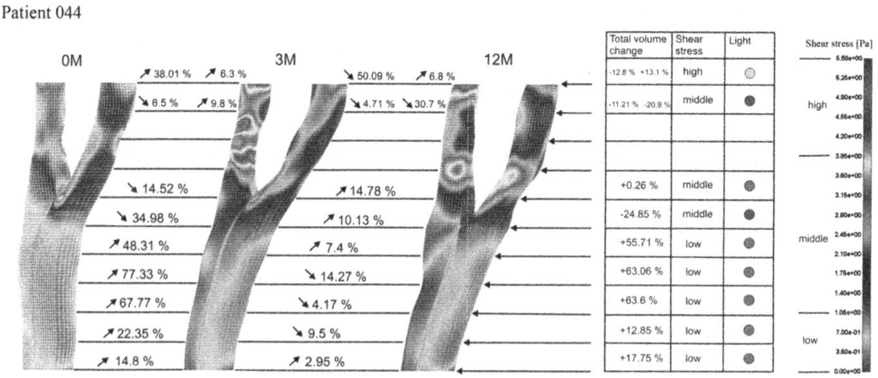

Patient 044

0M 3M 12M

↗ 38.01 % ↗ 6.3 % ↘ 50.09 % ↗ 6.8 %
↘ 6.5 % ↗ 9.8 % ↘ 4.71 % ↘ 30.7 %
↘ 14.52 % ↗ 14.78 %
↘ 34.98 % ↗ 10.13 %
↗ 48.31 % ↗ 7.4 %
↗ 77.33 % ↘ 14.27 %
↗ 67.77 % ↘ 4.17 %
↗ 22.35 % ↘ 9.5 %
↗ 14.8 % ↗ 2.95 %

Total volume change	Shear stress	Light
-12.8 % +13.1 %	high	
-11.21 % -20.9 %	middle	
+0.26 %	middle	
-24.85 %	middle	
+55.71 %	low	
+63.06 %	low	
+63.6 %	low	
+12.85 %	low	
+17.75 %	low	

Shear stress [Pa]
high
middle
low

Fig. 7.54 Correlation of cross-sections changes with wall shear stress for patient #1

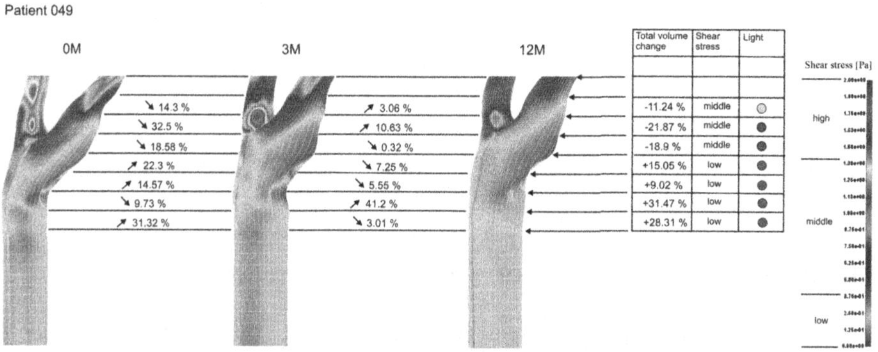

Patient 049

0M 3M 12M

↘ 14.3 % ↗ 3.06 %
↘ 32.5 % ↗ 10.63 %
↘ 18.58 % ↘ 0.32 %
↗ 22.3 % ↘ 7.25 %
↗ 14.57 % ↘ 5.55 %
↘ 9.73 % ↗ 41.2 %
↗ 31.32 % ↘ 3.01 %

Total volume change	Shear stress	Light
-11.24 %	middle	
-21.87 %	middle	
-18.9 %	middle	
+15.05 %	low	
+9.02 %	low	
+31.47 %	low	
+28.31 %	low	

Shear stress [Pa]
high
middle
low

Fig. 7.55 Cross-section areas versus time (0, 3 and 12 months) for patient #2

material parameters in equations of Sect. 8.2. The fitted numerical parameters are given in Table 7.13.

Fifty patients with carotid atherosclerotic disease are analyzed with MRI. Plaque components, i.e., lipid and fibrous tissue were manually delineated by two experienced MR readers.

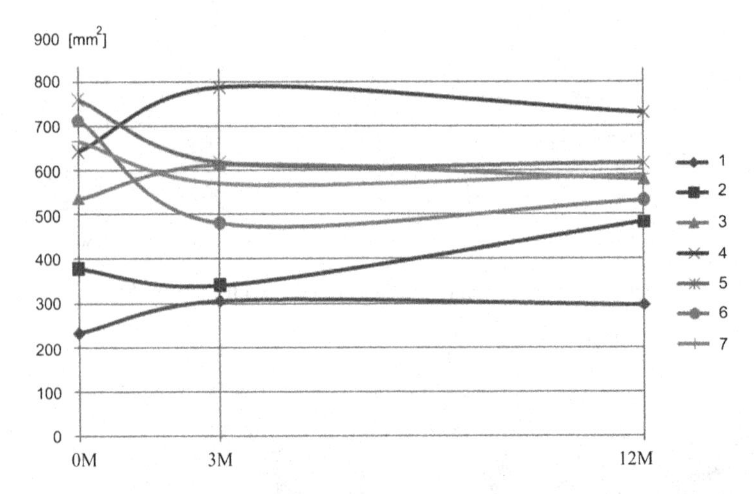

Fig. 7.56 Correlation of cross-section areas changes with wall shear stress for patient #2

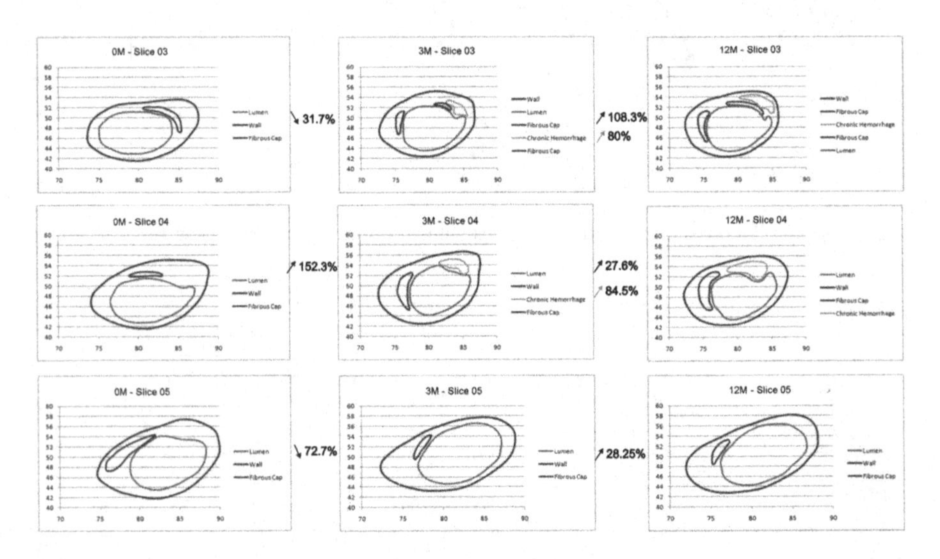

Fig. 7.57 Different material properties for a specific patient #1 for follow-up study

Elastic material property of the wall for property was assumed. Mass transfer equations for wall are considered to be stationary due very slow process of plaque formation and development. We implemented a fitting procedure for plaque volume growing which takes into account change of the coordinates, shear stress, as well as effective wall stress data from fluid–structure interaction calculation. It looks that low shear stress is not only factor for local plaque development. Wall stress analysis as well as inflammation model with cell level transformation and progression should be taken into account.

We examined patient data at three time points for the carotid artery, zero (baseline), 3 and 12 months in order to make fitting of the parameter model for

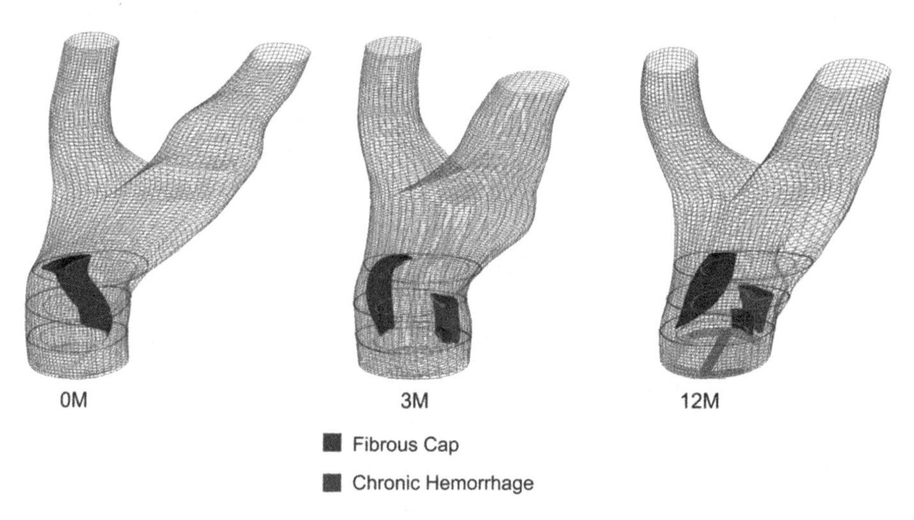

Fig. 7.58 3D presentation of plaque composition and different material properties for patient #1

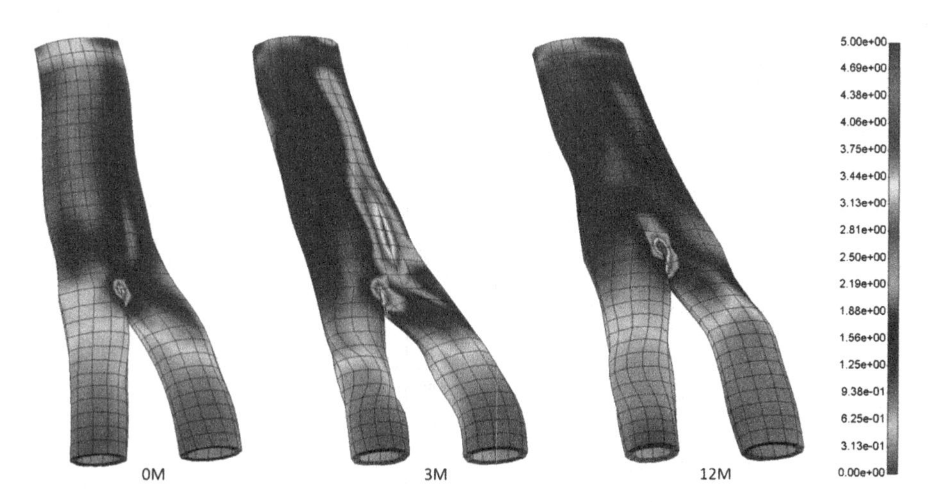

Fig. 7.59 Effective wall stress calculation for follow-up time baseline, 3, and 6 months for patient #1 (units 1 = 1e5 Pa)

specific patient. Three-dimensional reconstruction was performed from high resolution MRI system. Boundary conditions for the inlet velocity waveforms are measured from MR. Shear stress distribution mostly corresponds to the localization of the plaque volume progression. Fluid–structure interaction was implemented to analyze effective wall stress distribution. We fitted patient data for plaque volume progression with growth functions which depend from fluid shear stress and arterial wall effective stress.

Also we analyzed UCAM patient #8 and #9 for baseline and follow-up of 12 months. The WSS distribution results for these two patients are presented in Figs. 7.60 and 7.61.

Table 7.13 Values for clinical data

Lumen	$\rho = 1{,}000\ kg/m^3$	$\mu = 0.035$ [P]	$D_l = 3.8e^{-11}\ m^2/s$	Umean = 0.5 m/s	Pout = 100 mmHg	$Co = 2.5e^{-12}\ kg/m^3$
Intima			$D_w = 1.6e^{-11}\ m^2/s$	$r_w = -2.4e^{-4}$	Pmed = 100 mmHg	
Inflammation	$d_1 = 1e^{-7}\ m^2/s$	$d_2 = 1e^{-9}\ m^2/s$	$d_3 = 1e^{-7}\ m^2/s$	$k_1 = 2e^{-6}\ m^3/kg\ s$	$\lambda = 1\ s^{-1}$	$\gamma = 2.e\text{-}5\ s^{-1}$

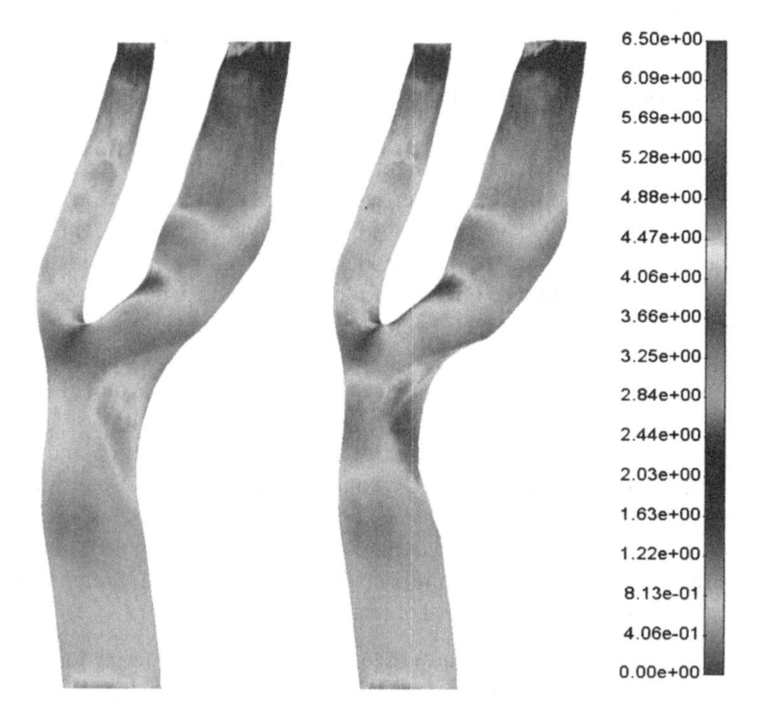

Fig. 7.60 UCAM patient #8, base line (*left*) and follow-up 12 months (*right*)

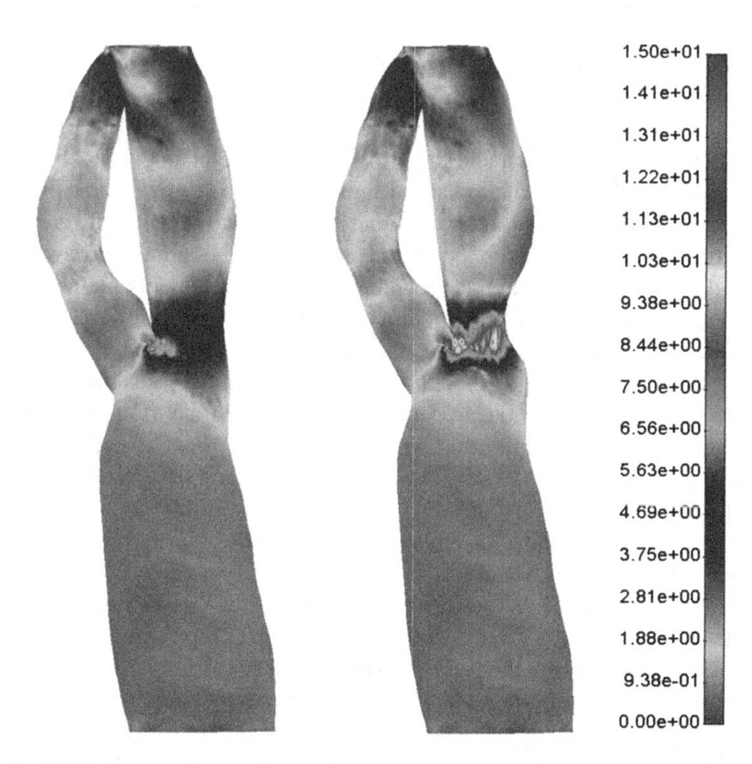

Fig. 7.61 UCAM patient #9, base line (*left*) and follow-up 12 months (*right*)

7.7 Discussion and Conclusions

Using partial differential equations, we provide a realistic three-dimensional model of the plaque initiation process. We perform numerical simulations in a two- and three-dimensional setting. Using in-house FE software from the University of Kragujevac a full three-dimensional model was developed. The mathematical model can be separated in three parts as following: a model for LDL transfer and oxidation (this model needs a careful computation of both the arterial tree geometry and the blood flow not only in the lumen but also in the arterial wall), a model for the inflammatory process that takes place in the intima and a model for lesion growth.

A full three-dimensional model was created for plaque formation and development, coupled with blood flow and LDL concentration in blood. Determination of plaque location and progression in time for a specific patient shows a potential benefit for future prediction of this vascular decease using computer simulation.

In this delivery a few continuum models based are described. Summarizing the results of the study, we have concluded that:

- A fluid-wall model for the transport of LDL from the arterial lumen to the arterial wall and inside the wall is developed. In this stage of the project we developed a single layer model while multilayered model is still developing. The convection terms are not neglected and the incremental-iterative procedure is applied.
- The models for plaque initiation and plaque progression are developed. These two models are based on partial differential equations with space and times variables and they describe the biomolecular process that takes place in the intima during the initiation and the progression of the plaque.
- The model for plaque formation and plaque progression despite some difficulties concerning the different time scales that are involved and the different blood velocities in the lumen and in the intima, its numerical treatment is developed by using decomposition techniques together with finite elements methods and by splitting the numerical scheme into three independent parts: blood flow and LDL transfer, inflammatory process and atheromatous plaque evolution. Using such a model, the initiation plaque modeling procedure, first comprehends the simulation of the plaque initiation procedure by performing a virtual and numerical computation of plaque initiation within a selected animal-specific artery. The main limitation of these models is that they have many parameters that are not easy to estimate.
- Nonlocal model for plaque progression is developed in order to build a macroscopic model that does not describe the inflammatory process in details but that focus on the evolution of the lumen's geometry. In this model, the nonlocal effects appear through a term that, in some sense, represents the membrane permeability. However, the very interesting features of this model, since the limitations are too important we will not use this model anymore.

- We proposed a model to describe the motion of mixtures of cell populations in a saturated medium with a constraint of the local density. We provided an adapted theoretical framework, based on a reformulation of the model as a gradient flow in a product space of densities, and proposed a discretization strategy which enjoys reasonable stability and accuracy properties.

References

1. P. Libby, Inflammation in atherosclerosis. *Nature,* 2002, 868–874.
2. J. Loscalzo& A.I. Schafer, *Thrombosis and Hemorrhage* (Third edition. Lippincott Williams & Wilkins, Philadelphia, 2003).
3. J.M. Tarbell, Mass transport in arteries and the localization of atherosclerosis, *Annual Review of Biomedical Engineering, 5*, 2003, 79–118.
4. P. Zunino, *Mathematical and numerical modeling of mass transfer in the vascular system.* (PhD thesis, Lausanne, Switzerland: EPFL, 2002).
5. A. Quarteroni, A. Veneziani, P. Zunino, Mathematical and numerical modeling of the solute dynamics in blood flow and arterial walls, *SIAM Journal of Numerical Analysis, 39*, 2002, 1488–1511.
6. M.R. Kaazempur-Mofrad, C.R. Ethier, Mass transport in an anatomically realistic human right coronary artery, *Ann Biomed Eng 29*, 2001, 121–127.
7. S. Wada, M. Koujiya, T. Karino, Theoretical study of the effect of local flow disturbances on the concentration of low-density lipoproteins at the luminal surface of end-to-end anastomosed vessels, *Med BiolEngComput 40*, 2002, 576–587.
8. D. K. Stangeby, C.R. Ethier, Computational analysis of coupled blood-wall arterial LDL transport, *J BiomechEng-T ASME 124*, 2002, 1–8.
9. N. Sun, N.B. Wood, A.D. Hughes, S.A.M. Thom, X.Y. Xu, Fluid-wall modelling of mass transfer in an axisymmetric stenosis: effects of shear dependent transport properties, *Ann Biomed Eng 34*, 2006, 1119–1128.
10. L. Ai, K. Vafai, A coupling model for macromolecule transport in a stenosed arterial wall, *Int J. Heat Mass Tran 49*, 2006, 1568–1591.
11. U. Olgac, V. Kurtcuoglu, V. Poulikakos, Computational modeling of coupled blood-wall mass transport of LDL: effects of local wall shear stress, *Am J. Physiol Heart CircPhysiol 294*, 2008, 909–919.
12. R. Ross, Atherosclerosis: a defense mechanism gone awry, *Am J Pathol., 143*, 1993, 987–1002.
13. N. Filipovic, N. Meunier, and M. Kojic, PAK-Athero, *Specialized three-dimensional PDE software for simulation of plaque formation and development inside the arteries*, University of Kragujevac, 34000 Kragujevac, Serbia, 2010.
14. Goldstein J., Anderson R., Brown M., "Coated pits, coated vesicles, and receptor-mediated endocytosis."*Nature*, 1979, Vol. 279, pp. 679–684.
15. Bratzler R. L., Chisolm G.M., Colton C. K., Smith K. A., Lees R. S., "The distribution of labeled low-density lipoproteins across the rabbit thoracic aorta in vivo." *Atherosclerosis*, 1977, Vol. 28, pp. 289–307.
16. Kojic, M., Filipovic, N., Stojanovic B., Kojic N., (2008) Computer modeling in bioengineering: Theoretical Background, Examples and Software, John Wiley and Sons, Chichester, England.
17. Brooks A.N., Hughes T.J.R., "Streamline upwind/Petrov-Galerkin formulations for convection dominated flows with particular emphasis on the incompressible Navier–Stokes equations."*Comput. Meths. Appl. Mech. Engrg.*, 1982, Vol. 32, pp. 199–259.

18. Filipovic N., Mijailovic S., Tsuda A., Kojic M. “An Implicit Algorithm Within The Arbitrary Lagrangian–Eulerian Formulation for Solving Incompressible Fluid Flow With Large Boundary Motions.” *Comp. Meth. Appl. Mech. Eng.*, 2006, Vol. 195, pp. 6347–6361.
19. Yuan S.W., Finkelstein A.B.., “Laminar pipe flow with injection and suction through a porous wall.” Transaction of ASME, 1956, Vol. 78, pp. 719–724.
20. Goh V. K., Lau C. P., Mohlenkamp S., Rumberger J. A., Achenbach A., Budoff M. J., Cardiovascular Ultrasound, 2010, 8:5.
21. Cheng C., Tempel D., Haperen V. R., Baan A. V. D., Grosveld F., Daemen Mat J.A.P., Krams R., Crom D.R., Atherosclerotic Lesion Size and Vulnerability Are Determined by Patterns of Fluid Shear Stress, Circulation Vol. 113, 2006, pp. 2744–2753
22. Himburg, H., Grzybowski, D., Hazel, A., LaMack, J. Li X. and Friedman M., Spatial comparison between wall shear stress measures and porcine arterial endothelial permeability. Am J Physiol Hear Circ Pysiol 286, 1916–1922, 2004.
23. Birukova A.A. , K. G. Birukov, K. Smurova, D. Adyshev, K. Kaibuchi, I. Alieva, J. G. Garcia and A. D. Verin. Novel role of microtubules in thrombin-induced endothelial barrier dysfunction. *Faseb J.* Dec 2004;18(15):1879–1890.
24. Tai S. C., G. B. Robb and P. A. Marsden. Endothelial nitric oxide synthase: a new paradigm for gene regulation in the injured blood vessel. *ArterioscleThrombVasc Biol.* Mar 2004;24 (3):405–412.
25. Vargas C. B., F. F. Vargas, J. G. Pribyl and P. L. Blackshear. Hydraulic conductivity of the endothelial and outer layers of the rabbit aorta. *Am J Physiol.* Jan 1979;236(1):H53–60.
26. Rappitsch G., K. Perktold, Pulsatile albumin transport in large arteries: a numerical simulation study, J. Biomech. Eng. 118 (1996) 511–519.
27. Wada, S., Karino, T., (2000) Computational study on LDL transfer from flowing blood to arterial walls. In: Yamaguchi, T. (Ed.), Clinical Application of Computational Mechanics to the Cardiovascular System. Springer, Berlin, 157–173.
28. Moore J.A., C.R. Ethier, Oxygen mass transfer calculations in large arteries, J. Biomech. Eng. 119 (1997) 469–475.
29. Stangeby D.K., C.R. Ethier, Coupled computational analysis of arterial LDL transport—effects of hypertension, Comput. Meth. Biomech. Biomed. Eng. 5 (2002) 233–241.
30. Kedem O., Katchalsky A. (1958): Thermodynamic analysis of the permeability of biological membranes to non-electrolytes. Biochim. Biophys. Acta 27, 229–246
31. Kedem O., Katchalsky A. (1961): A physical interpretation of the phenomenological coefficients of membrane permeability. J. Gen. Physiol. 45, 143–179
32. Chavent G., (2010) Nonlinear Least Squares for Inverse Problems, Nonlinear Least Squares for Inverse Problems Theoretical Foundations and Step-by-Step Guide for Applications, Springer, second print, New York.
33. Oberdan Parodi, Themis Exarchos, Paolo Marraccini, Federico Vozzi, Zarko Milosevic, Dalibor Nikolic, Antonis Sakellarios, Panagiotis Siogkas, Dimitris I. Fotiadis, Nenad Filipovic, 2012, Patient-specific prediction of coronary plaque growth from CTA angiography: a multiscale model for plaque formation and progression, IEEE Trans Inf Technol Biomed. 16 (5):952–65.
34. Sadat U, Teng Z, Young VE, Zhu C, Tang TY, Graves MJ, Gillard JH (2010) Impact of plaque haemorrhage and its age on structural stresses in atherosclerotic plaques of patients with carotid artery disease: an MR imaging-based finite element simulation study. Int J Cardiovasc Imaging DOI 10.1007/s10554-010-9679-z.
35. Yang C, Tang D and Atluri S (2010) Three-Dimensional Carotid Plaque Progression Simulation Using Meshless Generalized Finite Difference Method Based on Multi-Year MRI Patient-Tracking Data, CMES 57: 51–76
36. Nelder J. and Mead R 1965. “A simplex method for function minimization”, Computer Journal **7** (4): 308–313.

Chapter 8
Particle Dynamics and Design of Nano-drug Delivery Systems

Tijana Djukic

8.1 Introduction

Particles have been increasingly investigated in recent years as "smart" delivery systems, which can be applied in biomedical imaging, but also as therapeutical agents in cardiovascular and oncological treatments [1–3]. Such drug delivery systems consist of nanoparticles that can be loaded with contrast agents or drug molecules (monoclonal antibodies (mAbs) or small-molecule agents). Particles are sufficiently small to be injected at the systemic level and transported through the circulatory system to various organs and body districts. The size of particles ranges from few tens of nanometers to hundreds of nanometers [4, 5] to few microns [6]. Also, they can have various shapes, including spherical, spheroidal [7], or other more complex shapes [8]. The majority are made by polymeric or lipid materials, but can also be made of silica, gold, or iron oxide.

There are two strategies currently considered in the field of delivery of nanoparticles to solid tumors. The strategy that has been traditionally considered is the passive strategy, based on the enhanced permeability and retention effect (EPR) [9] – nanoparticles that extravasate through fenestrations found in the tumor vasculature are entrapped in the extracellular matrix and transported from the vascular compartment to the inner region of the tumor mass. As an alternative to the passive strategy, an active delivery strategy, which is gaining more and more interest recently, is based on the targeting of the tumor vasculature through ligand-receptor specific interactions. In this case particles are designed to be able to sense the difference between normal and tumor endothelium and "search" for biological and biophysical specificities such as overexpression of disease-specific receptor molecules [10] or the appearance of abnormally large inter- or intra-endothelial gaps [11]. In vascular targeting, nanoparticles should be able to attach to a specific

T. Djukic (✉)
Faculty of Engineering, R&D Center for Bioengineering, Kragujevac, Serbia
e-mail: tijana@kg.ac.rs

G. Rakocevic et al. (eds.), *Computational Medicine in Data Mining and Modeling*,
DOI 10.1007/978-1-4614-8785-2_8, © Springer Science+Business Media New York 2013

part of a blood vessel and release their payload, i.e., drug molecules or smaller particulate formulations specifically designed for further transport through the tumor mass.

In this type of drug delivery systems, it is very important to achieve particle margination, i.e., the capability to move towards the endothelium and sense the mentioned biological and biophysical diversities. An effective way to stimulate particle margination could be the control of their size, shape, and density [12].

In microcirculation, in proximity to the walls of small vessels, the flow is mainly governed by viscous forces. If there is no effect of gravitational or magnetic forces, light spherical particles tend to move parallel to the walls without crossing the streamlines [13, 14]. On the other hand, heavy particles, made of silica, gold, or iron oxide, would drift laterally under the influence of gravity [15]. However, nonspherical particles have a more complicated behavior. Hence, the conclusion was made that both wall proximity and particle inertia have a dramatic influence on particle dynamics [16, 17].

Numerical modeling of particle motion is important since it can facilitate the analysis of influence of various parameters relevant in design of nanoparticles, such as size, shape, and surface characteristics. Finite element method enables the development of adequate particle tracking models. But this method requires very fine meshes and very small time step to obtain precise simulation results. On the other hand, discrete particle methods are suitable since they can provide a more detailed analysis of particle trajectories and interaction forces between fluid and particles, as well as among the particles themselves. One of these discrete methods is the lattice Boltzmann method, which is considered in this chapter.

This chapter is organized as follows: in Sect. 8.2 the basics of lattice Boltzmann method are explained, including theoretical background and implementation details, such as discretization procedure and definition of boundary conditions. Section 8.3 explains the model that was used for simulations of solid–fluid interaction. Examples and results of simulations are the subject of Sect. 8.4. Section 8.5 concludes the chapter.

8.2 Lattice Boltzmann Method

Lattice Boltzmann method belongs to the class of problems named Cellular Automata (CA). This means that the physical system can be observed in an idealized way, so that space and time are discretized, and the whole domain is made up of a large number of identical cells [18]. Special form of CA, the so-called lattice gas automata (LGA) [19], describes the dynamics of particles that move and collide in discrete time-space domain. The advantages of this method are simple implementation, the stability of the solution, easy assignment of boundary conditions, and natural parallelization. However, this method has many drawbacks, like statistical error, special averaging procedures necessary to obtain macroscopic quantities, and others. In order to remove the aforementioned disadvantages of LGA, the new improved method was developed that can simplify the simulations of fluid flow. This method is lattice Boltzmann

(LB) method. The main goal during the development of the new LB method was to create a new improved model that can simplify the simulations of fluid flow. Using LB method, with certain limitations, it is possible to obtain the solution of Navier–Stokes equation, which means that it can be used to simulate fluid flow. Special propagation function is defined, so that it depends on the state of neighboring cells and it has an identical form for all cells. The state of all cells is updated synchronously, through a series of iterations, in discrete time steps. This way a greater numerical accuracy and efficiency is obtained in LB method.

8.2.1 Theoretical Background

The Boltzmann equation is a partial differential equation that describes the behavior and movement of particles in space and is valid for continuum. The basic quantity in Boltzmann equation is single distribution function f. The distribution function is defined in such a way that $f(\mathbf{x},\mathbf{v},t)$ represents the probability for particles to be located within a space element $d\mathbf{x}\, d\mathbf{v}$ around position $(\mathbf{x},\mathbf{v})$, at time t, where $\mathbf{x}$ and $\mathbf{v}$ are the spatial position vector and the particle velocity vector, respectively.

In the presence of an external force field $\mathbf{g}$, distribution function balance equation – Boltzmann equation – has the following form:

$$\frac{\partial f}{\partial t} + \mathbf{v}\frac{\partial f}{\partial \mathbf{x}} + \frac{\mathbf{g}}{m}\cdot\frac{\partial f}{\partial \mathbf{v}} = \Omega \tag{8.1}$$

where Ω is the collision integral or the collision operator, and it represents the changes in the distribution function due to the interparticle collisions.

The collision operator Ω is quadratic in f and is represented using a very complex expression. Hence, a simplified model is introduced, initially proposed by Bhatnagar, Gross, and Krook [20]. An assumption is made that the effect of the collision between particles is to drive the fluid towards a local equilibrium state. This model is known as the single relaxation time approximation or the Bhatnagar-Gross-Krook (BGK) model. Operator Ω is defined as follows:

$$\Omega = -\frac{1}{\tau}\left(f - f^{(0)}\right) \tag{8.2}$$

where τ is the relaxation time (the average time period between two collisions) and $f^{(0)}$ is the equilibrium distribution function, the so-called Maxwell-Boltzmann distribution function, which is given by:

$$f^{(0)}(\mathbf{x},\mathbf{v},t) = \frac{\rho(\mathbf{x},t)}{\left(2\pi\theta(\mathbf{x},t)\right)^{D/2}}\exp\left(-\frac{\left(\mathbf{u}(\mathbf{x},t)-\mathbf{v}\right)^2}{2\theta(\mathbf{x},t)}\right) \tag{8.3}$$

In this expression D is the number of physical dimensions ($D = 3$ for three-dimensional domain); $\theta = k_B T/m$; $k_B = 1,38 \cdot 10^{-23}\frac{J}{K}$ is the Boltzmann constant;

T is the absolute temperature, expressed in Kelvins (K); m is the particle mass (in the sequel, it is assumed that the mass of a single particle is $m = 1$).

Finally, BGK model of the continuous Boltzmann equation is given by:

$$\frac{\partial f}{\partial t} + \mathbf{v}\frac{\partial f}{\partial \mathbf{x}} + \mathbf{g}\frac{\partial f}{\partial \mathbf{v}} = -\frac{1}{\tau}\left(f - f^{(0)}\right) \tag{8.4}$$

If Eq. (8.4) is transformed according to the procedure described in [21, 22], the familiar Navier–Stokes equation for the incompressible fluid is obtained:

$$\frac{\partial \rho}{\partial t} + \frac{\partial}{\partial \mathbf{x}}(\rho \mathbf{u}) = 0 \tag{8.5}$$

$$\rho\frac{d\mathbf{u}}{dt} + \frac{\partial}{\partial \mathbf{x}}(p\mathbf{I} - 2\mu_S \mathbf{S}) = \rho \mathbf{g} \tag{8.6}$$

where $\mathbf{S}$ is the strain rate tensor and pressure p in LB method is introduced into the system of equations through the ideal gas law:

$$p = \rho\theta = \rho\frac{k_B T}{m} \tag{8.7}$$

In LB simulations there are two important approximations that need to be taken into account. First some new quantities have to be introduced. The characteristic length of the observed domain is denoted with L, and c_s denotes the characteristic speed of particles, which is also often called speed of sound in lattice units. This characteristic speed can be expressed as:

$$c_s \approx \sqrt{k_B \frac{T}{m}} = \sqrt{\theta} \tag{8.8}$$

Mean free path (the average path of a particle between two collisions) is denoted with $l = c_s \tau$. Knudsen number is defined as the ratio between mean free path and characteristic lengthscale of the considered system:

$$Kn = \frac{l}{L} \tag{8.9}$$

The entire approach and BGK model are valid only in the limit of small Knudsen number.

The Mach number is defined as the ratio between the characteristic fluid velocity and the "speed of sound" c_s:

$$Ma = \frac{|\mathbf{u}|}{c_s} \tag{8.10}$$

During the derivation procedure higher order members in some expressions are neglected, due to the introduction of an approximation that LB simulations are performed in the limit of small Mach number. This needs to be taken into account when defining the characteristic fluid velocity in simulations.

8.2.2 Discretization Procedure and Implementation Details

The original BGK Boltzmann equation is continuous in space domain and is related to continuous velocity field. This form is not suitable for numerical implementation. In order to develop a program that numerically solves this equation on a computer, it has to be previously discretized. But, the discretization needs to be conducted carefully, ensuring that the Navier–Stokes equations can still be derived from the newly obtained equations, in order to preserve the possibility to apply this method on fluid flow simulations.

The discretization procedure is conducted in two steps. First, the velocity field is discretized by applying the Gauss-Hermite quadrature rule [23, 24], and all integrals are transformed to weighted sums. Then the obtained equations are discretized in time and space domain by evaluating the integrals using the trapezoidal rule.

For a function $r(\mathbf{v})$, Gaussian quadrature seeks to obtain the best estimate of the integral $\int \omega(\mathbf{v}) r(\mathbf{v}) d\mathbf{v}$, by choosing the optimal set of abscissae $\boldsymbol{\xi}_i$, $i = 1, 2, \ldots, q - 1$, such that:

$$\int \omega(\mathbf{v}) r(\mathbf{v}) d\mathbf{v} \cong \sum_{i=1}^{n} \omega_i r(\boldsymbol{\xi}_i) \tag{8.11}$$

where ω_i, $i = 1, 2, \ldots, n$ is a set of constant weight coefficients.

It is an interesting fact that the discretization with a relatively small number of abscissae – 9 for two-dimensional domain (denoted by D2Q9) and 27 for three-dimensional domain (denoted by D3Q27) – is enough to correctly describe dynamics of isothermal incompressible fluid flow. The value of defined constant c_s depends on the discrete velocity set. For models D2Q9 and D3Q27, it is taken that $c_s^2 = 1/3$.

Figure 8.1 illustrates the layout of velocity abscissae for two mentioned cases. These directions are in the same time the directions of the distribution function. Coordinates of unit vectors of velocity abscissae and appropriate weight coefficients are listed in Table 8.1 (for D2Q9 case) and Table 8.2 (for D3Q27 case).

Finally, the equation that represents the LB numerical scheme and that is used in all the solvers based on LB method is given by:

$$f_i(\mathbf{x} + \boldsymbol{\xi}_i, t + 1) - f_i(\mathbf{x}, t) = -\frac{1}{\overline{\tau}} \left(f_i(\mathbf{x}, t) - f_i^{eq}(\mathbf{x}, t) \right) + \left(1 - \frac{1}{2\overline{\tau}} \right) \mathbf{F}_i \tag{8.12}$$

where $\overline{\tau}$ is the modified relaxation time (given by $\overline{\tau} = \tau + \frac{1}{2}$) that was introduced to provide better numerical stability of the solution and to enable explicit time steps

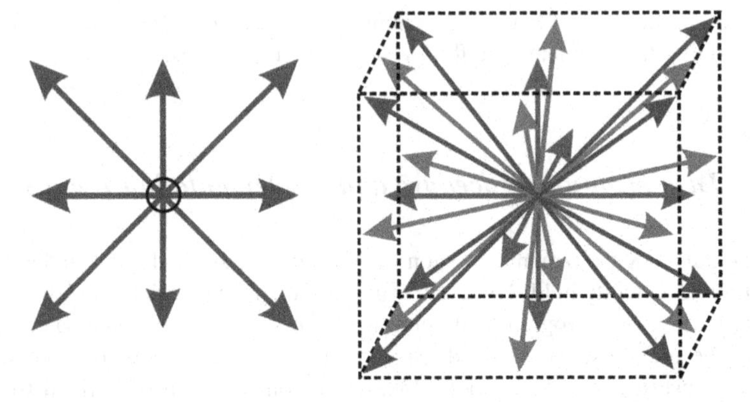

Fig. 8.1 Layout of velocity abscissae for two-dimensional D2Q9 lattice mesh (*left*) and three-dimensional D3Q27 lattice mesh (*right*)

Table 8.1 Velocity abscissae and weight coefficients for D2Q9 lattice mesh

Direction index	Weight coefficient	Unit vector
$i = 0$	$\omega_i = 4/9$	$\boldsymbol{\xi}_i = (0,0)$
$i = 1, 2, 3, 4$	$\omega_i = 1/9$	$\boldsymbol{\xi}_i = (\pm 1, 0)$; $\boldsymbol{\xi}_i = (0, \pm 1)$
$i = 5, 6, 7, 8$	$\omega_i = 1/36$	$\boldsymbol{\xi}_i = (\pm 1, \pm 1)$

Table 8.2 Velocity abscissae and weight coefficients for D3Q27 lattice mesh

Direction index	Weight coefficient	Unit vector
$i = 0$	$\omega_i = 8/27$	$\boldsymbol{\xi}_i = (0,0,0)$
$i = 1, \ldots, 6$	$\omega_i = 2/27$	$\boldsymbol{\xi}_i = (\pm 1, 0, 0)$; $\boldsymbol{\xi}_i = (0, \pm 1, 0)$; $\boldsymbol{\xi}_i = (0, 0, \pm 1)$
$i = 7, \ldots, 14$	$\omega_i = 1/216$	$\boldsymbol{\xi}_i = (\pm 1, \pm 1, \pm 1)$
$i = 15, \ldots, 26$	$\omega_i = 1/54$	$\boldsymbol{\xi}_i = (\pm 1, \pm 1, 0)$; $\boldsymbol{\xi}_i = (0, \pm 1, \pm 1)$; $\boldsymbol{\xi}_i = (\pm 1, 0, \pm 1)$

and thus a more appropriate time discretization. In numerical simulations, the whole calculation process is conducted using the modified relaxation time. In the Eq. (8.12) $\mathbf{F}_i$ represents the discretized force term, expressed as:

$$\mathbf{F}_i = \omega_i \rho \left(\frac{\boldsymbol{\xi}_i - \mathbf{u}}{c_s^2} - \frac{(\boldsymbol{\xi}_i \cdot \mathbf{u}) \cdot \boldsymbol{\xi}_i}{c_s^4} \right) \cdot \mathbf{g} \tag{8.13}$$

When LB method is implemented on a computer, Eq. (8.12) is solved in two steps – collision and propagation step. Two values of the distribution function can be defined, f_i^{in} and f_i^{out}, that represent the values of the discretized distribution function before and after the collision, respectively. The mentioned steps are expressed as follows:

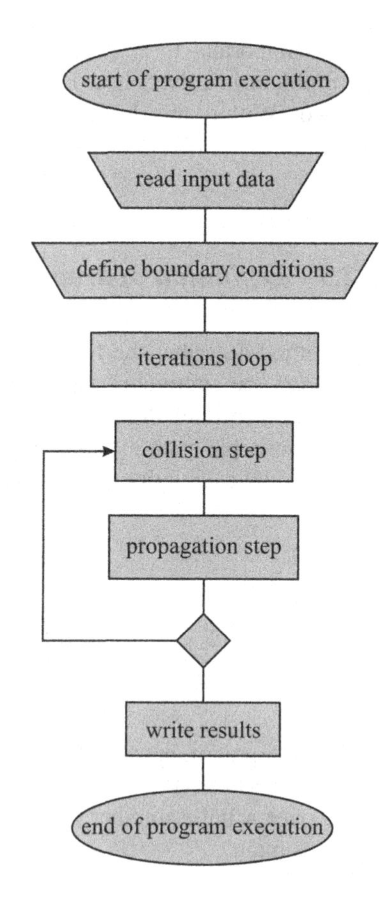

Fig. 8.2 Algorithm of execution of the program based on LB method

Collision step:

$$f_i^{out}(\mathbf{x},t) = f_i^{in}(\mathbf{x},t) - \frac{1}{\tau}\left(f_i^{in}(\mathbf{x},t) - f_i^{(0)}(\rho,\mathbf{u})\right) + \left(1 - \frac{1}{2\tau}\right)\mathbf{F}_i \tag{8.14}$$

Propagation step:

$$f_i^{in}(\mathbf{x} + \boldsymbol{\xi}_i, t+1) = f_i^{out}(\mathbf{x},t) \tag{8.15}$$

Each one of these steps must be applied to the whole system (to all particles) before the next one starts. The step that considers collisions is a completely local operation, and therefore, the equilibrium distribution function is calculated in every lattice cell individually, in every time step, and in terms of density ρ and macroscopic velocity $\mathbf{u}$. The step that considers propagation communicates with only few closest neighbors.

The program for the simulation of fluid flow, based on LB method is written in programming language C++. The algorithm of the program is schematically shown in Fig. 8.2.

It is important to emphasize that there must exist a loop of iterations, since the problem is solved in a large number of time steps, until the steady state is reached. Within this loop two steps defined with Eqs. (8.14) and (8.15) are carried out.

8.2.3 Definition of Macroscopic Quantities

LB method describes the fluid on a molecular level, and the characteristics of the fluid from the continuum aspect are implicitly contained in the model. The basic quantity determined in LB simulation (the distribution function) is used to calculate the macroscopic quantities that describe the fluid flow. Density and velocity can be evaluated by calculating the following integrals:

$$\rho(x,t) = \int_{\mathbf{v}} f(\mathbf{x},\mathbf{v},t)d\mathbf{v} \tag{8.16}$$

$$\mathbf{u}(\mathbf{x},t) = \frac{1}{\rho(\mathbf{x},t)}\int_{\mathbf{v}} f(\mathbf{x},\mathbf{v},t)\mathbf{v}d\mathbf{v} \tag{8.17}$$

The integration is carried out on the whole velocity space.

After discretization, fluid density and velocity can be calculated as weighted sums over a finite number of discrete abscissae that were used to discretize the space domain:

$$\rho = \sum_{i=0}^{q-1} f_i \tag{8.18}$$

$$\overline{\mathbf{u}} = \frac{1}{\rho}\sum_{i=0}^{q=1} \xi_i f_i = \frac{1}{\rho}(\rho\mathbf{u} - \rho\mathbf{g}/2) \tag{8.19}$$

From this equation it can be seen that the velocity that is calculated in terms of distribution function in LB simulations does not represent the "physical velocity." In order to evaluate the physical velocity (which is one of the most important characteristics of the flow observed on the macroscopic level), it is necessary to use the following expression:

$$\mathbf{u} = \overline{\mathbf{u}} + \frac{\mathbf{g}}{2} \tag{8.20}$$

This should be taken into consideration when analyzing the simulation results and when developing solvers based on LB method.

Dynamic viscosity in LB simulations (called "lattice viscosity") is calculated as follows:

$$\mu_s = c_s^2 \rho \left(\overline{\tau} - \frac{1}{2} \right) \tag{8.21}$$

The equation that defines the relation between fluid pressure and fluid density is given by:

$$p = c_s^2 \rho \theta = c_s^2 \rho \frac{k_B T}{m} \tag{8.22}$$

Stress tensor can be calculated using expression:

$$\mathbf{P} = c_s^2 \rho \mathbf{I} - \frac{\left(\overline{\tau} - \frac{1}{2} \right)}{c_s^4} \mathbf{S} \tag{8.23}$$

8.2.4 Boundary Conditions

All previous derivations did not take into account the boundary conditions (BC). Still, in order to obtain valid results and to correctly simulate fluid flow, it is necessary to define the boundary conditions appropriately. Typical types of boundary conditions (periodical, bounce-back, and predefined pressure and velocity Dirichlet BC) that are used to set up simulations in examples section will be briefly explained in the sequel of this chapter.

8.2.4.1 Periodical Boundary Condition

This is the simplest type of boundary condition. Practically, one can observe the boundary as if the inlet and the outlet are joined together. In the practical implementation, this BC is implemented within the propagation step. For all the nodes that are on the boundary of the domain, the components of the distribution function that should propagate outside of the domain boundary are being "redirected" such that these values are transferred to the nodes that are located on the other (opposite) boundary of the domain.

8.2.4.2 Bounce-Back Boundary Condition

This boundary condition is very simple for implementation and that is one of the reasons for its wide popularity. It is most commonly applied when solving problems with complex boundaries, such as the flow through a porous media [25]. If a certain node of the mesh is marked as a solid node, i.e., as an obstacle, the components of the distribution function in this node are copied from the components with opposite abscissae unit vectors.

However, bounce-back boundary condition has its drawbacks that are discussed in literature [26, 27]. The main drawback is that in case of a wrong implementation, some instabilities and the misbalance in the continuity equation may occur. A detailed analysis showed that applying this BC the second-order accuracy is achieved but only when the boundary between solid and fluid domain is located between the nodes of the mesh [28]. If the solid–fluid boundary is made of straight lines, then it is desirable to use this approach. But if complex geometries with curved boundaries are considered or when solid is placed inside fluid domain and solid–fluid interaction is simulated, it is more efficient to use a different approach, such as immersed boundary method, which will be discussed in the next chapter.

8.2.4.3 Pressure and Velocity Boundary Conditions

When the boundary conditions are observed in general, after the propagation step, those nodes that are on the domain boundary will contain certain information about the distribution function that are incoming from the wall, i.e., that are nonphysical. The main objective of the propagation is to transfer the information from one node to its closest neighboring nodes. Therefore it is evident that this information for boundary nodes should be obtained from a node inside the wall. Since the nodes inside the wall are not simulated, missing distribution function components must be recomputed using a different approach. Also it is necessary to keep in mind that the entire concept of derivation of Navier–Stokes equations must be valid in the whole system, i.e., in all nodes, including boundary nodes. The discussion about conserving the continuity equation can be found in literature [29].

When the simulations of fluid flow are performed, it is most common to define the value of velocity and pressure (that is the so-called Dirichlet boundary condition) or to define the derivatives of these quantities, i.e., the fluxes of certain quantities (that is the so-called Neumann boundary condition). It should be noted that in LB method the density is defined, instead of pressure, since these two quantities are related with the equation of state (8.22). Using these macroscopic values, it is possible to calculate the missing components of the distribution function coming from the wall.

There are several types of velocity and pressure BCs that were proposed in literature – the boundary condition proposed by Inamuro et al. [27], the Zou/He approach [30], the regularized method [31], and the finite difference method, based on an idea of Skordos [32]. In this implementation of LB method, the regularized boundary condition was used. It is assumed that the pressure or velocity is directly prescribed, i.e., the Dirichlet boundary condition is considered. Of course, since only one macroscopic quantity is predefined, the relation between two quantities – velocity and pressure – is also determined according to literature [22, 30]. The recalculation of the distribution function is performed, and afterwards both the collision and propagation steps are performed on all lattice nodes, including boundary nodes. It was only necessary to correct the unknown components that were incoming from the nodes inside the wall. Obviously the implementation of this boundary condition accurately recovers not only the velocity and density but also the stress tensor.

8.3 Modeling Solid–Fluid Interaction

There are certain types of problems that require the simulation of two or more physical systems that are in interaction. One of them is in the field of fluid flow simulation, and it is analyzed in this chapter – particles (regarded as solid bodies) moving through a fluid domain. In this case an external force exerted from the fluid is acting on the solid, causing solid movement or deformations and vice versa – solid is having certain influence on the fluid flow. That practically means that solid and fluid are forming a coupled mechanical system. In order to simulate a system like this, it is necessary to simulate both domains simultaneously, i.e., to model solid–fluid interaction. There are two approaches in modeling solid–fluid interaction: loose and strong coupling. In loose coupling approach the solid and fluid domains are solved separately, and all the necessary parameters obtained in one solver are passed to the solver for the other domain. In strong coupling both domains are simulated in the same time, like it were a single mechanical system. For certain problems it is easier to use loose coupling, due to easier implementation. However there are some drawbacks of this approach, like the problem of time integration. Since the physical characteristics of solid and fluid are different, it is not always possible to use the same time step for numerical solving. On the other hand, strong coupling is more applicable when it is necessary to accurately and precisely predict the movement of solid body inside fluid domain. But this approach also has its drawbacks. The solver that simultaneously solves both domains is much more complex and slower, which was expected considering the increased number of equations in the system. Examples in this chapter were simulated using strong coupling approach, and therefore, in the sequel of this section, theoretical basics of this approach will be discussed.

The basic idea of full interaction approach is to solve the complete domain (both fluid flow and particle motion) in every time step. This will provide that all quantities (both for particles and fluid) are changing simultaneously. The approach that was successfully applied for problems of particle movement through fluid domain [33] is used here to simulate motion of nanoparticles together with fluid. This method is called immersed boundary method (abbreviated IBM), and it was first developed and presented by Peskin [34]. This method uses a fixed Cartesian mesh to represent the fluid domain, so that the fluid mesh is composed of Eulerian points. This description is in accordance with LB representation of fluid domain, so it is evident that this approach can be easily applied if fluid flow is simulated using LB method. As far as particles (solid bodies in general definition of IBM) are concerned, IBM represents solid body as an isolated part of fluid, with a boundary represented by a set of Lagrangian points. The basic idea is to treat the physical boundary between two domains as deformable with high stiffness [35]. Fluid is acting on the solid, i.e., on the boundary surface, through a force that tends to deform the boundary. However, in the bounding area this deformation yields to a

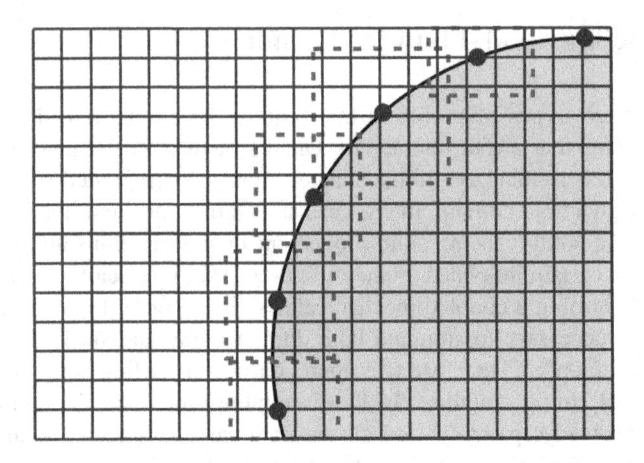

Fig. 8.3 The mutual influence of points from solid and fluid domain

force that tends to restore the boundary to its original shape. These two forces have to be in equilibrium. Practically, using the law of action and reaction (Newton's third law), the force exerted from the fluid and acting on the solid is acting on the fluid near the boundary too and is distributed through a discrete delta function. The entire solid–fluid domain is solved using Navier–Stokes equations, with external force term. There are several ways to determine this force representing the interaction between solid and fluid. Some of them are direct forcing term [36], enhanced version of direct forcing scheme [37], penalty formulation [33], momentum exchange method [38], and calculation based on velocity correction [39]. The latter is used in this study to simulate the full interaction between fluid and nanoparticles.

The following quantities are defined: $\mathbf{X}_B^l(s,t)$ represent the coordinates of Lagrangian boundary points; $l = 1, 2,.., m$, where m is the number of boundary points; $\mathbf{F}(s,t)$ is the boundary force density, exerted from the fluid, acting on the immersed object; $\mathbf{g}(\mathbf{x},t)$ is the fluid external force density; and $\delta(\mathbf{x} - \mathbf{X}_B^l(s,t))$ is a Dirac delta function.

All quantities that are required to simulate solid–fluid interaction are calculated using interpolation from the boundary points. Since different discretization of solid and fluid domain is possible, the problem of interpolation with diverse discretization has to be considered. This is solved such that for each boundary point belonging to the solid, the influence of a greater number of points from the fixed Cartesian mesh from fluid domain is considered, like it is shown in Fig. 8.3. For the interpolation of quantities, the Dirac delta function is used. This function is approximated using the following expression:

$$\delta(\mathbf{x} - \mathbf{X}_B(s,t)) = D_{ij}\left(\mathbf{x}_{ij} - \mathbf{X}_B^l\right) = \delta\left(x_{ij} - X_B^l\right)\delta\left(y_{ij} - Y_B^l\right) \tag{8.24}$$

The function $\delta(r)$ is defined in literature [34] as:

$$\delta(r) = \begin{cases} \frac{1}{4h}\left(1 + \cos\left(\frac{\pi r}{2}\right)\right), & |r| \leq 2 \\ 0, & |r| > 2 \end{cases} \tag{8.25}$$

where h is the mesh spacing between two nodes (points) of Eulerian fixed mesh (in this study, since LB method is used, $h = 1$).

The velocity in Lagrangian points is interpolated from the surrounding Eulerian points in fluid, and this can be expressed as:

$$\frac{\partial \mathbf{X}_B(s,t)}{\partial t} = \mathbf{u}(\mathbf{X}_B(s,t),t) = \int_{\Omega} \mathbf{u}(\mathbf{x},t)\delta(\mathbf{x} - \mathbf{X}_B(s,t))d\mathbf{x} \tag{8.26}$$

In IBM in every time step, the fluid and solid velocities have to be adjusted so that the nonslip condition is satisfied on the boundary. The approach used in this study ensures the equality of velocities by introducing a fluid velocity correction in boundary points, like it was proposed in literature [39]. The velocity correction can be expressed as:

$$\delta\mathbf{u} = \frac{1}{2\rho}\mathbf{g}(\mathbf{x},t) \tag{8.27}$$

The fluid velocity correction is set as unknown quantity that is determined based on the solid velocity correction at boundary points:

$$\delta\mathbf{u}(\mathbf{x},t) = \int_{\Gamma} \delta\mathbf{u}_B(\mathbf{X}_B,t)\delta(\mathbf{x} - \mathbf{X}_B(s,t))ds \tag{8.28}$$

In this equation $\delta\mathbf{u}_B(\mathbf{X}_B,t)$ represents the solid velocity correction at boundary points.

If two-dimensional problem is considered, using Eq. (8.24), the fluid velocity correction can be written as:

$$\delta\mathbf{u}(\mathbf{x}_{ij},t) = \sum_l \delta\mathbf{u}_B^l(\mathbf{X}_B^l,t)D_{ij}(\mathbf{x}_{ij} - \mathbf{X}_B^l)\Delta s_l \tag{8.29}$$

where Δs_l is the arc length of the boundary element (between two boundary points).

In order to satisfy the nonslip boundary condition, the fluid velocity at every boundary point must be equal to the velocity of the immersed body in that boundary point.

If the following changes of variables are introduced (to simplify the expressions):

$$\delta_{ij}^B = D_{ij}(\mathbf{x}_{ij} - \mathbf{X}_B^l)\Delta s_l \tag{8.30}$$

$$\delta_{ij} = D_{ij}(\mathbf{x}_{ij} - \mathbf{X}_B^l)\Delta x\Delta y \tag{8.31}$$

then velocity corrections can be calculated solving the following equation written in matrix form:

$$\mathbf{A} \cdot \mathbf{X} = \mathbf{B} \tag{8.32}$$

where the matrices are given by:

$$\mathbf{X} = \left\{ \delta\mathbf{u}_B^1, \delta\mathbf{u}_B^2, \ldots, \delta\mathbf{u}_B^m \right\}^T \tag{8.33}$$

$$\mathbf{A} = \begin{bmatrix} \delta_{11} & \delta_{12} & \cdots & \delta_{1n} \\ \delta_{21} & \delta_{22} & \cdots & \delta_{2n} \\ \vdots & \vdots & \ddots & \vdots \\ \delta_{m1} & \delta_{m2} & \cdots & \delta_{mn} \end{bmatrix} \cdot \begin{bmatrix} \delta_{11}^B & \delta_{12}^B & \cdots & \delta_{1m}^B \\ \delta_{21}^B & \delta_{22}^B & \cdots & \delta_{2m}^B \\ \vdots & \vdots & \ddots & \vdots \\ \delta_{n1}^B & \delta_{n2}^B & \cdots & \delta_{nm}^B \end{bmatrix} \tag{8.34}$$

$$\mathbf{B} = \begin{pmatrix} \mathbf{u}_B^1 \\ \mathbf{u}_B^2 \\ \vdots \\ \mathbf{u}_B^m \end{pmatrix} - \begin{bmatrix} \delta_{11} & \delta_{12} & \cdots & \delta_{1n} \\ \delta_{21} & \delta_{22} & \cdots & \delta_{2n} \\ \vdots & \vdots & \ddots & \vdots \\ \delta_{m1} & \delta_{m2} & \cdots & \delta_{mn} \end{bmatrix} \begin{pmatrix} \overline{\mathbf{u}}_1 \\ \overline{\mathbf{u}}_2 \\ \vdots \\ \overline{\mathbf{u}}_n \end{pmatrix} \tag{8.35}$$

In these expressions $\overline{\mathbf{u}}$ is the fluid velocity calculated from LB simulation ($\overline{\mathbf{u}} = \frac{1}{\rho} \sum_i \xi_i f_i$), m is the number of Lagrangian (boundary) points, and n is the number of surrounding Eulerian points that are used in the Dirac delta interpolation function.

Solving the system of Eq. (8.32), the unknown solid velocity corrections in boundary points $\delta\mathbf{u}_B^l$ are calculated. Using these values, it is straightforward to calculate the fluid velocity corrections, applying Eq. (8.29) and then to calculate the total fluid velocity.

Using velocity corrections, it is also easy to evaluate the forces exerted from the fluid, acting on the immersed body (using Eq. (8.27) and the third Newton law). The force exerted from the fluid that is acting on the immersed body in one boundary Lagrangian point is given by:

$$\mathbf{F}\left(\mathbf{X}_B^l\right) = 2\rho\delta\mathbf{u}_B^l \tag{8.36}$$

The overall influence of the fluid on the immersed body is expressed through a force and torque that are given by:

$$\mathbf{F}_R = -\sum_l \mathbf{F}\left(\mathbf{X}_B^l\right)\Delta s_l \tag{8.37}$$

$$\mathbf{M}_R = -\sum_l \left(\mathbf{X}_B^l - \mathbf{X}_R\right) \times \mathbf{F}\left(\mathbf{X}_B^l\right)\Delta s_l \tag{8.38}$$

where $\mathbf{X}_R$ is the vector of coordinates of the center of mass of the observed immersed object. The minus sign in both equations is the consequence of the Newton's third law (the law of action and reaction).

If it is necessary to consider the influence of gravity force on the solid object, another term is added in the expression for total force:

$$\mathbf{F}_R = \left(1 - \frac{\rho_f}{\rho_p}\right) m\mathbf{G} - \sum_l \mathbf{F}(\mathbf{X}_B^l)\Delta s_l \tag{8.39}$$

Here m is the mass of solid object (particle), $\mathbf{G}$ is the gravity force acceleration, and ρ_f and ρ_p are the fluid and particle density, respectively.

When force and torque acting on the particle are known, the equations of motion can be written for the immersed body:

$$m \frac{d\mathbf{v}_C}{dt} = \mathbf{F}_R \tag{8.40}$$

$$I_0 \frac{d\boldsymbol{\omega}}{dt} = \mathbf{M}_R \tag{8.41}$$

where m and I_0 are the particle mass and moment of inertia, respectively, and $\mathbf{v}_C$ and $\boldsymbol{\omega}$ are the velocity and angular velocity of the particle, respectively.

Integrating these equations the position and orientation of particle are obtained. Using this data the position of particle inside fluid domain is updated, and the entire procedure is repeated until particle collides with one of the boundaries of the domain. Figure 8.4 shows a schematic diagram of the strong coupling approach in modeling nanoparticles motion using immersed boundary method.

8.4 Numerical Results

Specialized and in-house developed software that numerically simulates fluid flow using the principles of LB method is used to simulate several problems of particle movement through fluid domain. Obtained results were compared with the analytical solutions or with results obtained using other methods that were found in literature.

All examples are related to two-dimensional (plane) problems and that is why the lattice denoted by D2Q9 was used for the simulations. For every example several characteristic quantities need to be defined. First, it is necessary to define resolution N that represents the number of nodes along one direction (most commonly it is the y axis direction). The domain is defined using characteristic lengths l_x and l_y, while "physical" lengths, denoted by L_x and L_y, are calculated as:

$$L_x = l_x \cdot N \tag{8.42}$$

$$L_y = l_y \cdot N \tag{8.43}$$

Figure 8.5 shows one domain, and the coordinate axes and mentioned lengths are labeled.

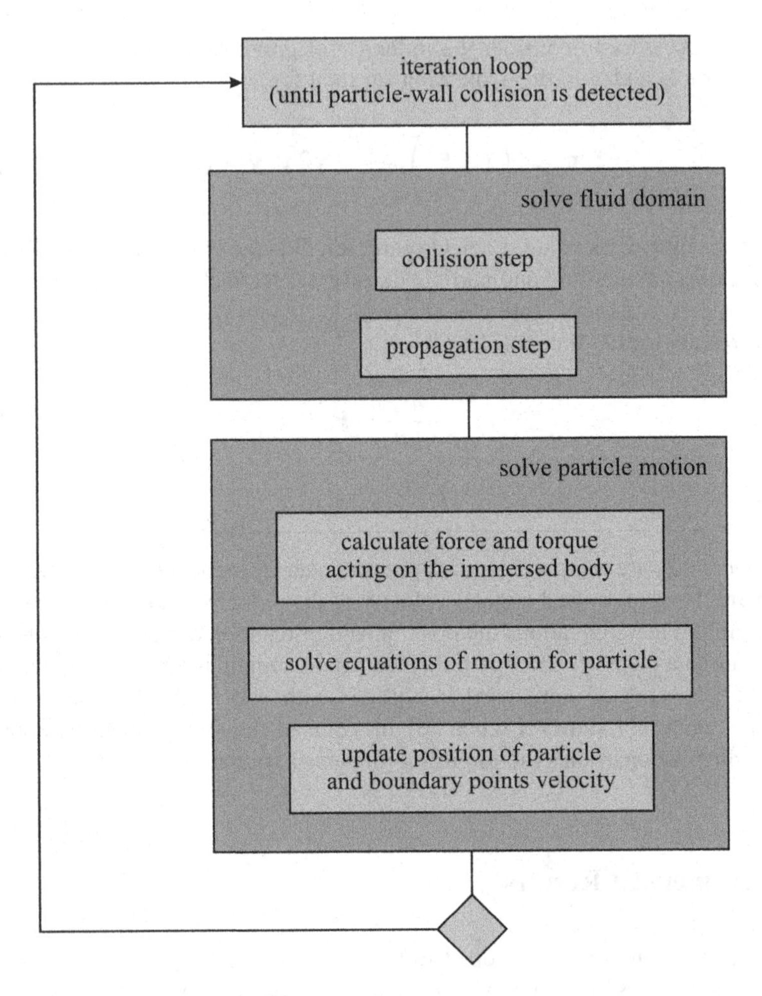

Fig. 8.4 Algorithm of strong coupling approach

In LB method a special calculation principle is used, such that all macroscopic quantities that are important for fluid flow simulation are defined in a so-called system of lattice units. Therefore it is first necessary to evaluate the values of these quantities in a dimensionless form and then to transform the dimensionless quantities to the real physical quantities for a specific example.

When defining characteristic quantities, it is important to define also the characteristic velocity (in lattice units), denoted by u_{lb}. The appropriate selection of this value ensures the satisfaction of condition to keep the LB simulations in the limit of

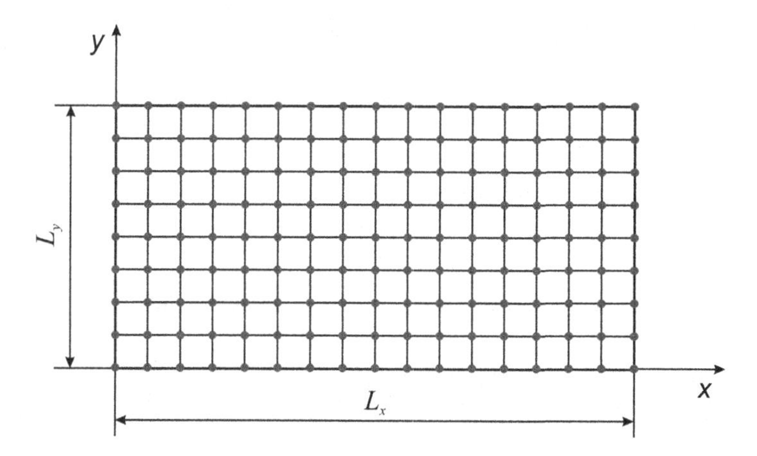

Fig. 8.5 Lattice mesh – two-dimensional domain

small Mach number, i.e., the fluid is retained weakly compressible. This practically means that the value of u_{lb} must satisfy the following inequation:

$$u_{lb} < \frac{1}{c_s} \tag{8.44}$$

The relation between kinematic fluid viscosity and Reynolds number is given by:

$$\mathrm{Re} = \frac{|\mathbf{u}|L}{\nu} \tag{8.45}$$

Most commonly in LB simulations, the Reynolds number is prescribed, and then using Eq. (8.45), the fluid viscosity in system of lattice units is evaluated, in order to obtain the relaxation time.

Besides standard parameters that should be defined in LB simulations, when the strong coupling approach is used, it is necessary to define the number of Lagrangian boundary points, denoted by N_L.

8.4.1 Drag Force on a Circular Particle

A circular particle is fixed inside fluid domain. The initial and boundary conditions are defined like in the previous example. The position of the particle is defined with coordinates of the center of mass (x_c,y_c). The geometrical data is shown in Fig. 8.6. The boundary condition is the prescribed velocity on the bottom and top wall equal to zero (the velocity is prescribed using the regularized BC). As initial condition the velocity profile and the outflow condition on the outlet is prescribed. The main goal is to calculate the drag force over the particle, exerted from the fluid.

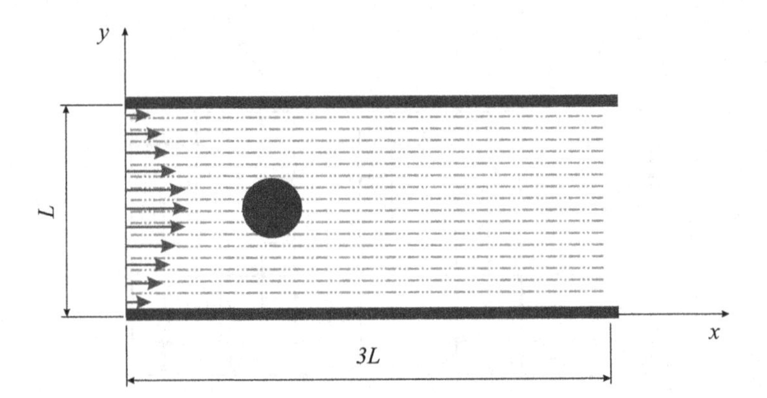

Fig. 8.6 Example 1 – geometrical data

Table 8.3 Data necessary for LB solver; Example 1

Quantity	Value
N	50
l_x	3
l_y	1
u_{lb}	0.02
Re	10
x_c	25
y_c	25
N_L	50

Drag force over the circular particle, in a laminar flow has been calculated analytically [40] and is given by:

$$F = 6\pi\mu RV\left(1 + \frac{3}{8}\mathrm{Re}\right) \tag{8.46}$$

where R is the radius of the particle.

Data defined in LB solver is shown in Table 8.3.

Figure 8.7 shows the fluid velocity field obtained using numerical simulation.

Results obtained using LB solver are compared to the mentioned analytical solution, as well as with results obtained using DPD method (abbreviated from Dissipative Particle Dynamics) [41, 42]. The comparison of results is shown in Fig. 8.8.

It should be noted that the dependent variable on this diagram is given by:

$$\Omega = F/6\pi\mu V\left(1 + \frac{3}{8}\mathrm{Re}\right) \tag{8.47}$$

As it can be seen from Eq. (8.46), this ratio is equal to the particle radius R.

Good agreement of results with the analytical solution is obtained, for different values of the particle radius.

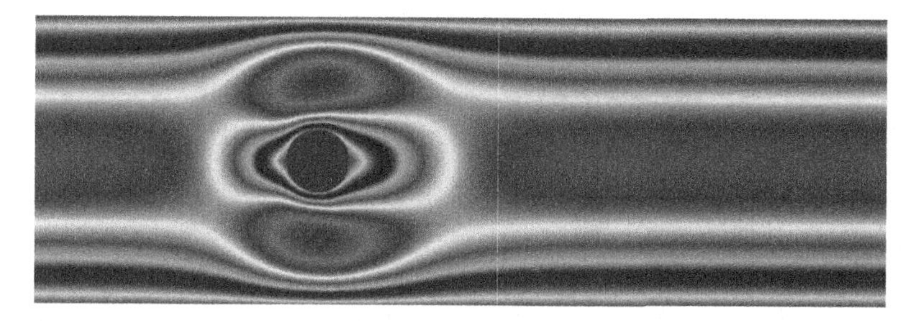

Fig. 8.7 Example 1 – fluid velocity field

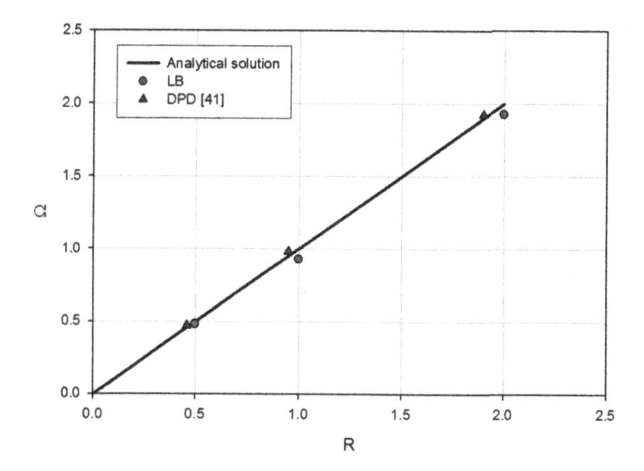

Fig. 8.8 Example 1 – comparison of results for drag force

8.4.2 Pure Rotation of Elliptical Particle

Elliptical particle is immersed in the fluid domain. As initial condition in the fluid domain, the velocity profile is prescribed on the inlet. The boundary condition is set such that the velocity of the lower wall (for $y = 0$) is equal to zero, while the velocity of the upper wall (for $y = L_y$) is equal to u_{lb} – forming the so-called shear flow. Position of the particle is defined with coordinates of the center of mass (x_c,y_c), and since here elliptical particle is considered, the major and minor semi-lengths a and b are also defined. The particle is fixed (it cannot move freely through the fluid domain) but is free to rotate. Since in this example it is necessary to solve the equations of motion for the particle, it is necessary to define the particle density when defining simulation parameters. Using particle density, the mass and moment of inertia of the particle are straightforwardly calculated. Geometrical data for this example is shown in Fig. 8.9.

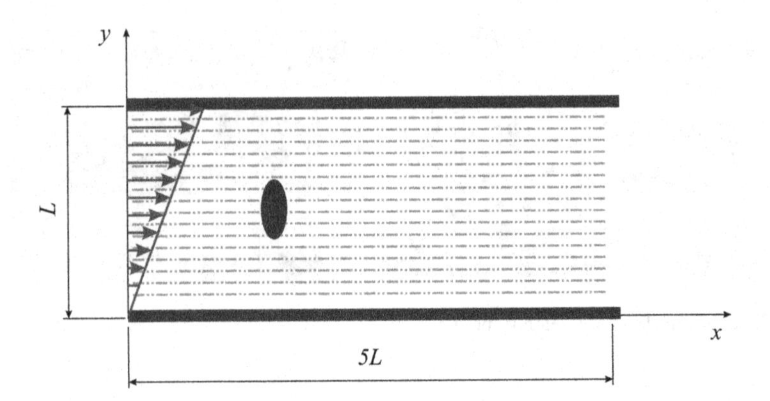

Fig. 8.9 Example 2 – geometrical data

Table 8.4 Data necessary for LB solver; Example 2

Quantity	Value
N	50
l_x	5
l_y	1
u_{lb}	0.01
Re	10
ρ_p	1.25 ρ
x_c	25
y_c	25
a	10
N_L	50

Ratio of particle semi-lengths is given by:

$$\gamma = \frac{b}{a} = 0.5 \tag{8.48}$$

The initial angle between the major semiaxis and vertical axis is equal to $\frac{\pi}{2}$.

Data defined in LB solver is shown in Table 8.4.

Figure 8.10 shows the obtained fluid velocity field.

The analytical solution for this problem was determined by Jeffery [43]. Angular velocity ω, expressed in terms of the angle of rotation θ, is given by:

$$\omega = \frac{1}{2}\left(1 + \frac{1-\gamma^2}{1+\gamma^2}\cos(2\theta)\right) \tag{8.49}$$

The angle of rotation can be expressed as:

$$\theta = \arctan\left(\frac{1}{\gamma}\tan\left(\frac{\gamma t}{1+\gamma^2}\right)\right) \tag{8.50}$$

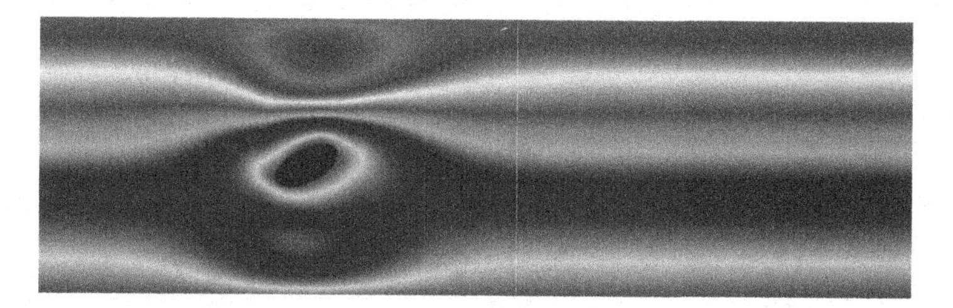

Fig. 8.10 Example 2 – fluid velocity field

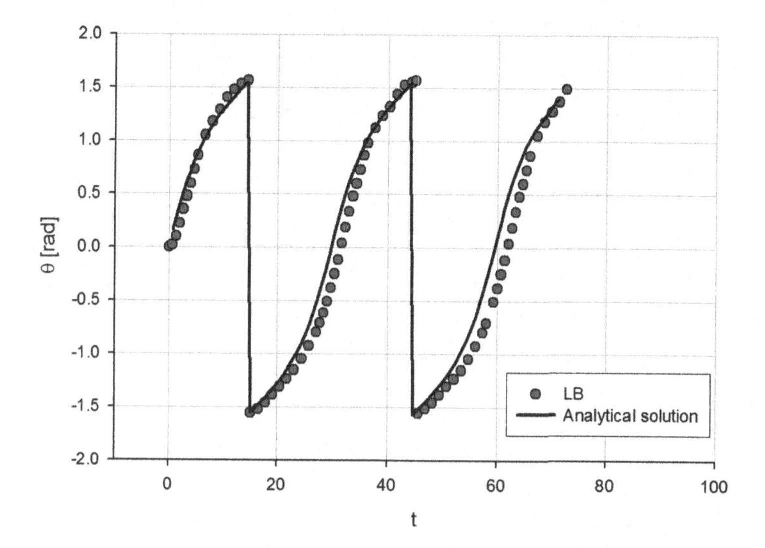

Fig. 8.11 Example 2 – comparison of results for angle of rotation

The variation of angle of rotation θ with time is shown in Fig. 8.11. The analytical solution is compared with the results of numerical simulation, obtained using LB solver. Good agreement of results is achieved.

Figure 8.12 shows the variation of angular velocity with time. The analytical solution is compared with the results obtained using LB solver and results obtained using DPD method (presented in literature [41, 42]).

8.4.3 *Simulation of Movement of Elliptical Particle*

In this example the movement of elliptical particle in fluid domain is considered. All parameters are identical to the ones in example 2. The boundary and initial conditions are also identical. Figure 8.13 shows the geometrical data for this example. It is important to emphasize that the initial angle between the major semiaxis of the elliptical particle and the vertical axis is $\theta_0 = \frac{\pi}{2}$ and the initial position of the particle along y axis is $y_0 = 1.32a$, where a is the major semi-length of the elliptical particle.

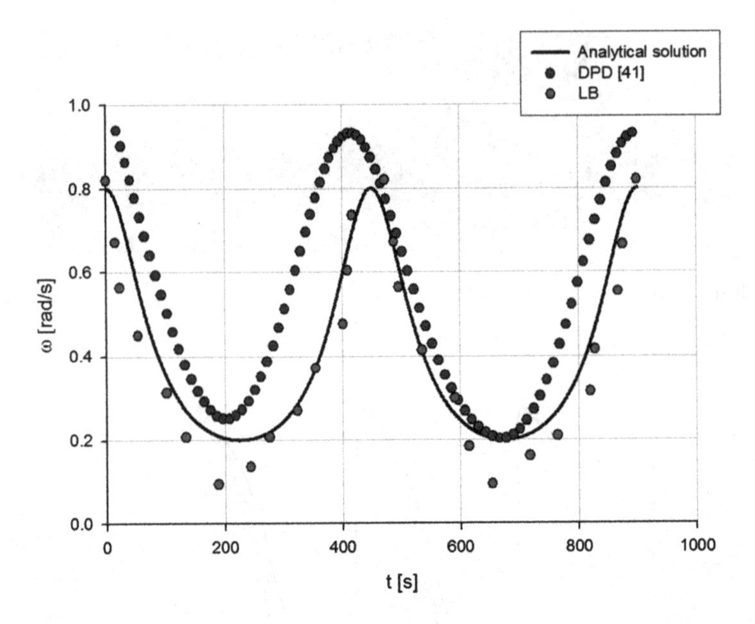

Fig. 8.12 Example 2 – comparison of results for angular velocity

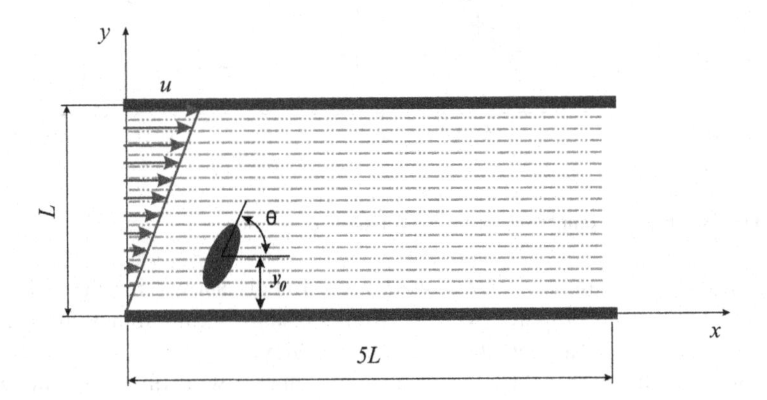

Fig. 8.13 Example 3 – geometrical data

The Stokes number has a great influence on the particle trajectory. Stokes number is defined by:

$$St = \frac{\rho_p b^2 S}{\mu} \tag{8.51}$$

where ρ_p is the particle density.

The relation between Stokes and Reynolds number is given by:

$$St = \frac{\rho_p}{\rho} \mathrm{Re} \tag{8.52}$$

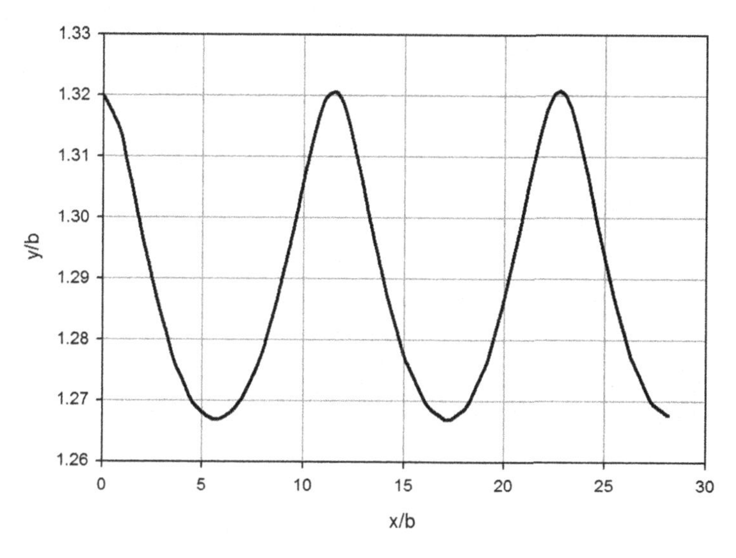

Fig. 8.14 Example 3 – elliptical particle trajectory for St = 1.25

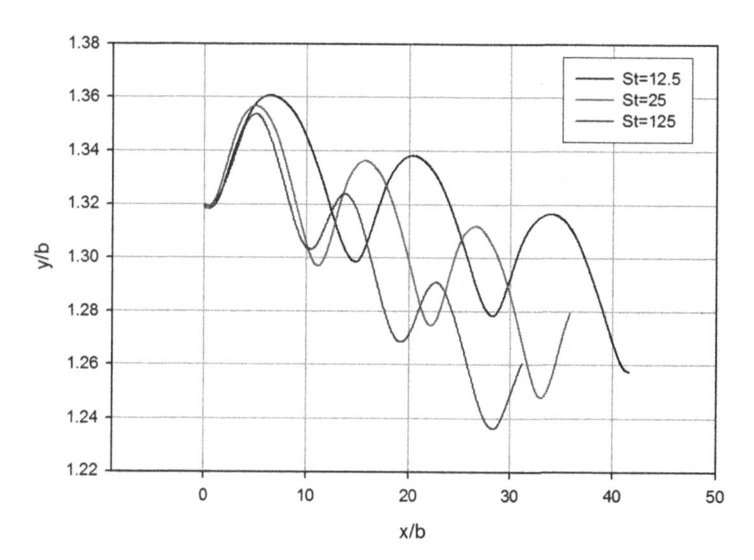

Fig. 8.15 Example 3 – elliptical particle trajectory for diverse values of the Stokes number

When Stokes number is equal to 1.25, the movement of inertialess elliptical particle is oscillatory, with a zero lateral drift, like it is shown in Fig. 8.14. This drift velocity grows with the Stokes number. If Reynolds number and fluid density are kept constant and particle density is varied, particle inertia will break the symmetry of motion and particle will begin to drift laterally, with an oscillatory translational and rotational motion. Figure 8.15 shows the particle trajectories for diverse Stokes number.

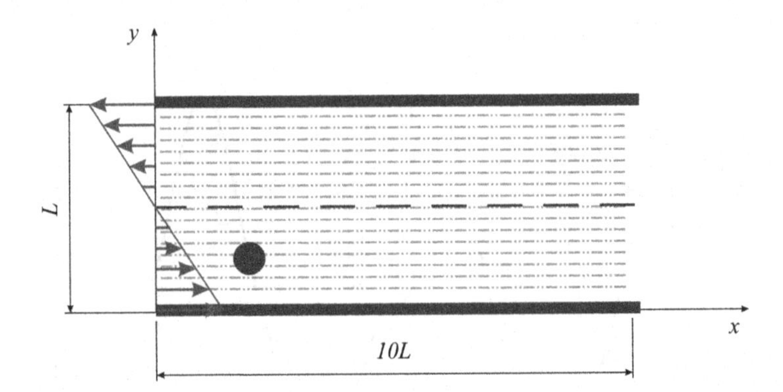

Fig. 8.16 Example 4 – geometrical data

Table 8.5 Data necessary for LB solver; Example 4

Quantity	Value
N	80
l_x	10
l_y	1
u_{lb}	0.0375
Re	40
x_c	50
y_c	20
r	10
N_L	50

In the design of nano-drug delivery systems, these observations should be considered. Both spherical and nonspherical particles with appropriate mass and geometrical inertia will drift on their way through blood vessels and hence will be able to sense abnormalities in vessel walls.

8.4.4 Simulation of Movement of a Circular Particle in Linear Shear Flow

A circular particle is immersed in fluid domain and is free to move. As boundary condition the velocity of upper and lower wall is prescribed to be equal to u_{lb}, and the walls are moving in opposite directions. As initial condition the velocity profile on the inlet is defined, the so-called double shear flow. The geometrical data is shown in Fig. 8.16. In this example it is considered that solid and fluid densities are equal. The initial position of the particle is defined with coordinates of the center of mass (x_c, y_c), and the particle radius r is also defined. It should be noted that the initial position of the particle is defined such that the particle is located at $\frac{L}{4}$ above the bottom wall.

Data defined in LB solver is shown in Table 8.5.

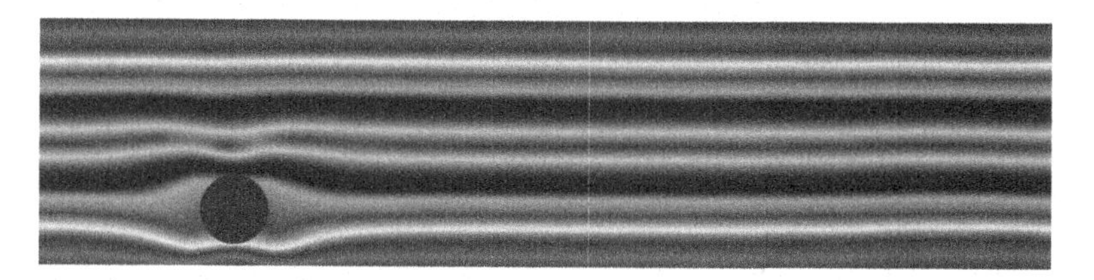

Fig. 8.17 Example 4 – fluid velocity field

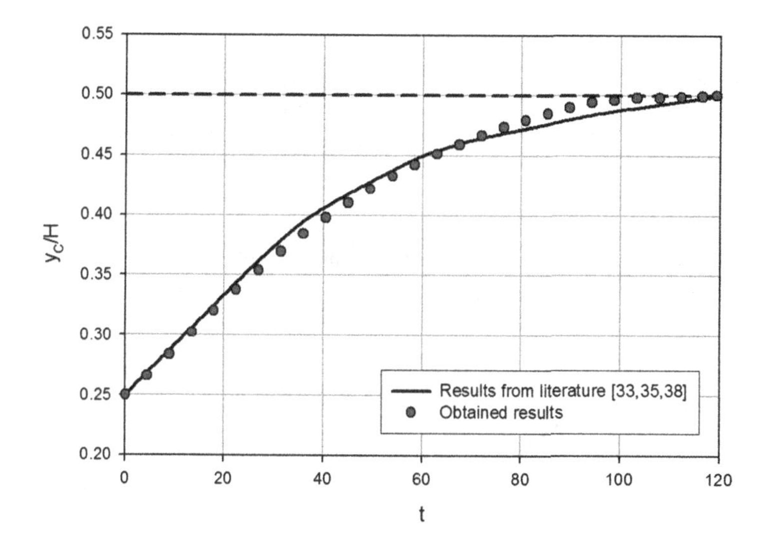

Fig. 8.18 Example 4 – comparison of results for particle position in y axis direction

This problem was first investigated by Feng et al. [44] (and simulated using finite element method), and they concluded that no matter how the particle is initially positioned, it always tends to migrate to the centerline of the channel. Later this same problem has been studied by others and simulated using other methods, including LB method and diverse approaches for solid–fluid interaction. Here the obtained results were compared to the results presented in literature [33, 35, 38].

Figure 8.17 shows the fluid velocity field during particle movement.

Figures 8.18, 8.19, and 8.20 show comparison of results obtained using the developed LB solver with results found in literature, for movement of particle in y axis direction and the components of translational velocities in x and y axis direction. On all three diagrams, the dimensionless values obtained in simulations are shown on both axes.

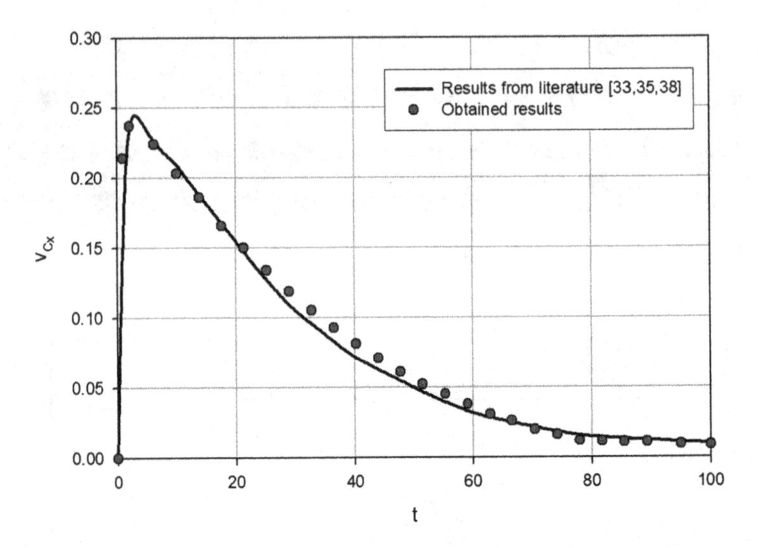

Fig. 8.19 Example 4 – comparison of results for x component of particle velocity

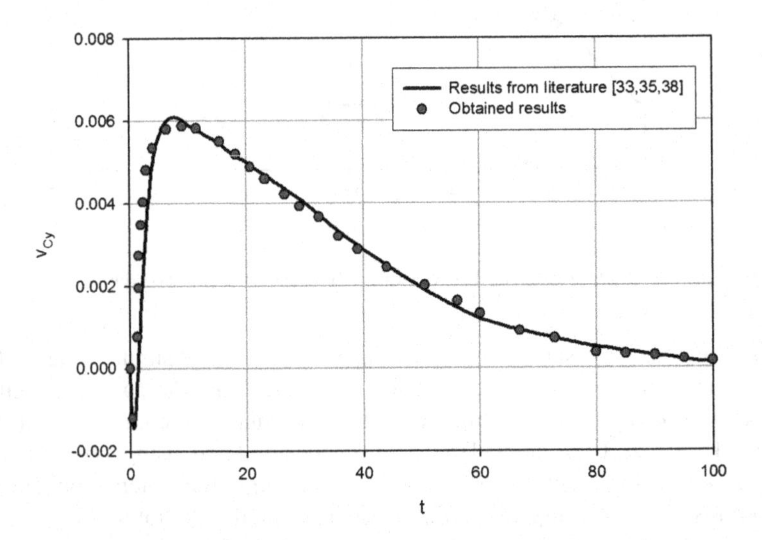

Fig. 8.20 Example 4 – comparison of results for y component of particle velocity

8.4.5 Particle Sedimentation in Viscous Fluid

This example illustrates the movement of particle due to gravity force. At the beginning of the simulation, the circular particle and fluid are in static state (the initial velocity of the particle as well as fluid velocity in all lattice nodes are equal to zero). Particle is located at the middle of fluid domain along horizontal axis and at $\frac{2}{3}$

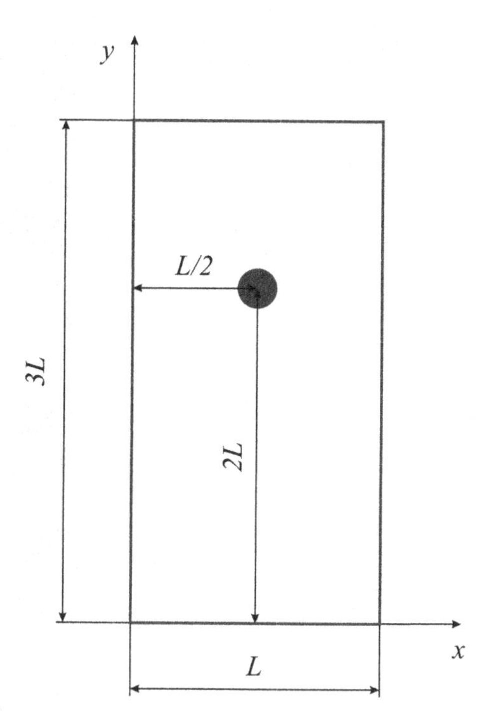

Fig. 8.21 Example 5 – geometrical data

Table 8.6 Data necessary for LB solver; Example 5

Quantity	Value
N	100
l_x	3
l_y	1
u_{lb}	0
Re	40
ρ_p	1.25 ρ
x_c	50
y_c	200
r	6.25
N_L	50

of height along vertical axis (as shown in Fig. 8.21). Due to gravity force, the particle is falling down, i.e., moving through the fluid.

This problem has also been extensively discussed in literature, so all parameters necessary for LB simulation are set in such a way to enable easy comparison of the obtained results with results found in literature. Data defined in LB solver is shown in Table 8.6.

Figure 8.22 shows the fluid velocity field during the simulation (while the particle is sedimenting in viscous fluid).

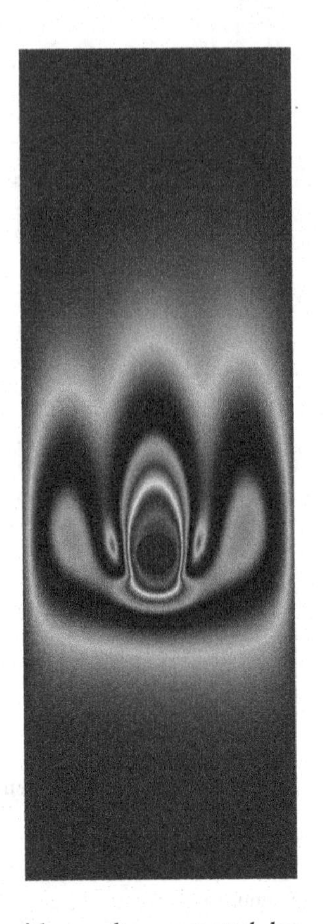

Fig. 8.22 Example 5 – fluid velocity field

Results obtained using LB solver are compared with results presented by Wu et al. [35] (the simulation was performed using LB method), as well as with results presented by Wan and Turek [45] (the simulation was performed using finite element method). The variations of four quantities are observed – the position of particle in y axis direction, y component of particle velocity, Reynolds number, and translational kinetic energy. Reynolds number in a specified moment in time can be evaluated using the following expression:

$$\mathrm{Re} = \frac{\rho_p r \sqrt{v_{Cx}^2 + v_{Cy}^2}}{\mu} \tag{8.53}$$

The expression for the translational kinetic energy in a specific moment in time is given by:

$$E_t = \frac{1}{2} m \left(v_{Cx}^2 + v_{Cy}^2 \right) \tag{8.54}$$

As it can be seen from Figs. 8.23, 8.24, 8.25, and 8.26, the obtained results compare very well with results found in literature.

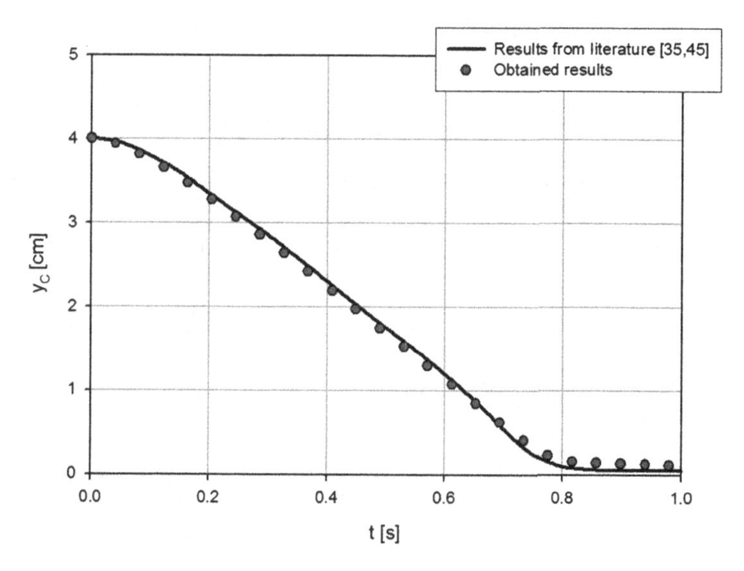

Fig. 8.23 Example 5 – comparison of results for particle position in y axis direction

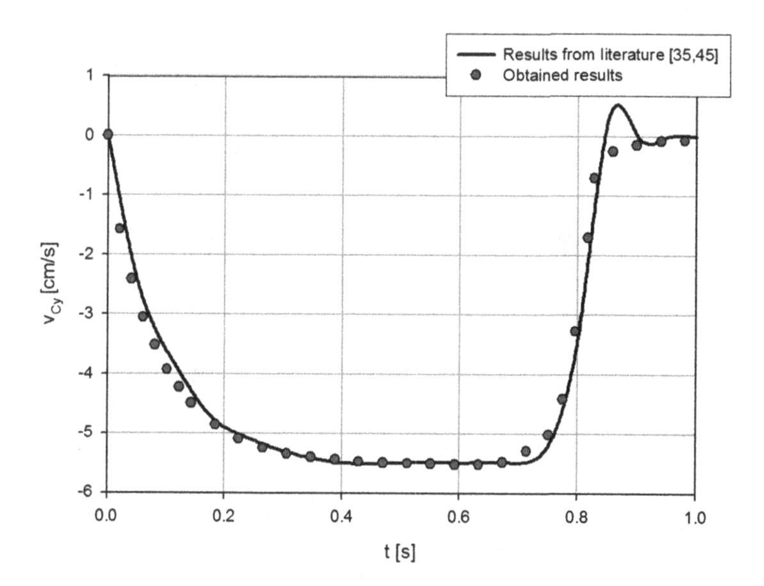

Fig. 8.24 Example 5 – comparison of results for y component of particle velocity

8.4.6 Simulation of Movement of Circular Particle Through a Stenotic Artery

The efficiency of the strong coupling approach in simulations of solid–fluid interaction becomes obvious when the fluid domain has complex boundaries. One such example is the simulation of movement of particle through a stenotic artery (artery

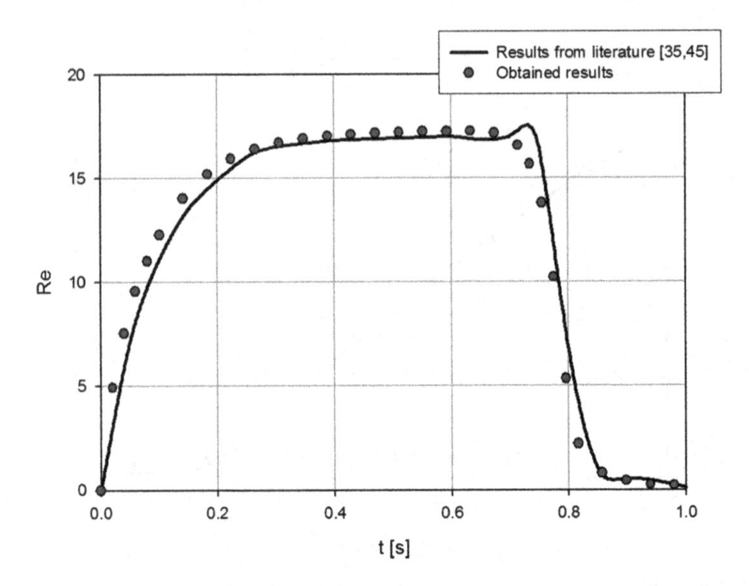

Fig. 8.25 Example 5 – comparison of results for Reynolds number

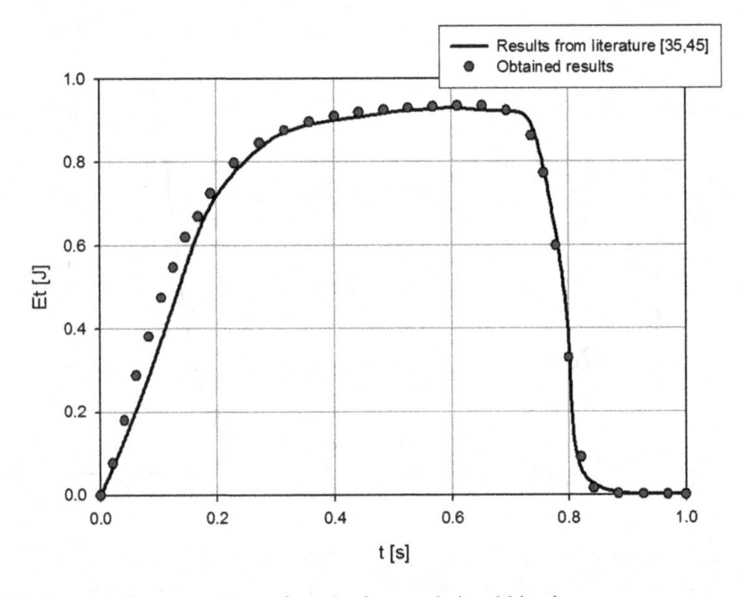

Fig. 8.26 Example 5 – comparison of results for translational kinetic energy

with constriction). Li et al. [46] have analyzed the pulsatile flow in a mildly or severely stenotic artery, and Wu and Shu simulated the motion of particles through stenotic artery in one of their papers [35]. Therefore here the results found in literature will be compared with results obtained using LB solver.

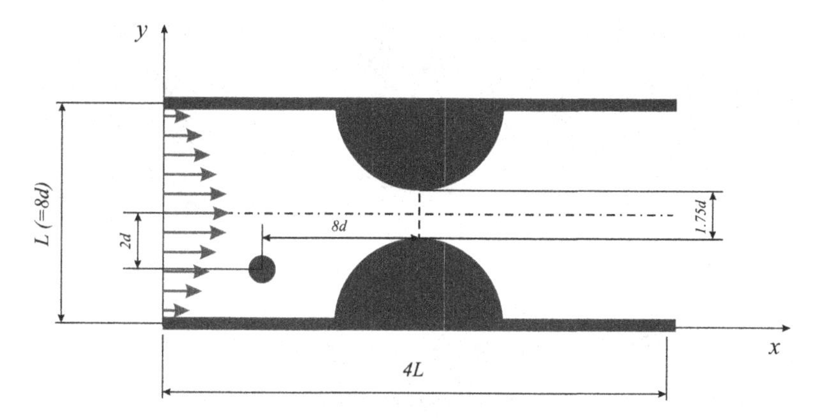

Fig. 8.27 Example 6 – geometrical data

Table 8.7 Data necessary for LB solver; Example 6

Quantity	Value
N	64
l_x	4
l_y	1
u_{lb}	0.01
Re	40
ρ_p	1.25 ρ
x_c	24
y_c	48
r	4
N_L	50

All the necessary geometrical parameters are shown in Fig. 8.27, and the data defined in LB solver is shown in Table 8.7. In the fluid domain as initial condition, the pressure difference on the inlet and outlet is prescribed, and the boundary condition is set such that the velocity of the upper and lower wall is equal to zero. The stenosis (and generally the complex boundary) can be implemented in two ways – applying the bounce-back or using immersed boundary method (it is implemented similarly to the solid–fluid interaction when particles are immersed in fluid domain). Here the first approach is used, which leads to a certain loss of accuracy but is numerically more efficient, due to a smaller number of calculations that are required.

Fluid velocity field for the specific moment in time when particle is passing through the stenotic artery is shown in Figs. 8.28 and 8.29 represents a schematic diagram of particle trajectory (in several steps) from the initial position to the final position symmetrically on the other side of the stenosis.

The comparative diagrams for x and y components of particle velocity for this example are shown in Figs. 8.30 and 8.31 (results obtained using LB solver and results obtained by Wu and Shu [35] are plotted).

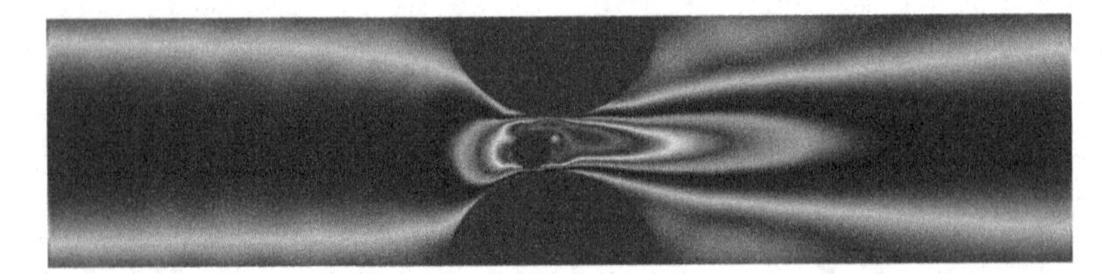

Fig. 8.28 Example 6 – fluid velocity field

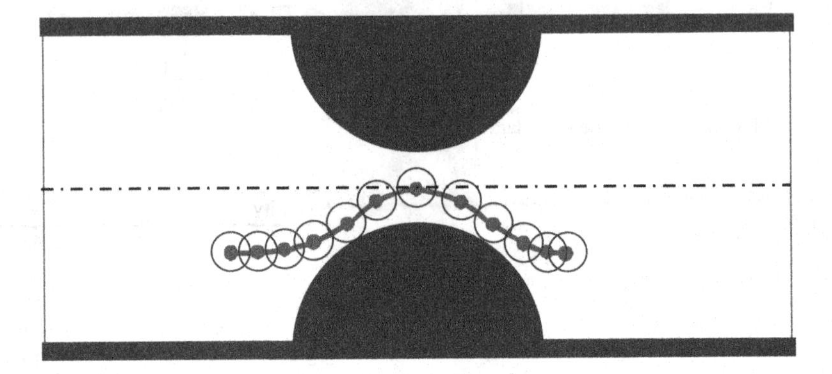

Fig. 8.29 Example 6 – trajectory of the particle through a stenotic artery

8.4.7 Simulation of Movement of Two Circular Particles Through a Stenotic Artery

This example has exactly the same boundary and initial conditions like example 6 and was simulated based on a similar example presented in literature [35]. The only difference between this example and example 6 is that instead of one particle, the movement of two particles is simulated. Also, it is necessary to implement the interaction force between two particles. In this case, the force acting on particle j exerted from particle i is given by:

$$\mathbf{F}^{col} = \begin{cases} 0 & , X_R^{i,j} > r_i + r_j + \zeta \\ 2.4\varepsilon \cdot \left[2\left(\dfrac{r_i + r_j}{X_R^{i,j}}\right)^{14} - \left(\dfrac{r_i + r_j}{X_R^{i,j}}\right)^{8} \right] \cdot \dfrac{\mathbf{X}_R^i - \mathbf{X}_R^j}{(r_i + r_j)^2} & , X_R^{i,j} \leq r_i + r_j + \zeta \end{cases} \tag{8.55}$$

where ζ is the threshold and is set to the distance between two lattice nodes (in this case it is considered that $\zeta = 1$), r_i and r_j are the radii of particles, $\mathbf{X}_R{}^i$ and $\mathbf{X}_R{}^j$ are the position vectors of centers of mass of particles, and ε and $X_R{}^{i,j}$ are defined using following equations:

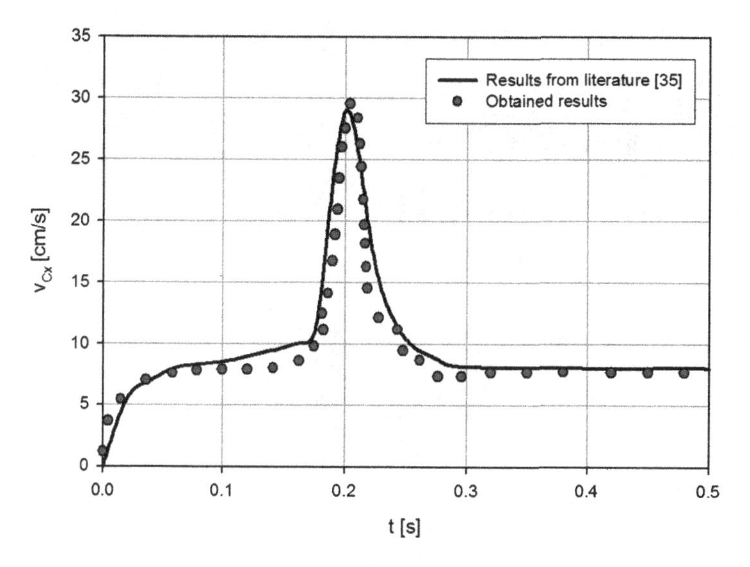

Fig. 8.30 Example 6 – comparison of results for x component of particle velocity

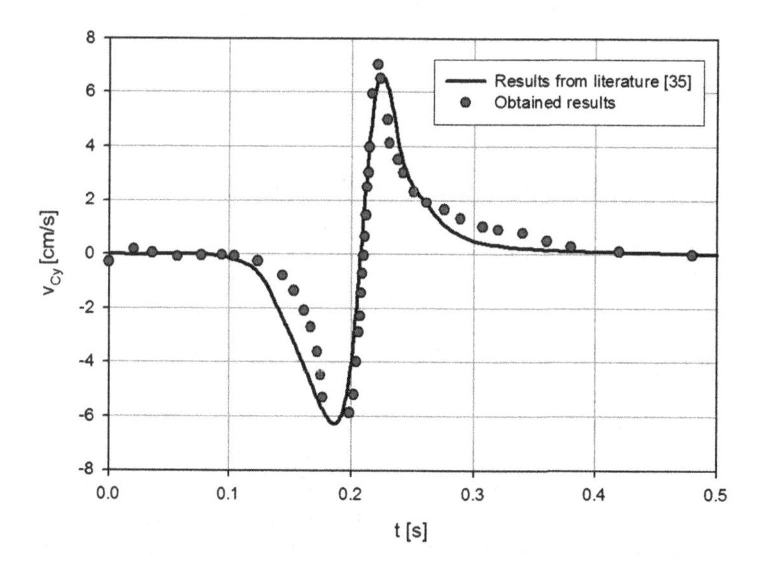

Fig. 8.31 Example 6 – comparison of results for y component of particle velocity

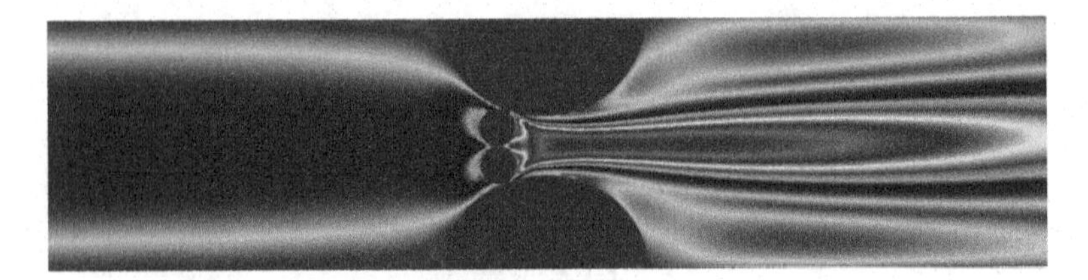

Fig. 8.32 Example 7.1 – fluid velocity field

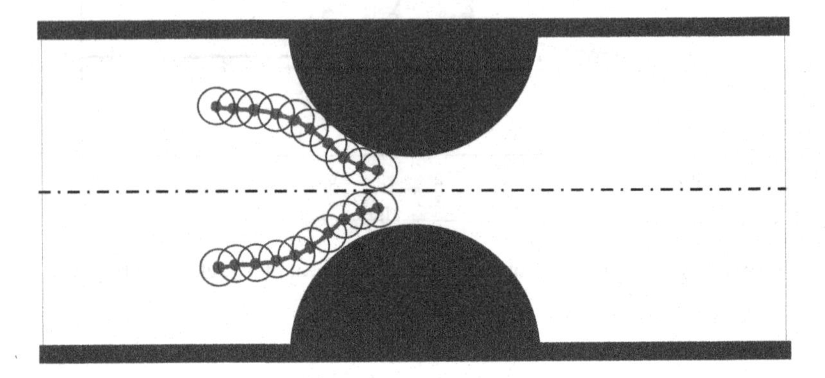

Fig. 8.33 Example 7.1 – trajectories of particles

$$\varepsilon = \left(\frac{2r_i r_j}{r_i + r_j}\right)^2 \tag{8.56}$$

$$X_R^{i,j} = \left|\mathbf{X}_R^i - \mathbf{X}_R^j\right| \tag{8.57}$$

When the interaction force is calculated, it is added in the expression for total force acting on every particle individually (Eq. (8.39)) and then the equations of motion are solved. The total force is now given by:

$$\mathbf{F}_R = \left(1 - \frac{\rho_f}{\rho_p}\right) m\mathbf{G} + \mathbf{F}^{col} - \sum_l \mathbf{F}\left(\mathbf{X}_B^l\right)\Delta s_l \tag{8.58}$$

Considering that the free space in the artery (the gap between two protuberances) is less than $2d$ (as it is shown in Fig. 8.27), it is obvious that both particles cannot pass the throat side by side. If both particles are initially placed symmetrically to the centerline of the artery, they will start moving towards the throat but will stay stuck at the entrance and block the throat. Figure 8.32 shows the fluid velocity field, and Fig. 8.33 shows the schematic diagram of particles' trajectories for this case.

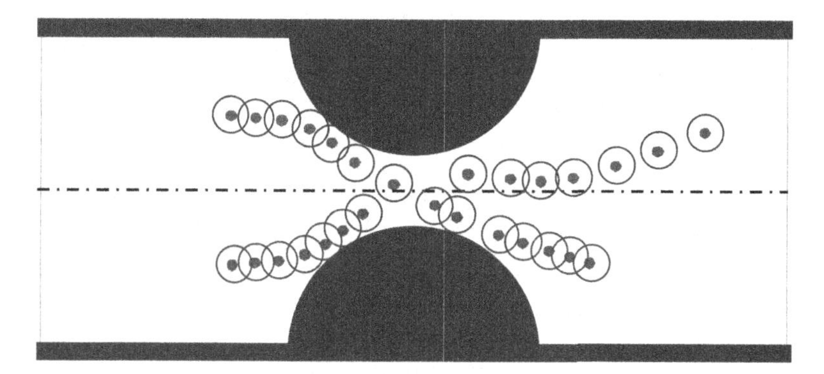

Fig. 8.34 Example 7.2 – trajectories of particles

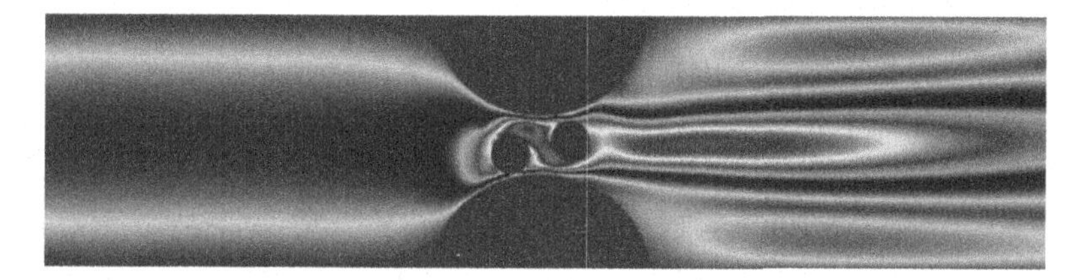

Fig. 8.35 Example 7.2 – fluid velocity field

If one of the particles is initially moved for a very small distance towards the centerline (e.g., for only 1 lattice unit), the particles will be able to pass the throat and move on through the artery. This is shown schematically in Fig. 8.34, while Fig. 8.35 shows the fluid velocity field in the moment of passing through the throat.

8.4.8 Simulation of Movement of a Circular Particle Through an Artery with Bifurcation

This is another example of the simulation of movement of particle through a fluid domain with complex boundary. In this case it is an artery with bifurcation. Geometrical data and all necessary geometrical parameters are shown in Fig. 8.36.

As boundary condition the velocities on all walls are prescribed to be equal to zero. Since the domain boundaries are not regular, it is possible to use two approaches to model the walls – bounce-back BC or immersed boundary method. In this simulation the first approach was used (like in examples 6 and 7), due to better efficiency, and the lower numerical accuracy does not have a strong effect on the final result. As initial condition the prescribed Poiseuille velocity profile is used, together with outflow condition on the outlet branches. Data defined in LB solver is shown in Table 8.8.

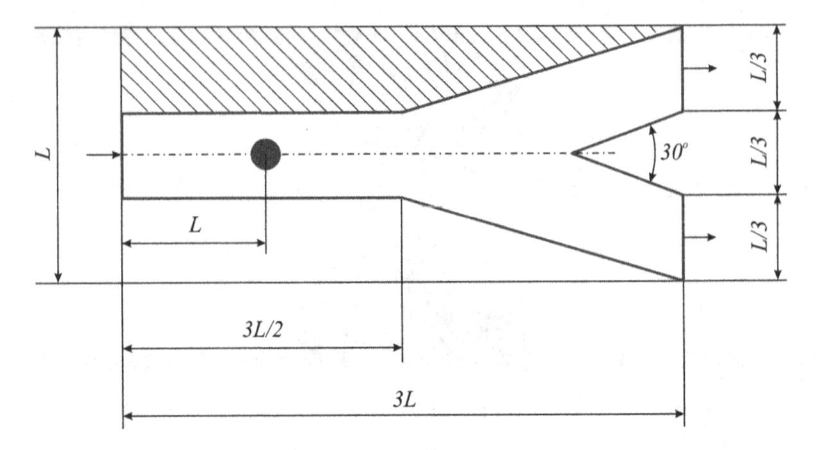

Fig. 8.36 Example 8 – geometrical data

Table 8.8 Data necessary for LB solver; Example 8

Quantity	Value
N	60
l_x	3
l_y	1
u_{lb}	0.01
Re	40
ρ_p	1.25 ρ
x_c	60
y_c	30
r	3
N_L	20

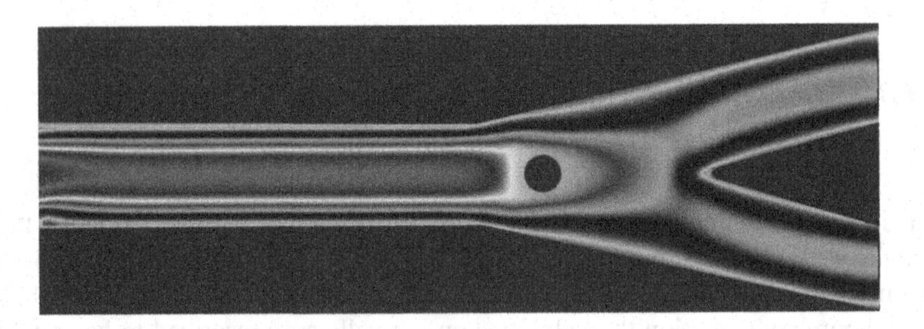

Fig. 8.37 Example 8 – fluid velocity field

When the particle starts moving, it comes across the bifurcation. Strong repulsive force is acting on the particle so that it rebounds from the wall and continues to move through the upper branch, until the end of the domain. Figure 8.37 shows the fluid velocity field during the motion of particle through the main branch of the artery, and Fig. 8.38 shows schematically the trajectory of the particle.

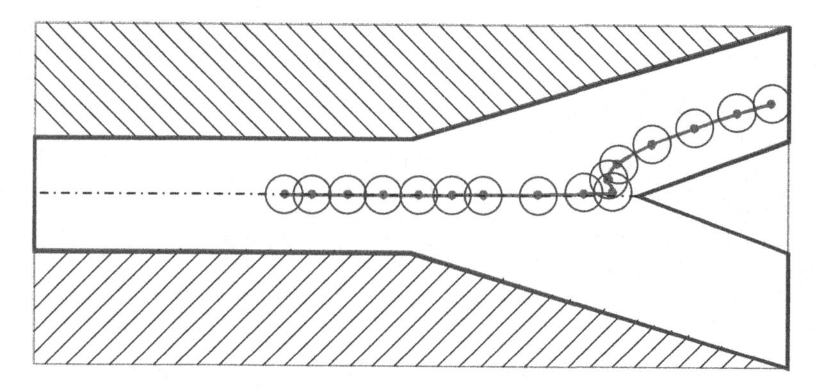

Fig. 8.38 Example 8 – trajectory of the particle

8.5 Conclusion

Drug delivery system should be designed such that nanoparticles are able to drift across the streamlines of blood flow in vessels and to interact with the vessel walls. In order to analyze the dynamics of arbitrarily shaped particles in fluid flow, it is necessary to use numerical simulations. By choosing an appropriate balance between size, shape, geometrical inertia, and density, the rate of particles that are interacting with walls can be customized. This will enable the formation of an efficient drug delivery system, with particles capable of sensing biological and biophysical abnormalities in endothelial cells.

In this chapter lattice Boltzmann method was used to simulate motion of particles through fluid domain, and a specific type of particle-fluid interaction was modeled. Movement of both spherical and nonspherical particles was analyzed in diverse complex geometrical fluid domains. Agreement between the LB method and analytical, FEM, and other solutions found in literature demonstrates that this method and the developed software can be successfully used to model complex problems of fluid flow and fluid-particle interaction in microcirculation, especially in the fields of particle transport and margination to the vessel walls, bio-imaging, and drug delivery.

References

1. Peer, D., Karp, J.M., Hong, S., Farokhzad, O.C., Margalit, R. and Langer, R., 2007. Nanocarriers as an emerging platform for cancer therapy. Nat. Nanotechnol. 2, 751–60.
2. Ferrari, M., 2005. Cancer Nanotechnology: Opportunities and Challenges. Nature Rev. Cancer. 5, 161–171.
3. LaVan, D.A., Mcguire, T., Langer, R., 2003. Small-scale systems for in vivo drug delivery. Nat. Biotech. 21(10): 1184–1191.

4. Choi, Y.S., Thomas, T., Kotlyar, A., Islam, M.T., Baker, J.R., 2005. Synthesis and Functional Evaluation of DNA-Assembled Polyamidoamine Dendrimer Clusters for Cancer Cell-Specific Targeting. Chemistry & Biology, 12:35–43, DOI 10.1016/j.chembiol.2004.10.016
5. Duncan, R. 2003. The dawning era of polymer therapeutics. Nat. Rev Drug Discov. 2: 347–360.
6. Cohen, M.H., Melnik, K., Boiarski, A.A., Ferrari, M., Martin, F.J., 2003. Microfabrication of silicon-based nanoporous particulates for medical applications, Biomedical Microdevices, 5: 253–259.
7. Dillen, van T., van Blladeren, A., Polman, A., 2004. Ion beam shaping of colloidal assemblies, Materials Today: 40–46.
8. Rolland, J.P., Maynor, B.W., Euliss, L.E., Exner, A.E., Denison, G.M., DeSimone, J., 2005. Direct fabrication and harvesting of monodisperse, shape specific nano-biomaterials. J. M.J. Am. Chem. Soc. 127: 10096–10100.
9. Greish, K., Enhanced permeability and retention effect for selective targeting of anticancer nanomedicine: are we there yet ?, Drug Discovery Today: Technologies, vol. 9, issue 2, pp. 161–166, 2012.
10. Neri, D. and Bicknell, R. 2005. Tumour vascular targeting, Nat. Cancer. 570, 436–446.
11. Jain, R.K., 1999. Transport of molecules, particles, and cells in solid tumors. Annu. Rev. Biomed. Eng., 1, 241–263.
12. Decuzzi, P., Ferrari, M., 2006. The adhesive strength of non-spherical particles mediated by specific interactions, Biomaterials, 27(30):5307–14.
13. Goldman, A.J., Cox, R.J. and Brenner, H., 1967, Slow viscous motion of a sphere parallel to a plane wall I. Motion through a quiescent fluid, Chem. Eng. Sci., 22, 637–651.
14. Bretherton, F.P., 1962., The motion of rigid particles in a shear flow at low Reynolds number, Journal of Fluid Mechanics, 14, 284–304.
15. Decuzzi, P., Lee, S., Bhushan, B. and Ferrari, M., 2005. A theoretical model for the margination of particles within blood vessels. Ann. Biomed. Eng., 33, 179–190.
16. Gavze, E. and Shapiro, M., 1997. Particles in a shear flow near a solid wall: Effect of nonsphericity on forces and velocities. International Journal of Multiphase Flow, 23, 155–182.
17. Gavze, E. and Shapiro, M., 1998. Motion of inertial spheroidal particles in a shear flow near a solid wall with special application to aerosol transport in microgravity, Journal of Fluid Mechanics, 371, 59–79.
18. S. Wolfram, *Cellular Automaton Fluids 1: Basic Theory*.: J. Stat. Phys., 3/4:471–526, 1986.
19. D. H. Rothman and S. Zaleski, *Lattice Gas Cellular Automata. Simple models of Complex Hydrodynamics*. England: Cambridge University Press, 1997.
20. P. L. Bhatnagar, E. P. Gross, and M. Krook, "A model for collision processes in gases. i. small amplitude processes in charged and neutral one-component systems," *Phys. Rev. E*, vol. 77, no. 5, pp. 511–525, 1954.
21. T. Đukić, Modelling solid–fluid interaction using LB method, *Master thesis,* Kragujevac: Mašinski fakultet, 2012.
22. O. P. Malaspinas, *Lattice Boltzmann Method for the Simulation of Viscoelastic Fluid Flows*. Switzerland: PhD dissertation, 2009.
23. V. I. Krylov, *Approximate Calculation of Integrals*. New York: Macmillan, 1962
24. P.J. Davis and P. Rabinowitz, *Methods of Numerical Integration*. New York, 1984.
25. M.C. Sukop and D.T. Jr. Thorne, Lattice Boltzmann Modeling - An Introduction for Geoscientists and Engineers. Heidelberg: Springer, 2006.
26. M.A. Gallivan, D.R. Noble, J.G. Georgiadis, and R.O. Buckius, "An evaluation of the bounce-back boundary condition for lattice Boltzmann simulations," Int J Num Meth Fluids, vol. 25, no. 3, pp. 249–263, 1997.
27. T. Inamuro, M. Yoshina, and F. Ogino, "A non-slip boundary condition for lattice Boltzmann simulations," Phys. Fluids, vol. 7, no. 12, pp. 2928–2930, 1995.

28. I. Ginzbourg and D. d 'Humières, "Local second-order boundary method for lattice Boltzmann models," J. Statist. Phys., vol. 84, no. 5–6, pp. 927–971, 1996.
29. B. Chopard and A. Dupuis, "A mass conserving boundary condition for lattice Boltzmann models," Int. J. Mod. Phys. B, vol. 17, no. 1/2, pp. 103–108, 2003.
30. Q. Zou and X. He, "On pressure and velocity boundary conditions for the lattice Boltzmann BGK model," Phys. Fluids, vol. 9, no. 6, pp. 1592–1598, 1997.
31. J. Latt and B. Chopard, "Lattice Boltzmann method with regularized non-equilibrium distribution functions," Math. Comp. Sim., vol. 72, no. 1, pp. 165–168, 2006.
32. P.A. Skordos, "Initial and boundary conditions for the lattice Boltzmann method," Phys. Rev. E, vol. 48, no. 6, pp. 4823–4842, 1993.
33. Z. Feng and E. Michaelides, "The immersed boundary-lattice Boltzmann method for solving fluid-particles interaction problem," Journal of Computational Physics, vol. 195, no. 2, pp. 602–628, 2004.
34. C. S. Peskin, "Numerical analysis of blood flow in the heart," Journal of Computational Physics, vol. 25, no. 3, pp. 220–252, 1977.
35. J. Wu and C. Shu, "Particulate flow simulation via a boundary condition-enforced immersed boundary-lattice Boltzmann scheme," Commun. Comput. Phys., vol. 7, no. 4, pp. 793–812, 2010.
36. Z. Feng and E. Michaelides, "Proteus: A direct forcing method in the simulations of particulate flows," Journal of Computational Physics, vol. 202, no. 1, pp. 20–51, 2005.
37. M. Uhlmann, "An immersed boundary method with direct forcing for the simulation of particulate flows," J. Comput. Phys., vol. 209, no. 2, pp. 448–476, 2005.
38. X. D. Niu, C. Shu, Y. T. Chew, and Y. Peng, "A momentum exchange-based immersed boundary-lattice Boltzmann method for simulating incompressible viscous flows," Phys. Lett. A, vol. 354, no. 3, pp. 173–182, 2006.
39. J. Wu and C. Shu, "Implicit velocity correction-based immersed boundary-lattice Boltzmann method and its application," J. Comput. Phys., vol. 228, no. 6, pp. 1963–1979, 2009.
40. A. T. Chwang and T. Y. Wu, "Hydromechanics of low-Reynolds-number flow, Part 4, Translation of spheroids," J. Fluid Mech., vol. 75, no. 4, pp. 677–689, 1976.
41. N. Filipovic, M. Kojic, P. Decuzzi, and M. Ferrari, "Dissipative Particle Dynamics simulation of circular and elliptical particles motion in 2D laminar shear flow," Microfluidics and Nanofluidics, vol. 10, no. 5, pp. 1127–1134, 2010.
42. N. Filipovic, V. Isailovic, T. Djukic, M. Ferrari, and M. Kojic, "Multi-scale modeling of circular and elliptical particles in laminar shear flow," IEEE Trans Biomed Eng, vol. 59, no. 1, pp. 50–53, 2012.
43. G. B. Jeffery, "The motion of ellipsoidal particles immersed in a viscous fluid," Proc. R. Soc. Lond. A, vol. 102, no. 715, pp. 161–180, 1922.
44. J. Feng, H. H. Hu, and D. D. Joseph, "Direct simulation of initial value problems for the motion of solid bodies in a Newtonian fluid. Part 2, Couette and Poiseuille flows," J. Fluid Mech., vol. 277, pp. 271–301, 1994.
45. D. Wan and S. Turek, "Direct numerical simulation of particulate flow via multigrid FEM techniques and the fictitious boundary method," *Int. J. Numer. Meth. Fluids*, vol. 51, no. 5, pp. 531–566, 2006.
46. H. Li, H. Fang, Z. Lin, S. Xu, and S. Chen, "Lattice Boltzmann simulation on particle suspensions in a two-dimensional symmetric stenotic artery," *Phys. Rev. E*, vol. 69, no. 3, p. 031919, 2004.

Chapter 9
Computational Modeling of Ultrasound Wave Propagation in Bone

Vassiliki T. Potsika, Maria G. Vavva, Vasilios C. Protopappas, Demosthenes Polyzos, and Dimitrios I. Fotiadis

9.1 Introduction

Computational modeling is considered as a significant tool used in engineering applications as well as for the virtual investigation of various problems that may be encountered in clinical settings. In the field of bone characterization, several clinical and preclinical studies have demonstrated the diagnostic capabilities of ultrasound since its characteristics are related to long bone's mechanical and geometrical properties. However, the recent introduction of computer simulations has extended our understanding of the underlying wave propagation phenomena opening thus new horizons in the quantitative evaluation of bone pathologies such as fracture healing and osteoporosis.

Computational simulations offer significant advantages as compared to in vitro and in vivo experiments since (a) the effect of parameters related to bone material and geometrical properties on wave propagation can be independently determined, (b) it is easier to discriminate the different types of propagating waves, and (c) experimental artifacts and problems regarding the properties of the examined bone specimens can be avoided.

The development of powerful computers and simulation tools has reduced the computational time and cost, rendering computational techniques more and more attractive. In addition, the significant availability of high-resolution two-dimensional (2D) and three-dimensional (3D) images of bone structure derived using

V.T. Potsika (✉) • V.C. Protopappas • D.I. Fotiadis
Unit of Medical Technology and Intelligent Information Systems, Department of Materials Science and Engineering, University of Ioannina, GR 45 110 Ioannina, Greece
e-mail: vpotsika@cc.uoi.gr; vprotop@mech.upatras; dfotiadis@cs.uoi.gr

M.G. Vavva • D. Polyzos
Department of Mechanical Engineering and Aeronautics, University of Patras, GR 26 500 Patras, Greece
e-mail: marvavva@gmail.com; polyzos@mech.upatras.gr

G. Rakocevic et al. (eds.), *Computational Medicine in Data Mining and Modeling*,
DOI 10.1007/978-1-4614-8785-2_9,

microcomputed tomography (μCT) and scanning acoustic microscopy (SAM) has paved the way for the development of more realistic computational models.

The scope of this chapter is to review the relevant computational studies on ultrasound wave propagation in bone reported in the literature and present the current status of knowledge in the field. The chapter is organized as follows: In the next section some introductory points regarding the bone physiology and its pathologies are presented. Section 9.3 is devoted to the ultrasonic methods that have been proposed in the literature for examining long bones. Thereafter we present the computational studies of wave propagation in intact and osteoporotic bones by classifying into two broad categories, i.e., ultrasound simulations in (a) cortical and (b) trabecular bone models. Finally, an overview of the computational studies for the simulation of ultrasound propagation in models of healing bones is presented.

9.2 Bone Physiology and Pathologies

9.2.1 Bone Structure

Bone is a heterogeneous, porous, and anisotropic material with a complex structure. In order to fully evaluate bone's mechanical properties, it is important to give insight into the mechanical properties of its component phases and the structural relationship between them at the various levels of hierarchical structural organization.

From a macrostructural point of view, bone is consisted of the periosteum, bone tissue, bone marrow, blood vessels, and nerves. Bone tissue is composed of the cortical or compact bone and the cancellous or trabecular bone. The main differentiation between these two types of bone is their degree of porosity or density.

Trabecular bone is found in the inner parts of bones and has a significantly porous structure. The porosity in trabecular bone ranges between 50 % and 95 %, usually found in cuboidal bones, flat bones, and at the ends of long bones [1]. In the microstructure level, trabecular bone is consisted of three-dimensional cylindrical structures, called trabeculae, with a thickness of about 100 μm and a variable arrangement form [2]. This porous network of trabecular bone includes pores filled with marrow which produces the basic blood cells and is consisted of blood vessels, nerves, and various types of cells.

Cortical bone composes the external surface of all bones and has a porosity of about 5–10 %. From a microscopic point of view, it is consisted of cylindrical structures called osteons or Haversian systems with a diameter of about 10–500 μm [2] formed by cylindrical lamellae surrounding the Haversian canal (size of 3–7 μm) [3]. The boundary between the osteon and the surrounding bone is called the cement line. The most significant type of porosity, known as vascular porosity, is formed by the Haversian canals (aligned with the long axis of the bone) and the Volkmann canals (transverse canals connecting Haversian canals) with capillaries and nerves [1]. Experimental studies have shown that the shear modulus of wet single femoral osteons is higher than that of the whole bone which suggests that microstructural effects are significant [4, 5].

In the nanostructure level, the basic components are collagen and hydroxyapatite which are made of collagen molecules organized in fibrils [2]. Inorganic components (hydroxyapatite crystals) are mainly responsible for the compression strength and stiffness, while organic components (collagen fibers, proteoglycans, osteocalcin) are responsible for tension properties [1]. The hydroxyapatite crystals are located in the interfibrillar spaces. Mineralized fibers are aligned to form bone lamellae of typical thickness of a few micrometers. The orientation of the fibers depends on the lamellae and may change within lamellar sublayers [2, 3].

It is evident that bone's material and structural properties differ according to the examined hierarchical level. Thus, development of realistic models necessitates consideration of bone's microstructural effects as well as porosity and anisotropy properties. Computational simulation of ultrasound wave propagation in such bone models can provide valuable information about the complicated wave guidance and scattering phenomena that occur in deeper cortical layers due to the complex structure of bone and is difficult to be investigated experimentally.

9.2.2 Bone Pathologies: Osteoporosis and Fracture Healing

9.2.2.1 Osteoporosis

Osteoporosis is a skeletal disease in which the density and quality of bone are reduced, leading to weakness of the skeleton and increased risk of fracture, particularly of the spine, wrist, hip, pelvis, and arm [6]. The main cause of osteoporosis is hormonal deficiency, and thus the most frequent disease is postmenopausal osteoporosis [6]. At least 40 % of postmenopausal women over the age of 50 and 15–30 % of men will sustain one or more fragility fractures [6]. The bone resorption starts from the endosteal (inner) surface, leading to thinning of the cortex and trabecularization of the inner cortical layer [7]. It has been shown that the cortical thickness is related to decreasing total bone strength and increasing fracture risk [8]. The diagnosis of osteoporosis has been mostly based on bone mineral density measurements using dual-energy X-ray absorptiometry (DEXA) scan, which however is an inconvenient and expensive method.

Quantitative ultrasound (QUS) has been used for decades for the assessment of the skeletal status and the monitoring of osteoporosis [7, 9]. The effectiveness of QUS techniques has been evaluated in a large number of studies [6]. The relative low cost and portability make QUS a challenging noninvasive and non-radiating technique which can enhance the prediction of the risk of fractures. Nevertheless, further numerical and experimental research is needed as QUS technologies are not yet widely accepted due to technical immaturity and to the lack of standardization between different technical approaches and among various manufacturers [6].

9.2.2.2 Bone Fracture Healing

Bone fracture healing is a complex and dynamic regenerative process, which involves a series of cellular and molecular events that result in a combination of intramembranous and endochondral bone formation. The healing process advances in stages of callus formation and consolidation and is eventually completed within some months. Worldwide, millions of fractures occur annually due to a common injury or as a result of osteoporosis, and the human suffering is immense, while the financial costs for the monitoring and the treatment are staggering. There are two types of bone healing with significant differences in the evolution of the healing process, i.e., primary and secondary healing. Secondary healing is a complex regenerative process which involves the progressive formation, differentiation, and ossification of the fracture callus tissue in order to restore the original material properties and structural integrity of intact bone. This is the most common healing type. In primary healing rigid stabilization is required with or without compression of the bone ends, and lamellar bone is directly formed without the formation of callus tissue.

Although the vast majority of fractures are treated successfully, complications such as delayed union and nonunion are frequently encountered [10]. Both types of complications require further conventional or surgical procedures extending thus the already prolonged treatment period. In clinical practice, the assessment of fracture healing is performed by serial clinical and radiographic examinations. However, both techniques strongly depend on the orthopedic surgeon's experience and clinical judgment, which renders the development of more objective and quantitative means of the evaluation of bone fracture healing more than necessary. Several noninvasive biomechanical methods for monitoring fracture healing have been reported in the literature. The most popular biomechanical methods include the attachment of strain gauges to external fixation devices for measuring the axial or bending deformation, the use of vibrational testing, and the acoustic emission technique [11–16].

QUS has been used for the assessment and monitoring of the fracture healing process for over five decades. Ultrasonic techniques evaluate the alterations that occur within the callus tissue from the early inflammation stage, the endochondral and intramembranous ossification stages, up to the bone remodeling stage. As the waves propagate across the fracture site, the callus properties gradually evolve during bone healing which is reflected in the ultrasound velocity. Animal and clinical studies have shown that ultrasound velocity changes as the healing progresses which suggests that QUS can be used for monitoring purposes as well as for the early detection of complications and the accurate estimations of the callus tissue [17–25].

9.3 Ultrasound Configurations and Measuring Parameters

The most common ultrasound techniques that have been used for bone characterization are the through and the axial transmission. The through-transmission technique is more suitable for the assessment of trabecular bone. Typically two transducers a transmitter and a receiver are placed on opposite sides of the skeletal site to be measured. This technique has been mainly applied to the measurement of different skeletal sites such as finger phalanxes and heel and more recently at the forearm or the proximal femur at the hip [6].

Axial transmission is the most suitable method for the assessment of cortical bone and has been mainly applied for obtaining measurements from long bones such as tibia and radius [26, 27]. A set of transmitter(s) and receiver(s) are placed along the bone axis at a constant or variable distance either in contact with the skin (percutaneous application) [26] or through implantation directly onto the bone surface (transosseous application) [28]. As opposed to through-transmission techniques, the transducer setup is much easier and can therefore be applied to a greater number of skeletal sites. In particular, the axial-transmission technique is preferable in the case of fracture healing as this setup can be positioned more easily on each side of the injured region.

The most common measuring parameters are the velocity and the attenuation of the first arriving signal (FAS). In the case of fracture healing, the average propagation velocity is the component of the propagation velocities in intact bone and callus. Thus, the value of the velocity depends on the distance between the transducers for a specific fracture gap, whereas in the case of osteoporosis the velocity is practically independent of the transducers' distance. In addition, several studies have indicated that ultrasound wave attenuation can be used as an effective means of monitoring bone healing. The numerical evaluation of the ultrasound attenuation in bone is mainly based on the measurement of the first half cycle of the FAS. Different criteria have been proposed to detect the FAS at the receiver including extrema, zero crossings, or threshold-based time criteria. However, the FAS propagates as a lateral wave for wavelengths comparable to or smaller than the cortical thickness reflecting thus only the periosteal region of the bone.

To this end, the study of the propagation of guided waves in bone has drawn the interest of several researchers. Guided waves are sensitive to material and geometrical changes and can provide valuable information for the ultrasound propagation features occurring at deeper layers. Different numerical approaches have been presented for the study of guided waves based on the Lamb wave theory. According to one technique, ultrasound signals are recorded at various transmitter–receiver distances to produce grayscale plots of amplitude as a function of time and distance. From the (r, t) diagrams propagating waves were visualized, and, by fitting a line to the peaks within a wave packet, velocities were measured. Other authors have used the wavenumber–frequency diagrams ((k, f) diagrams) obtained from the simulated (r, t) diagrams by applying the 2D fast Fourier transform (2D FFT). Analytically derived Lamb dispersion curves are superimposed on the (k, f) diagram in order to detect the dominant modes. Another technique, which is mainly used in fracture healing, is based on time–frequency (t, f) analysis by using the reassigned smoothed pseudo Wigner–Ville (RSPWV) distribution function. Frequency–group velocity (f, c_g) dispersion curves are computed for the plate model based on the Lamb wave theory. At each healing stage the (f, c_g) dispersion curves are superimposed to the (t, f) representations in order to estimate the evolution of the dispersion modes during healing. The (t, f) analysis provides an effective means of representing guided waves in bones and has several advantages over the analysis of an (r, t) diagram. The derivation of an (r, t) diagram usually requires a manual procedure for the collection of multiple waveforms, while (t, f) analysis can represent multiple guided modes using only a broadband excitation [17]. Therefore, the acquisition of an (r, t) diagram increases the measuring time and impedes the applicability of the method in cases where the accessibility to the skeletal site of interest is limited (e.g., when the transducers are rigidly attached or implanted).

More recent numerical studies have used multiple scattering theories in order to estimate the frequency-dependent wave dispersion and attenuation coefficient in media with porous nature such as trabecular bone and callus. Nevertheless, the investigation of the scattering and absorption mechanisms induced by the interaction of ultrasound with the complex bone microstructure is a new and promising research field.

9.4 Computational Modeling of Wave Propagation in Intact and Osteoporotic Bones

The development of computational bone models has played a key role in bone characterization in general, but especially for the diagnosis and evaluation of osteoporosis, it has facilitated the investigation of new configurations and measuring indices. Numerical simulations provide a significant supplementary for the investigation of the underlying mechanisms of ultrasound propagation in bone that cannot be interrogated by clinical and animal experiments.

9.4.1 Computational Methods

The different computational algorithms that have been mainly presented so far in the literature for simulating wave propagation in bones are as follows: (a) the finite element method (FEM), (b) the finite-difference method (FDM), and (c) the boundary element method (BEM). The Elastodynamic Finite Integration Technique (EFIT) has been also used in a more recent study [29].

The FEM is a numerical technique which divides a structure into many elements, called finite elements, to approximate a solution over the domain of interest. Specifically, a complex system of points (nodes) forms a grid, called mesh. The mesh contains the material and geometrical properties which determine the deformation of a structure under certain loading conditions. Approximate solutions of partial differential equations as well as of integral equations are derived. The FDM is also a numerical method that solves differential equations using finite-difference equations to approximate derivatives. The FDM is the most popular method due to the ease of implementation and efficiency and has been particularly popular in osteoporosis studies. According to the BEM the dimensionality of the solution is reduced by one as only the boundaries of a structure have to be discretized. Thus, less time and memory requirements are needed. Finally, the EFIT has been used to simulate both linear and nonlinear interaction between a propagated wave and the boundaries of the structure.

Accurate computational models and efficient simulation of elastic wave propagation necessitate (a) careful assignment of material properties to the model, (b) selection of parameters related to the ultrasound configurations with cautiousness, as well as (c) detailed discretization of the spatial and temporal domains.

9.4.2 2D Studies on Cortical Bone

The first computational studies of ultrasound wave propagation aiming at cortical bone assessment were based on two-dimensional models. The initial objective of the first studies on the field was to perform FAS velocity measurements so as to investigate the potential of FAS to (a) reflect structural changes in cortical thickness, (b) offer quantitative criteria for the diagnosis of osteoporosis, and (c) interpret the findings from previous animal and clinical studies. More recently, several authors have studied the propagation of guided modes in order to interpret ultrasound propagation phenomena occurring at deeper bone layers.

The first two studies on ultrasound wave propagation on 2D bone-mimicking plates are published in 2002 [30, 31] aiming at the evaluation of osteoporosis. The bone was simply modeled as a linear elastic 2D acrylic plate. Since osteoporosis is characterized by reduced cortical thickness, ultrasound simulations were performed for varying plate thicknesses. In another subsequent 2D study [32], the bone models were enhanced by assuming realistic geometry derived from X-ray computed tomography reconstructions of human radius. Solution to the 2D ultrasound wave propagation problem was given using the FDM. Velocity measurements were also performed for various transmitter–receiver distances. It was found that when the plate thickness, d, is larger than the wavelength in bone, λ_{bone}, the FAS wave corresponds to a lateral wave which propagates at the bulk longitudinal velocity of bone. When $d/\lambda_{bone} \leq 1$, the velocity of the FAS wave decreases with decreasing plate thickness. For very thin plates, $d/\lambda_{bone} \leq 0.4$, the FAS wave propagates as the lowest-order symmetric plate mode [10]. When $0.4 < d/\lambda_{bone} < 1$, the nature of

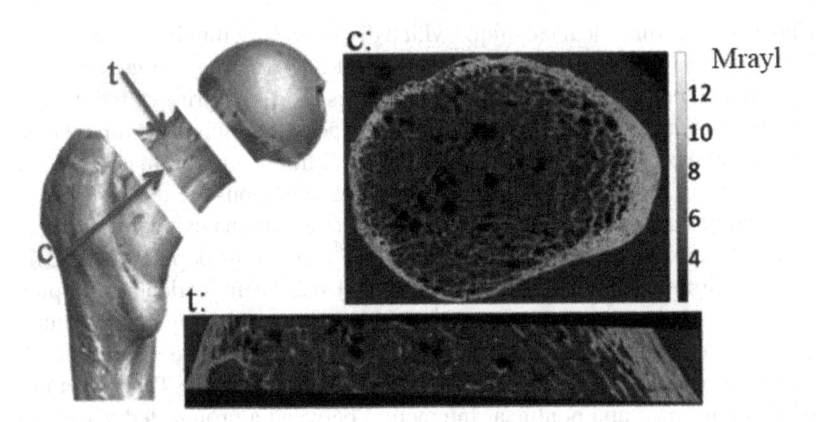

Fig. 9.1 Sample preparation scheme of the femoral neck disks. The disks were extracted between greater trochanter and femoral neck (1). c_{33} and c_{11} were obtained from scan of cross sections (*c*) and transverse sections (*t*) [34]

the FAS has not been understood yet, but it may be interpreted as an interference between the lateral wave and some plate modes [31].

In two recent 2D studies reported in the literature [33, 34], bone models were enhanced by considering porosity and anisotropy issues. Numerical simulations were performed by using either a hybrid spectral/FE method [33] or the finite-difference time-domain (FDTD) method [34]. In [34] the bone models' geometrical and material properties were derived from scanning acoustic microscopy images of human femoral neck. Cortical bone was assumed to be transversely isotropic with elastic properties derived from scans of cross sections and transverse sections as shown in Fig. 9.1. Seven 2D models were created which account for the presence or absence of trabecular bone, cortical porosity, and homogeneous or heterogeneous matrix elasticity [34]. The authors concluded that the porosity plays a significant role in the FAS wave propagation with the relative change of the propagation time to be highly correlated to the relative change of porosity and tissue elasticity c_{33}. In addition, it was reported in [33] that the FAS velocity decreases with increasing porosity.

However, as it was previously mentioned for wavelengths smaller than the cortical thickness, the FAS wave behaves as a lateral wave which propagates in the subsurface of bone, reflecting thus only the periosteal region. Therefore FAS velocity measurements cannot capture the material and structural alterations that occur in deeper cortical layers. To this end, 2D computational studies have been also performed to investigate the propagation of guided waves in bone which are sensitive on both the material and the geometrical properties [30–32]. This measurement method is based on the assumption that the propagation of Lamb modes in a plate is in consistency with the type of waves propagating in the cortical layer. Ultrasound signals were recorded at various transmitter–receiver distances to produce grayscale plots of amplitude as a function of time and distance. From the distance–time diagrams (r, t), the observation of at least two distinct wave modes was enabled: (a) the fast first arriving mode which corresponds to the first symmetric Lamb mode (S0) and (b) a slower guided wave contribution corresponding to the propagation of the first antisymmetric mode (A0). Nevertheless, bones are surrounded by marrow and soft tissues which provide a leakage path for the ultrasonic energy and give rise to additional guided modes. In order to identify the propagation of these modes, computational simulations on immersed plates have also been performed [35]. It was shown that the A0 mode remained the dominant mode being unaffected by the presence of the soft tissues.

Two more recent computational bone studies [36, 37] investigated guided and Rayleigh wave propagation in 2D bone models with microstructural effects. Microstructural effects were accounted by using the Mindlin Form II theory of gradient elasticity which introduces intrinsic parameters that correlate microstructure with macrostructure. In [36] by exploiting the simplest form of this theory (which incorporates two coefficients), it was shown that bone's microstructure has a significant effect on the dispersion of the propagating guided waves. In [37] the effect of bone's microstructure on Rayleigh wave propagation was investigated by adopting both the simple and the general Mindlin Form II gradient elastic theory (in which four constants are involved denoted as l_1, l_2 and h_1, h_2). Since the determination of the microstructural constants introduced in the stress analysis is an open issue when applying enhanced elastic theories to real problems, the constants were first assigned with values from closed form relations derived from a realistic model proposed by [38]. This model associates the internal length-scale values with the periodicity of geometrical and elastic properties of the osteons. BEM simulations were also performed for different combinations of the microstructural constants whose values were at the order of the osteon's size. Figure 9.2 presents the time–frequency diagrams of the generated surface wave for the different examined cases annotated by superimposing the theoretical dispersion curve of the first-order antisymmetric mode derived from the dipolar elastic theory. It can be seen that Rayleigh wave is dispersive only when microstructural effects are represented by different shear stiffness and inertia internal length-scale parameters. These studies demonstrated that bone's microstructure plays an important role in the propagation of guided modes and should be thus taken into consideration.

9.4.3 2D Studies on Trabecular Bone

Luo et al. [39] presented the first computational study aiming at the investigation of ultrasound wave propagation in trabecular bone. Micro-CT was used to scan a sample of calcaneal trabecular bone to obtain 15 3D data sets. 2D specimens derived from each one of the 3D data sets were then examined to evaluate their respective architectures and densities. Trabecular bone was modeled as a nonhomogeneous medium consisted of trabecular rods and blood. Simulations of ultrasound propagation were performed using the FDM. It was shown that both the density and architecture of trabecular bone have a significant effect on ultrasound propagation.

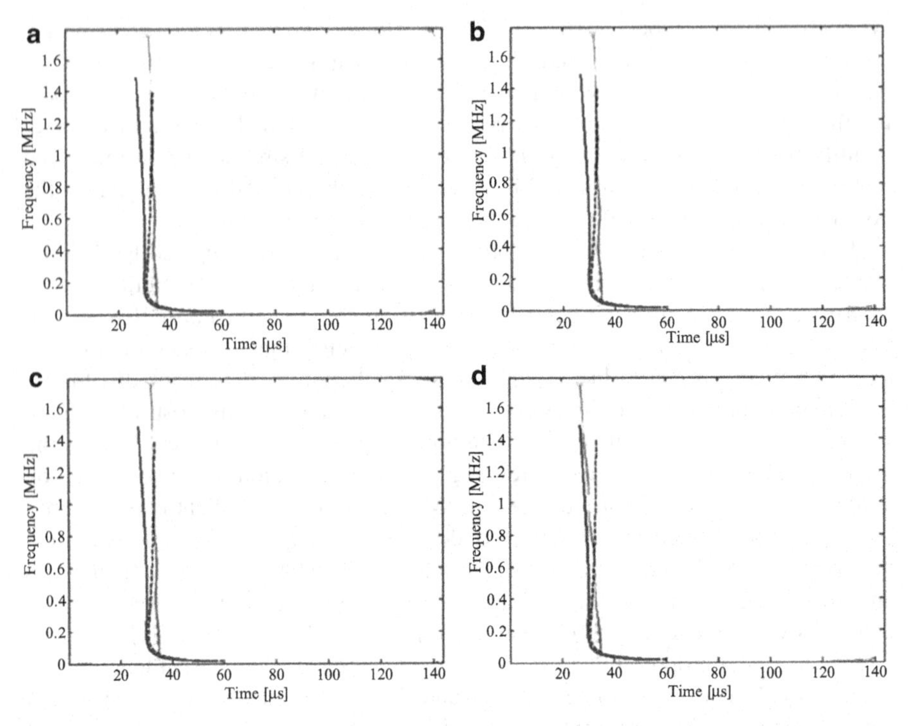

Fig. 9.2 (Color online) Time–frequency diagram of a Rayleigh waves evaluated for (**a**) $l_1 = h_1 = 1.04 \times 10^{-4}$ m, $l_2 = h_2 = 0.74 \times 10^{-4}$ m, (**b**) $h_1 = 1.04 \times 10^{-4}$ m, $l_1 = l_2 = h_2 = 0.74 \times 10^{-4}$ m, (**c**) $l_1 = 1.04 \times 10^{-4}$ m, $l_2 = h_1 = h_2 = 0.74 \times 10^{-4}$ m, (**d**) $l_2 = 1.04 \times 10^{-4}$ m, $l_1 = l_2 = h_2 = 0.74 \times 10^{-4}$ m. The solid line corresponds to first antisymmetric mode evaluated theoretically for $g = 1.04 \times 10^{-4}$ m and $h = 0.74 \times 10^{-4}$ m, whereas the dotted line to the classical elastic theory [37]

Another research group [40–43] used different 2D computational methods to simulate wave propagation in trabecular bones. Specifically, in [40] computational simulations of wave propagation were performed by using both the elastic and Biot's FDTD methods. Trabecular bone was modeled as a nonhomogeneous and anisotropic material composed of numerous randomly distributed trabecular rods and bone marrow. The porosity was also taken into account by changing the dimensions of the trabecular rods. Through-transmission measurements were performed by keeping a constant distance between the transducers. It was shown that by using the elastic FDTD model, neither the fast nor the slow wave could be analyzed due to (a) the insufficient trabecular frame and (b) the non-considering of the viscous loss caused by the pore fluid motion [40]. On the other hand, when the Biot's FDTD method was applied, both waves could be clearly observed. Moreover, it was found that the change in the amplitude ratio of the fast and slow waves is correlated with the porosity variation. In [42], the same research group extended its study to numerically investigate ultrasound propagation in trabecular bone in the directions parallel and perpendicular to the trabecular alignment. Both the popular viscoelastic FDTD method and the Biot's FDTD method for a fluid-saturated porous medium were used to simulate wave propagation. In that study two models were developed with different orientations of the trabecular rods. It was found that the propagation of the Biot's fast and slow longitudinal waves in the direction parallel to the trabecular alignment could be analyzed using Biot's FDTD method rather than the viscoelastic FDTD method, whereas the single wave propagation in the perpendicular direction could be more clearly analyzed using the viscoelastic FDTD method [40, 42]. These studies made clear that the porosity variation and the orientation of trabecular structure strongly affect the ultrasound propagation mechanisms in trabecular bone.

The relation between the ultrasound propagation direction and the trabecular alignment was also investigated in a more recent 2D study [44] by using an FE method for anisotropic porous media (Biot's model) coupled with two acoustic fluids. This method takes into account nonhomogeneous physical characteristics (e.g., porosity, rigidity, permeability) or irregularities of the medium [44]. Trabecular bone was assumed as a fluid-saturated orthotropic poroelastic medium immersed in an acoustic fluid. Through-transmission measurements were performed to investigate the propagation of the transmitted and reflected waves for different angles of incidence. It was shown that the characteristics of the transmitted and reflected waves depend strongly on the anisotropy of trabecular bone. In particular, it was found that when an ultrasound pulse propagates in the direction parallel to the main trabecular alignment, the numerical results include two separate transmitted and reflected waves. On the other hand, when the wave propagates approximately perpendicularly to the main trabecular direction, both the transmitted and reflected waves do not depend perceptibly on the propagation angle.

In a more recent study [29], phase velocity calculations in trabecular bone-mimicking phantoms were performed by using analytical multiple scattering and time-domain numerical approaches. Numerical calculations were derived using the Elastodynamic Finite Integration Technique (EFIT) in which linear elastodynamic equations in heterogeneous media are solved, while the analytical predictions were based on the approximation of Waterman and Truell's (WT2D) corrected model.

Through-transmission measurements were performed for different frequencies by using trabecular bone models with different volume fraction or diameters of the trabecular rods (called scatterers). The numerical and analytical results were in excellent agreement with the experimental findings of Wear [45]. However, the phase velocities computed theoretically were closer to the experimental data, although the numerical EFIT computations correspond to a closer experimental setup, while WT2D is based on a random distribution of scatterers [29].

Analytical studies in trabecular bone models have been also presented by Haiat et al. [46, 48]. In [46] a multiple scattering model was developed aiming at the estimation of the frequency dependence of the phase velocity and at the investigation of the physical mechanisms that lead to a negative velocity dispersion in trabecular bone. The 2D homogenization model derived from [47] was extended to account for viscoelastic absorption effects. Trabecular bone was modeled as a two-phase medium consisted of a viscoelastic matrix including randomly distributed viscoelastic infinite cylinders. It was shown that analytical phase velocity predictions were in agreement with the experimental findings in [45]. Also, it was suggested that scattering effects are responsible for the reported negative dispersion in trabecular bone, whereas the dependence of the absorption coefficient on frequency in the trabeculae and bone marrow leads to a dispersion increase. The same research group in [48] presented a new analytical model of trabecular bone that takes into consideration viscoelastic absorption effects in combination with independent scattering. There are two main differences between this model and the model developed in [46]: (a) the first one refers to the assumption of independent versus multiple scattering and (b) the second one is related to the modeling of the attenuation coefficient in the matrix and in the scatterers [48]. Specifically, in [48] a squared frequency dependence was assumed for the attenuation coefficient, while a linear dependence was reported in [46]. The results were compared with the experimental findings of [45] and the analytical results of [46]. It was found that for low frequencies the independent scattering model better approaches the experimentally measured phase velocities. On the other hand, for high values of scatterer diameter and volume fraction, the results predicted from the multiple scattering model were in better agreement with the experimental results. This was attributed to the significant effect of the multiple scattering phenomena.

9.4.4 3D Models of Cortical Bone

Although 2D computational bone models have played key role in the interpretation of real bone measurements, the incorporation of the 3D bone realistic geometry has provided new supplementary knowledge about the ultrasound wave propagation mechanisms.

The first 3D computational study was published in 2004 aiming to investigate the effect of bone's tubular geometry on FAS wave propagation [49]. Ultrasound measurements were first performed on 3D homogeneous isotropic hollow cylinders with properties equal to those of bone. Anisotropy issues were then taken into account. The influence of the cortical thickness was also examined by changing the inner radius of the model. Numerical simulations were performed for ultrasonic frequencies ranging from 500 kHz to 2 MHz using the FDM. The FAS wave was found not to be affected by the 3D geometry and curvature since its velocity values measured on tubular cortical shells were identical to those measured on bone plates of equal thickness. However anisotropy had a significant influence on the FAS velocity as a function of cortical thickness. It was also made clear that the lateral wave only reflects material and structural changes that occur at a thin periosteal layer with cortical depth approximately from 1 to 1.5 mm for 500 kHz to 2 MHz ultrasound waves.

To this end, two subsequent 3D studies aiming at the assessment of osteoporosis have investigated the propagation of the lowest-order antisymmetric mode

[50, 51]. In the first study [50], the real 3D geometry of bone diaphysis was modeled based on X-ray computed tomography reconstructions of human radius. Axial-transmission measurements were performed using an FDTD algorithm for different transmitter–receiver distances for frequencies from 300 to 500 kHz. Due to the strong circumferential curvature of the human radius, the first antisymmetric plate mode (A0) was replaced by a tubular fundamental flexural mode (F_{11}) in order to enhance the accuracy of the thickness prediction. In order to account for the effect of the soft tissues on guided wave propagation, the authors further extended their model by assuming bone to be a 3D bilayer tube system surrounded of a solid empty inner cavity and an outer radius of a fluid shell [51]. It was shown that the intensity of F_{11} decreases rapidly with increasing the thickness of the fluid shell, showing a corresponding decrease in the normal displacement associated with F_{11}. Thus, F_{11} could only be distinguished from other modes and noise when the thickness of the fluid shell was below 3 mm, indicating that thickness evaluation based on F_{11} is not possible for typical soft tissues surrounding human radius. On the other hand, it was observed that as the thickness of the fluid shell was increasing higher-order modes were dominant. This finding could potentially provide additional valuable information for cortical bone assessment.

The effect of the overlying soft tissues was also investigated in a more recent study [52] modeling the propagation of elastic waves in coupled media mimicking the bone. Bone's hard and soft phases were modeled using 3D elastic, homogeneous, and isotropic elements. The FEM was used for the simulation of the wave propagation problem. The estimation of the stress field showed a clear leakage of wave energy from the hard to the soft phase which could be identified by both the FAS and the second arriving signal (SAS). It was also found that the coupled media affect in a different manner the FAS and the SAS and the most significant changes occur when the soft tissues are initially introduced.

9.4.5 3D Models of Trabecular Bone

Most of the 3D computational models that have been reported in the literature in the context of trabecular bone assessment have been based on high-resolution computed tomography (CT) reconstructions of real bones. Bossy et al. [53] developed computational models of trabecular bone with geometrical and material properties derived from high-resolution synchrotron radiation microcomputed tomography (SR-μCT) of 31 human femur samples (Figs. 9.3, 9.4 and 9.5). Through-transmission ultrasound attenuation and velocity measurements were performed using the FDTD method. It was found that the attenuation varied linearly for frequencies up to 1.2 MHz and the velocity increased with increasing bone volume concentration. In the majority of the examined specimens, the velocity exhibited negative dispersion. It was also shown that two types of waves could be observed (a fast wave and a slow wave travelling slower than waves in water) when

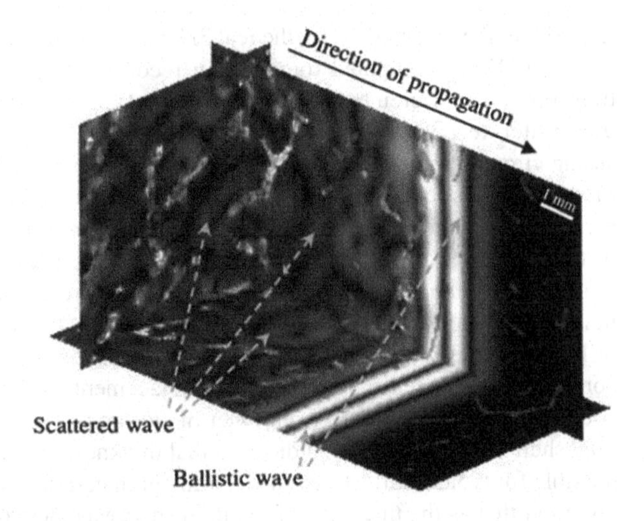

Fig. 9.3 3D snapshot obtained from FDTD calculations, illustrating the propagation of a quasi-plane wave through trabecular bone. The trabecular bone geometry was derived from high-resolution synchrotron computed micro-tomography [53]

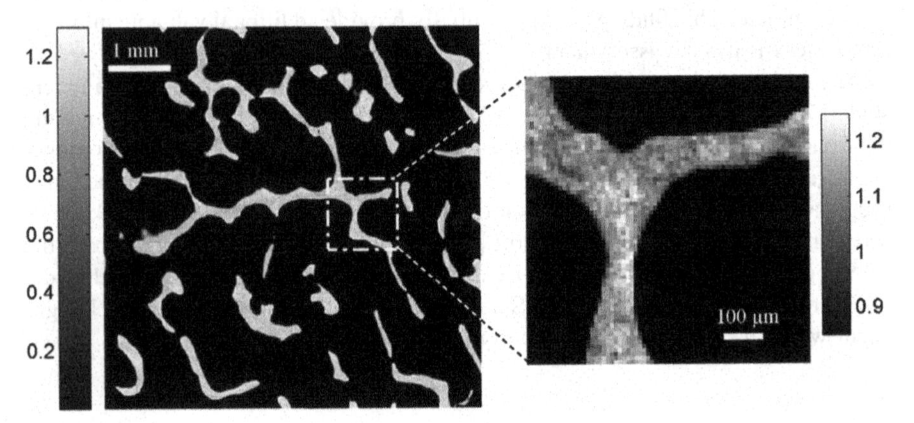

Fig. 9.4 Typical two-dimensional SR-μCT grayscale reconstruction. The pixel size on the images corresponds to the original 10 μm resolution. The grayscale bars indicate the bone tissue mineralization (g cm^{-3} of hydroxyapatite crystals) [53]

the ultrasonic wave propagates parallel to the main orientation of the trabeculae (Fig. 9.3) [53].

The same research group in [54] extended its work to evaluate wave attenuation in trabecular bone models based on SR-μCT of human femur. Numerical simulations were performed using the FDTD method, and the results were compared to experimental findings. Scattering was accounted in the numerical simulations, while absorption was neglected. It was found that numerical simulations can provide normalized broadband ultrasound attenuation (nBUA) values which approach the experimental ones, especially for specimens with low bone volume fraction. Also, it was reported that scattering is the factor that mainly affects nBUA for bone specimens with low bone volume fraction, whereas in

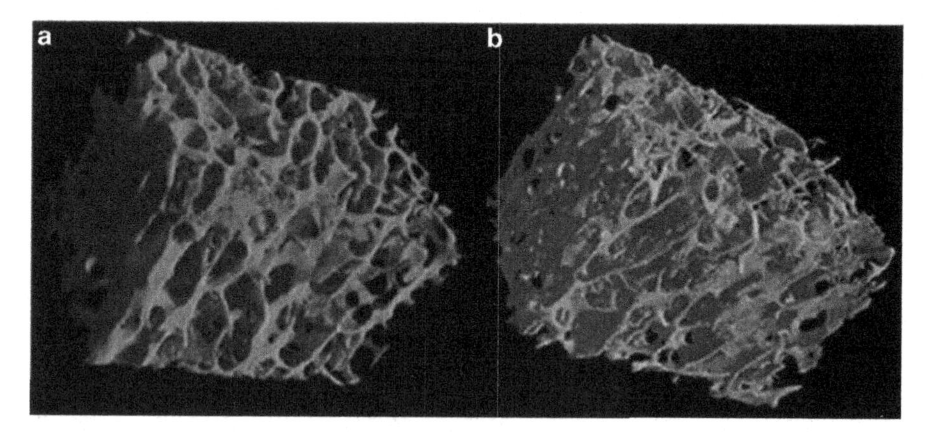

Fig. 9.5 Three-dimensional view of synchrotron micro-tomographic reconstruction of typical dense (**a**) and porous (**b**) trabecular samples [53]

denser specimens the role of other mechanisms such as absorption is significant and should be taken into account.

Wave attenuation in trabecular bone was also examined in [55, 56] by using the FDTD to simulate ultrasound propagation. In [55], three-dimensional X-ray CT data of actual bone samples were used to develop realistic trabecular bone models. Through-transmission measurements were performed using a plane emitter and receiver in order to investigate the generation mechanism and propagation behavior of the fast wave. The attenuation of the fast wave was found to be always more significant in the early state of propagation, while it gradually decreased along the wave propagation path. This phenomenon was attributed to the complicated mechanisms of fast wave propagation in trabecular bone. Also, Hosokawa et al. [56] investigated the effect of porosity on the propagation attenuation and velocity. The complicated pore structure of trabecular bone was examined using a 3D X-ray μCT image. A 3D trabecular bone model was developed consisting of spherical pores in a solid bone. Using a viscoelastic FDTD algorithm, ultrasound propagation through trabecular bone was simulated, and two different directions of propagation (parallel and perpendicular to the main trabecular orientation) were investigated. The porosity was shown to be correlated to wave attenuation and propagation velocity.

The effect of trabecular porosity was also investigated in [57] by examining the ultrasonic propagation of the fast wave in trabecular bone. FDTD numerical simulations were performed for computational models of trabecular bone developed using 34 μCT human femoral specimens. Nonviscous water was used to model the marrow, and bone was assumed to be isotropic, nonabsorbing, and homogeneous. The main trabecular alignment (MTA) and the degree of anisotropy (DA) were examined. DA values were found to range between 1.02 and 1.9, and the bone volume fraction (BV/TV) was varying from 5 % to 25 %. The influence of the BV/TV on the propagation of both the fast and the slow wave was examined, and a heuristic method was used to detect when the two wave modes are time separated. It was shown that both waves overlap in time when the propagation direction is perpendicular to the MTA, whereas when these directions are parallel, both waves are separated in time for samples with high DA and BV/TV values [57]. It was also shown that higher values of the DA correspond to lower values of the BV/TV.

The same research group in [58] investigated numerically the relationship between ultrasonic parameters and trabecular bone elastic modulus. The 3D FDTD numerical computations of wave propagation, the micro-finite element analysis, and the fabric tensor analysis were coupled to 3D segmented digital models of trabecular structure based on the human femoral specimens derived from [57]. Numerical simulations were performed in the three perpendicular directions for each sample and each direction [58]. Bone tissue was assumed to be isotropic, nonabsorbing, and homogeneous. This model neglected the absorption or viscous phenomena that occur during ultrasound propagation. It was shown that when the direction of ultrasound propagation is parallel to the main trabecular orientation, the predictive power of QUS parameters decreases and the fabric tensor analysis provides better results [58]. This decrease was attributed to the presence of two longitudinal wave modes. In addition, in all cases, the combination of BV/TV and fabric tensor was shown to be a more effective indicator than the QUS parameters. Specifically, the fabric tensor analysis was found to be significantly better than QUS parameters in the assessment of the Young's modulus when the direction of testing was parallel to the MTA.

9.5 Ultrasound Wave Propagation in Healing Bones

Ultrasound wave propagation has been also used for the monitoring of the fracture healing process but to a lesser extent than in osteoporosis. Computational modeling in this field has allowed for the examination of ultrasound interaction with a discontinuity in bone that is subsequently filled in by a dynamically changing material, and this has played a major role in following this repairing process and with the aim to devise quantitative criteria that describe its outcome. The most significant findings in this research area are presented herein by focusing on the modeling of the callus geometrical and material properties which evolve during the healing process.

The first computational study examining ultrasound wave propagation in healing bones was presented in [17]. Bone was modeled as a 2D isotropic plate, and the healing process was assumed as a 7-stage process with the callus material properties to vary according to the examined healing stage. Computational simulations were performed using the axial-transmission technique by placing the transducers direct onto the plate's upper surface. The receiver was progressively shifted by 0.5 mm steps with the center-to-center distance from the transmitter to increase from 20 to 35 mm. Numerical solution to the elastic problem was performed using the FD method. The callus was initially modeled by simply filling the fracture gap without

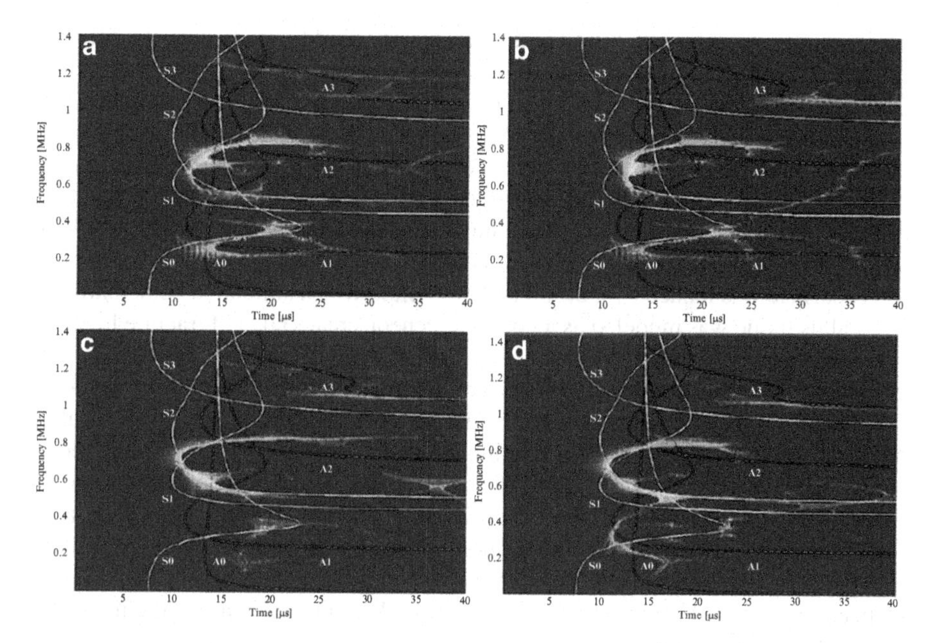

Fig. 9.6 The RSPWV distribution of the signals obtained from (**a**) first, (**b**) second, (**c**) third, and (**d**) fifth stage of healing [17]

considering its geometry. Thereafter simulations were performed by describing the callus geometry with two regions outside the plate borders corresponding to the periosteal and endosteal formation of callus. The axial transmission of ultrasound was simulated by two transducers (transmitter and receiver) placed in direct contact with the plate's upper surface [17]. The excitation frequencies of 500 kHz and 1 MHz were investigated. The results indicated that the FAS velocity decreases during the first and the second healing stages. However an increase was observed at later healing stages gradually approaching the values of intact bone. In addition, the FAS velocity at each stage was not influenced by the excitation frequency or the callus geometry. Although the FAS measurements could provide information for the monitoring of healing, it was made clear that it could not reflect the changes that occurred within the endosteal callus tissue during the healing stages. This limitation was addressed in this study by also investigating the propagation of guided waves as a different means of bone healing evaluation. Signal analysis was performed in the (t, f) domain. The RSPWV distributions of the signals obtained from different healing stages are shown in Fig. 9.6. Mode identification was performed by using the velocity dispersion curves derived from the Lamb wave theory. As shown in Fig. 9.6, the callus geometrical and material properties had a significant effect on the dispersion of the theoretical Lamb modes, with the S2 and A3 modes to dominate.

In a subsequent study [28], the same group further extended their work by addressing more realistic conditions which account for the effect of the soft tissues on guided wave propagation. Three different cases were examined which corresponded to different fluid-loading boundary conditions applied on the 2D models of bone. The obtained signals were analyzed in both time and (t, f) domain. In the first case, the bone was assumed to be immersed in blood which occupied the semi-infinite spaces of the upper and lower surfaces of the plate. In the second case, the bone model was assumed to have the upper surface loaded by a 2-mm thick layer of blood and the lower surface loaded by a semi-infinite fluid with properties close to those of bone marrow. The third case, involved a three-layer model in which the upper surface of the plate was loaded by a layer of blood, whereas the lower surface was loaded by a 2-mm layer of a fluid which simulated bone marrow. The callus tissue was modeled as a nonhomogeneous material, and fracture healing was simulated as a three-stage process. Axial-transmission measurements were carried out by placing a set of transducers in contact to the plate's upper surface, equidistant from the fracture callus. The FAS velocity was found to be practically not affected by the different fluid-loading boundary conditions since it exhibited a similar behavior for all the examined cases. More specifically it was found to decrease at the first healing stage and gradually increase during the second and the third healing stages. On the other hand, guided wave analysis clearly indicated that the application of realistic boundary conditions has a significant effect on the dispersion of guided waves and should be thus taken into account for the interpretation of real measurements.

The same research group in [59] also performed a feasibility study of an alternative ultrasonic configuration in which two of the pins of an already applied external fixation device are used as a means of ultrasound transmission and reception. In particular, the pins of an already applied external fixation device were used as a means of ultrasound transmission and reception. The effectiveness of the proposed technique in the monitoring of the fracture healing process was evaluated by performing velocity measurements on 2D models of intact and healing bones. Bone was modeled as a three-layer isotropic and homogeneous medium, and the FDM was used to simulate wave propagation. The fracture callus tissue was modeled as a nonhomogeneous material consisting of six distinct ossification regions. Two stainless steel pins of an external fixation were modeled and incorporated in the bone model. The pins were inserted into the first three layers of the bone model, and their center-to-center distance was 40 mm. In the case of the healing bone models, the pins were placed equidistant from the fracture callus, and the transmitter and receiver were attached to the extracorporeal tip of the first and the second pin, respectively. Different pin inclination angles were examined. Furthermore, axial-transmission measurements were also performed by using the percutaneous and the transosseous configurations. It was shown that the presence of the pins leads to higher velocity measurements since the waves also travel along the metal medium whose bulk velocity is higher than that of bone. In addition, in all the examined cases, the velocity was found to increase during healing, and this behavior was not influenced by any pin inclination angle.

The next series of computational studies in healing bones investigate the potential of amplitude and attenuation of the FAS to monitor the healing course. In the first study [60], the healing bone was modeled as a 2D isotropic plate with the size of the fracture gap to vary from 1 to 10 mm. Simulations of wave propagation were performed using the FDM that was used to simulate wave propagation in models. Axial-transmission measurements were performed for different distances between the transducers (40–80 mm). The attenuation data were estimated as a sound pressure level (SPL) according to the transducers' distance. The difference in sound pressure levels, denoted as a Fracture Transmission Loss (FTL), at a specific measurement position x was calculated as the difference between the SPL(x) of an intact bone and a fractured specimen. SPL was found to decrease as the distance between the transducers increases.

In the next two subsequent studies of the same group, the effects of different fracture geometries on ultrasound signal loss [61, 62] were investigated. Different healing stages were represented by incorporating different fracture geometries to the plate model. Initially, a simple transverse and oblique fracture filled with water was introduced to simulate the inflammatory stage. Then, a symmetric external callus surrounding a transverse fracture was modeled to represent an advanced stage of healing. Axial-transmission measurements were performed by using the FDM to simulate wave propagation. Human cortical bone was assumed to be an isotropic flat plate, and the transducers were positioned at a constant distance of 5 mm over the bone surface. The results made clear that as opposed to the intact plate model, a large net loss in the signal amplitude was produced for both the simple transverse and oblique geometries. Moreover, the introduction of the geometry of an external callus in the numerical simulations caused a remarkable reduction of the net loss of the signal amplitude. It was also found that the arrival time and the signal amplitude displayed a different variation depending on the receiver position and the fracture geometry for a constant gap width. In the case of the oblique fracture, a decrease in the extra time delay was observed and also an increase in the signal loss of the propagating wave as compared with the transverse fracture. The authors concluded that the FAS amplitude measurements could capture alterations in the callus geometrical and mechanical properties during fracture healing. However, the inhomogeneity of the callus tissue was not considered in this study.

In a more recent similar computational study, this was addressed by assuming callus as a nonhomogeneous medium consisted of six different tissue types with material properties evolving during healing [27]. Cortical bone was assumed to be a homogeneous and isotropic plate with a thickness of 3 mm. The model of the healing bone is shown in Fig. 9.7. Axial-transmission measurements were performed by using the FDM. FAS velocity and SPL measurements were performed for four cases corresponding to daily-changing models of callus as proposed in [62]. A 1-MHz point-source transducer was placed at 20 mm from the center of the fracture gap and a point-receiver transducer at 40 mm from the source. To simulate more realistic conditions, the transducers were placed at a distance of 4.5 mm above the surface of the bone plate. The FAS propagation time was found to decrease during healing, while the callus composition could not well explain the changes in energy attenuation. In all the examined cases, a loss in SPL was reported in the first days after fracture, while a different SPL trend was observed at later stages of healing depending on the model. Moreover, the

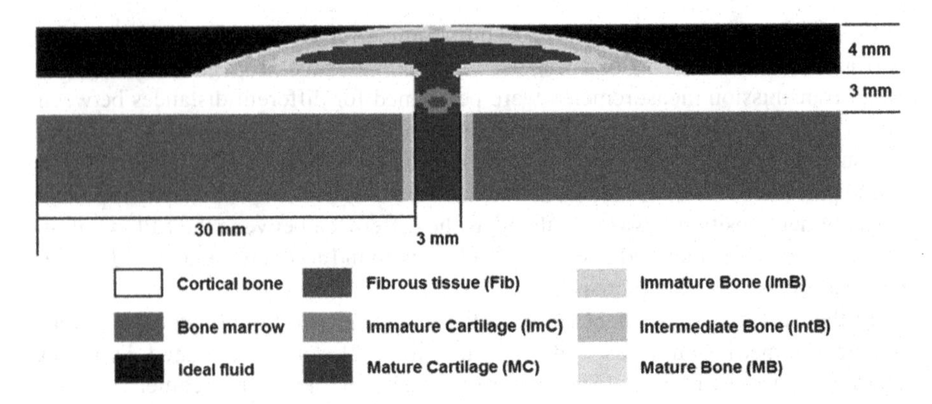

Fig. 9.7 Dimensions and callus tissue composition for the sixth day of bone healing [27]

propagation time was found to be sensitive only to superficial changes in the propagation path.

This study was further extended by also accounting for the influence of cortical bone mineralization on ultrasound axial-transmission measurements [19]. The authors first presented an experimental study in a cortical bovine femur sample with a 3-mm fracture gap. A cortical bone slice, which was extracted from another location in the bone sample, was submitted to a progressive demineralization process with ethylenediaminetetraacetic acid (EDTA) for 12 days. Axial-transmission measurements were performed with the demineralized slice placed into the fracture gap to mimic different stages of mineralization during the healing process. The calcium loss of the slice due to the demineralization process was recorded, and SAM was used to assess the mineralization degree of the bone slice. Thereafter, the experimental conditions were incorporated in computational simulations with the aim to develop a bone model of the time evolution of the callus mechanical properties. The FDM was used to perform axial-transmission measurements by placing the transducers at a distance of 28 mm, positioned 0.5 mm above the plate surface. Both the simulations and the experiments showed a significant and progressive increase in the propagation time during the first 4 days of the demineralization process. Although the simulated measurements were slightly larger than the experimental ones, they both exhibited a similar time-dependence trend. Furthermore, it was suggested that the ultrasound propagation time is affected by changes in local mineralization and could be used as an indicator of bone healing.

Nevertheless in all the aforementioned bone healing studies, the effect of callus porosity was not investigated. To this end, the last two computational studies in the field [63, 64] examine the propagation of ultrasound in healing bones by taking into account the porous nature of callus through the use of SAM images (Fig. 9.8). In particular, in [63] an FD code was used to carry out 2D numerical simulations of wave propagation in realistic computational models of healing long bones developed based on SAM images. Acoustic impedance images were used representing embedded longitudinal sections of 3-mm osteotomies in the right tibia of female Merino sheep [65]. Each SAM image corresponded to a representative healing

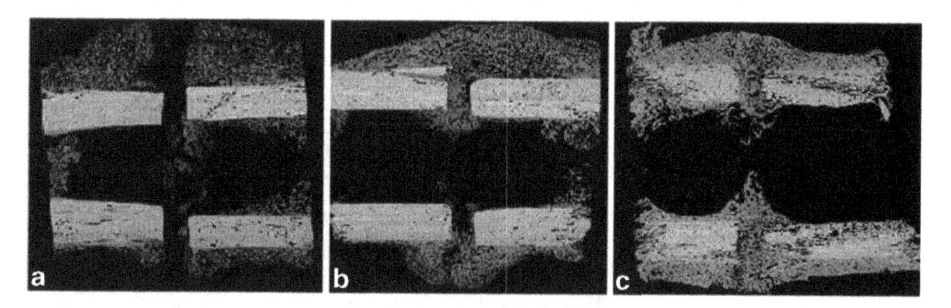

Fig. 9.8 SAM images representing the (**a**) third, (**b**) sixth, and (**c**) ninth postoperative week [63]

stage after 2, 3, 6, and 9 weeks of consolidation. From these maps, the geometry and material properties of cortical and mineralized callus tissues were directly transferred into the simulation model. The histogram was calculated for each one of the calibrated maps of the acoustic impedance. Subsequently, the equipartition of the pixels was performed into 14 material groups. The material properties for each group were defined using empirical relations. Axial-transmission measurements were conducted by placing one transmitter and one receiver on each side of the osteotomy directly onto the cortical bone surface. The center-to-center distance between the transducers was set to 20 mm. Additionally, the transducers were placed on segments of intact cortex in order to measure the velocity of intact bone. Two sets of measurements were performed in order to examine both the upper and the lower surface of the cortex. By plotting the FAS velocity against the excitation frequency, an increase in velocity values was observed in the range of 0.1–0.5 MHz, while for higher frequencies the measurements showed a tendency to reach plateau values. This indicated the dispersive nature of the FAS wave as well as that it changes mode of propagation throughout this frequency range. Guided wave analysis was also performed, and the material and geometrical changes in the callus tissue during healing were found to affect the features of the dominant dispersion modes. The same SAM images were used in [64], to estimate wave dispersion and attenuation in the callus tissue by using an iterative effective medium approximation (IEMA) [66] which is significantly accurate for highly concentrated elastic mixtures. The callus tissue was assumed to be a composite medium consisting of a matrix with spherical inclusions. In week 3, blood was considered as the matrix of the medium and osseous tissue as the material of the spherical inclusions, while the opposite assumption was made in weeks 6 and 9 as at later healing stages the presence of blood is more limited. Group velocity and attenuation estimations were carried out in the frequency range 24–1,200 kHz for different inclusions' diameters and volume concentrations depending on the healing stage. A negative dispersion was observed in all the examined cases, while the attenuation coefficient was found to increase with increasing frequency.

It was shown that the role of scattering, material dispersion, and absorption phenomena is more significant during the early healing stages enhancing wave dispersion and attenuation estimations.

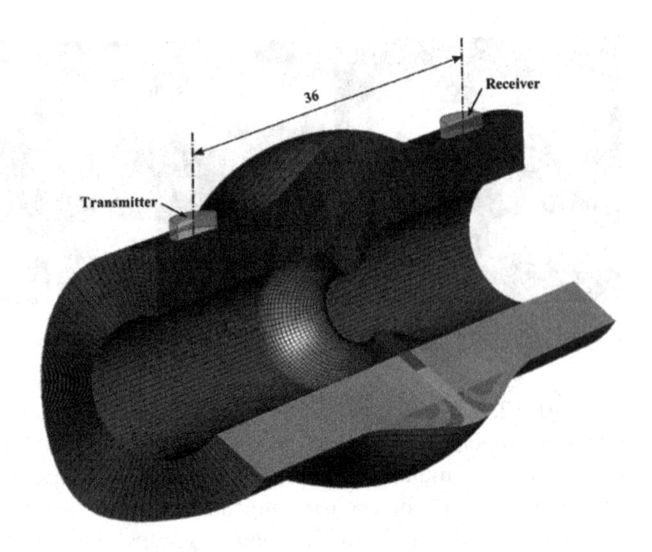

Fig. 9.9 The model of the diaphyseal segment of cortical bone incorporating the fracture callus (sagittal section). The transmitter–receiver configuration is also illustrated [67]

Despite the intensive work that has been performed in the previous 2D bone healing studies, only one model has been presented so far in the literature to account for the 3D irregular geometry of cortical bone (Fig. 9.9) [67]. The cortical bone was modeled as a linear elastic and homogeneous material. In a first series of simulations, it was considered isotropic, whereas in a second series it was considered transversely isotropic in order to achieve more realistic conditions. The model of the callus was similar to that presented in previous 2D computational studies of the same research group [28]. Bone healing was simulated as a three-stage process. Wave propagation in the intact bone model was first studied, and comparisons were made with a simplified geometry using analytical dispersion curves of the tube modes. Then, the influence of callus consolidation on the guided wave propagation was investigated during the healing process. The transmitter and receiver were placed equidistant from the fracture gap, and their center-to-center distance was 36 mm. Concerning the intact bone models, it was shown that the propagation features of the dominant modes were significantly influenced by the irregularity and anisotropy of bone. On the other hand, the FAS wave corresponded to a lateral wave, and the propagation velocity was not influenced by the different material symmetry assumptions. For the healing bone models, guided waves were found to be sensitive to material and geometrical changes that take place during the healing process. It was also demonstrated that the FAS velocity cannot reflect the changes that occur in the whole structure of the callus tissue [67].

9.6 Conclusions

The use of quantitative ultrasound for bone characterization has attracted the interest of many research groups worldwide as it can play significant role in the noninvasive and radiation-free evaluation of metabolic disorders such as osteoporosis as well as in the monitoring of the fracture healing process.

Computational modeling of ultrasonic wave propagation in bone has paved the way for the interpretation of experimental and in vivo findings and has given insight into the mechanisms of interaction of ultrasound with bone. In recent years, the availability of powerful numerical tools in combination with high-resolution 2D and 3D images (SAM, μCT) of the bone structure has facilitated the development of more realistic computational bone models.

While a 2D simulation can often be a starting point when addressing a new problem, the findings, when possible, should be verified with 3D simulations, especially in studies examining significantly nonhomogeneous and anisotropic media such as trabecular bone and fractured bones. Numerical modeling can be performed at several scales, as bone properties and geometrical features differ from the nanostructure to the macrostructure level, and a new trend is multi-scale modeling, i.e., models that extend across different scales. A computational study also depends on multiple parameters which should be carefully chosen to approximate similar conditions to the corresponding in vivo experiment. The ultrasound configuration, the excitation signal, the material properties, the geometric representation, the simulation algorithm, the creation of the mesh, and the size of the element are only a few of the parameters that have to be carefully determined.

The majority of the numerical studies dealing with ultrasound propagation in bone have been based on the FDM, whereas the FEM and the BEM, although popular in other engineering fields, have been applied to a more limited extent. The FDM method is such popular as it has been proved to be a simple and efficient method. Both the FDM and the FE methods depend on space discretization of a specific structure over some mesh. The main advantage of the BEM compared to the FEM and FDM is the reduction in the dimensionality of the problem by one, by discretizing only the boundaries surrounding the examined geometry. Concerning the EFIT, it is considered as a reliable tool to numerically simulate wave propagation in nonhomogeneous media. This method can provide different material scenarios for multiphase media to carry out ultrasonic scanning procedures similar to the experimental setup [29].

In the context of intact and osteoporotic bones, the FAS velocity and the propagation of guided waves have been used as the main indicators for bone assessment. The first studies used simple 2D, isotropic, and homogeneous geometries to model bone [30, 31, 37, 39–42], while tubular geometries were later developed [49–51, 53–58]. It has been shown that the relation between the wavelength and the thickness of the plate or tube plays a key role as the FAS wave under certain conditions propagates as a lateral wave and cannot reflect bone material and structural properties at deeper layers [10, 30, 49]. In addition, the porosity plays a significant role in the FAS wave propagation with the relative change of the propagation time to be highly correlated to the relative change of porosity and tissue elasticity [33, 34]. Anisotropy was also found to have a significant influence on the FAS velocity as a function of cortical thickness. Analysis of the propagation of guided waves has shown the dominance of two main modes: (a) a fast first arriving mode which corresponds to the first symmetric Lamb mode S0 and (b) a slower guided wave contribution corresponding to the propagation of the first antisymmetric mode A0. Computational simulations on immersed plates have also shown that the A0 mode remained the dominant mode being unaffected by the presence of the soft tissues [35]. More recent studies have indicated that bone's microstructure plays an important role in the propagation of guided modes [36, 37]. It should be also noticed that several authors investigating the interaction of ultrasound with the complex trabecular bone geometry have reported that the porosity variation and the orientation of trabecular structure strongly affect the ultrasound propagation mechanisms [40–42, 49, 57]. Ultrasonic attenuation as well as scattering and absorption effects have been estimated numerically and analytically by using realistic geometries based on μCT images [32]. It was suggested that scattering effects are responsible for the observed negative dispersion in trabecular bone, whereas the frequency dependence of the absorption coefficient in bone marrow and in trabecular rods can lead to a dispersion increase [32].

A series of computational studies have been also presented to investigate wave propagation during the fracture healing process. The FAS velocity and attenuation and guided wave analysis have been investigated as significant monitoring means of the healing progress. 2D and 3D computational studies have indicated a FAS velocity decrease at the first healing stages followed by a constant increase as healing progresses [17, 28, 63]. However, when the FAS wave corresponds to a lateral wave, its velocity is sensitive only to the properties of a small superficial region and cannot reflect the gradual restoration of the callus geometrical and material properties occurring at deeper layers of the callus tissue [10]. Other studies estimated the attenuation data by investigating the variation of the sound pressure level during the healing process [51, 60]. It was found that the SPL decreases as the distance between the transducers increases [60]. A loss in SPL was reported in the first days after fracture, while a different SPL variation was observed at later stages of healing depending on the examined bone model [27]. On the other hand, guided waves are sensitive to geometrical and material changes in the callus tissue during the healing process. However, several parameters should be examined carefully as the characteristics of the guided waves are significantly affected by the irregular geometry and anisotropy of the cortical bone and callus as well as by the presence of the soft tissues surrounding cortical bone [28, 67]. More recent studies investigated the evolution of the scattering effects at different healing stages induced by the porous nature of callus. More realistic conditions were applied by using realistic bone models based on SAM images. A negative dispersion was observed for all the examined healing stages, while the attenuation coefficient was found to increase exponentially with increasing frequency. It was shown that the role of scattering, material dispersion, and absorption phenomena is more significant during the early healing stages enhancing wave dispersion and attenuation estimations [63, 64].

In conclusion, the use of QUS for bone assessment is a promising application field as ultrasound systems are low cost, safe, easy to operate, and in some cases portable and wearable. The development of numerical bone models for the assessment of bone pathologies can provide supplementary information to experimental observations and give insight to phenomena that have not been explained yet. Nevertheless, several issues need to be further addressed, such as assignment of realistic mechanical properties and geometries, incorporation of modern imaging modalities that carry information at different scales, and use of advanced constitutive theories depending on the type of bone or the scale at which is examined. Thus, the results derived from numerical studies should always be interpreted with caution preferably in combination with experimental and clinical findings.

References

1. M. Doblaré, J. M. Garciá, and M. J. Gómez, "Modelling bone tissue fracture and healing: a review," Engineering Fracture Mechanics **71**, 1809–1840 (2004).
2. P. Laugier and G. Haïat, "Bone quantitative ultrasound," Springer Dordrecht Heidelberg London New York: Science+Business Media B.V., 4–5 (2011).
3. J.-Y. Rho, L. Kuhn-Spearing, and P. Zioupos, "Mechanical properties and the hierarchical structure of bone," Medical Engineering & Physics **20,** 92–102 (1998).
4. A. Ascenzi, P. Baschieri, and A. Benvenuti, "The torsional properties of single selected osteons," Journal of Biomechanics **27**(7), 875–884 (1994).
5. R. S. Lakes and J. F. C.Yang, "Micropolar elasticity in bone: rotation modulus," 18th Midwest Mechanics Conference, Developments in Mechanics (1983).
6. P. Laugier and G. Haïat, "Bone quantitative ultrasound," Springer Dordrecht Heidelberg London New York: Science+Business Media B.V. (2011).
7. P. Molero, P. H. F. Nicholson, V. Kilappa, S. Cheng, and J. Timonen, "Assessment of the cortical bone thickness using ultrasonic guided waves: modeling and in vitro study," **33**(2), 254–62 (2007).
8. P. J. Meunier, C. Roux, S. Ortolani, M. Diaz-Curiel, J. Compston, P. Marquis, C. Cormier, G. Isaia, J. Badurski, J. D. Wark, J. Collette, and J. Y. Reginster, "Effects of long-term strontium ranelate treatment on vertebral fracture risk in postmenopausal women with osteoporosis," Osteoporosis International **20**(10), 1663–1673 (2009).
9. A. Tatarinov, N. Sarvazyan, and A. Sarvazyan, "Use of multiple acoustic wave modes for assessment of long bones: Model study," Ultrasonics **43**(8), 672–680 (2005).
10. V. C. Protopappas, M. G. Vavva, D. I. Fotiadis, and K. N. Malizos, "Ultrasonic monitoring of bone fracture healing," IEEE Transactions on Ultrasonics Ferroelectrics and Frequency Control **55**, 1243–1255 (2008).
11. J. L. Cunningham, J. Kenwright, and C. J. Kershaw, "Biomechanical measurement of fracture healing," J. Med. Eng. Technol. **14**(3), 92–101 (1990).
12. Y. Nakatsuchi, A. Tsuchikane, and A. Nomura, "Assessment of fracture healing in the tibia using the impulse response method," J. Orthop. Trauma **10**(1), 50–62 (1996).
13. G. Nikiforidis, A. Bezerianos, A. Dimarogonas, and C. Sutherland, "Monitoring of fracture healing by lateral and axial vibration analysis," Journal of Biomechanics **23**(4), pp. 323–330 (1990).
14. Y. Hirasawa, S. Takai, W. C. Kim, N. Takenaka, N. Yoshino, and Y. Watanabe, "Biomechanical monitoring of healing bone based on acoustic emission technology," Clin. Orthop. Relat. Res. **402**, 236–244 (2002).
15. Y. Watanabe, S. Takai, Y. Arai, N. Yoshino, and Y. Hirasawa, "Prediction of mechanical properties of healing fractures using acoustic emission," J. Orthop. Res. **19**(4), 548–553 (2001).
16. L. Claes, R. Grass, T. Schmickal, B. Kisse, C. Eggers, H. Gerngross, W. Mutschler, M. Arand, T. Wintermeyer, and A. Wentzensen, "Monitoring and healing analysis of 100 tibial shaft fractures," Langenbecks Arch. Surg. **387**(3–4), 146–152 (2002).
17. V. C. Protopappas, D. I. Fotiadis, and K. N. Malizos, "Guided ultrasound wave propagation in intact and healing long bones," Ultrasound in Medicine and Biology **32**, 693–708 (2006).

18. G. Barbieri, C. H. Barbieri, N. Mazzer, and C. A. Pelá, "Ultrasound Propagation Velocity and Broadband Attenuation Can Help Evaluate the Healing Process of an Experimental Fracture," Journal of Orthopaedic Research **29**, 444–451 (2011).
19. C. B. Machado, W. C. de Albuquerque Pereira, M. Granke, M. Talmant, F. Padilla, and P. Laugier, "Experimental and simulation results on the effect of cortical bone mineralization measurements: A model for fracture healing ultrasound monitoring," Bone **48**, 1202–1209 (2011).
20. P. J. Gill, G. Kernohan, I. N. Mawhinney, R. A. Mollan, and R. McIlhagger, "Investigation of the mechanical properties of bone using ultrasound," Proc. Inst. Mech. Eng. **203**, 61–63 (1989).
21. E. Maylia and L. D. Nokes, "The use of ultrasonics in orthopaedics – a review," Technol. Health Care **7**(1), 1–28 (1999).
22. S. Saha, V. V. Rao, V. Malakanok, and J.A. Albright, "Quantitative measurement of fracture healing by ultrasound," in Biomed. Engin. I: Recent Developments, Pergamon Press, New York, 247–249 (1982).
23. K. N. Malizos, A. A. Papachristos, V. C. Protopappas, and D. I. Fotiadis, "Transosseous application of low-intensity ultrasound for the enhancement and monitoring of fracture healing process in a sheep osteotomy model," Bone **38**(4), 530–539 (2006).
24. V. C. Protopappas, D. A. Baga, D. I. Fotiadis, A. C. Likas, A. A. Papachristos, and K. N. Malizos, "An ultrasound wearable system for the monitoring and acceleration of fracture healing in long bones," IEEE Trans. Biomed. Eng. **52**(9), 1597–1608 (2005).
25. G. T. Anast, T. Fields, and I. M. Siegel, "Ultrasonic technique for the evaluation of bone fractures," Am. J. Phys. Med. **37**, 157–159 (1958).
26. E. Bossy, M. Talmant, and P. Laugier, "Three-dimensional simulations of ultrasonic axial transmission velocity measurement on cortical bone models," The Journal of the Acoustical Society of America **115**, 2314–2324 (2004).
27. C. B. Machado, W. C. de Albuquerque Pereira, M. Talmant, F. Padilla, and P. Laugier, "Computational evaluation of the compositional factors in fracture healing affecting ultrasound axial transmission measurements," Ultrasound in Medicine & Biology **36**, 1314–1326 (2010).
28. M. G. Vavva, V. C. Protopappas, L. N. Gergidis, A. Charalambopoulos, D. I. Fotiadis, and D. Polyzos, "The effect of boundary conditions on guided wave propagation in two dimensional models of healing bone", Ultrasonics **48**, 598–606 (2008).
29. M. Molero and L. Medina, "Comparison of phase velocity in trabecular bone mimicking-phantoms by time domain numerical (EFIT) and analytical multiple scattering approaches," Ultrasonics **52,** 809–814 (2012).
30. P. Nicholson, P. Moilanen, T. Kärkkäinen, J. Timonen, and S. Cheng, "Guided ultrasonic waves in long bones: modelling, experiment and application," Physiological Measurements **23**, 755–768 (2002).
31. E. Bossy, M. Talmant, and P. Laugier, "Effect of bone cortical thickness on velocity measurements using ultrasonic axial transmission: a 2d simulation study," Journal of the Acoustical Society of America **112**, 297–307 (2002).
32. P. Moilanen, M. Talmant, V. Bousson, P. H. F. Nicholson, S. Cheng, J. Timonen, and P. Laugier, "Ultrasonically determined thickness of long cortical bones: two-dimensional simulations of in vitro experiments," Journal of the Acoustical Society of America **122**, 1818–1826 (2007).
33. V.-H. Nguyen and S. Naili, "Simulation of ultrasonic wave propagation in anisotropic poroelastic bone plate using hybrid spectral/finite element method," International Journal for Numerical Methods in Biomedical Engineering **28**, 861–876 (2012).
34. D. Rohrbach, G. Grondin, P. Laugier, R. Barkmann, and K. Raum, "Evidence based numerical ultrasound simulations at the human femoral neck," Biomedizinische Technik, Rostock, conference proceeding (2010).

35. P. Moilanen, P. H. Nicholson, V. Kilappa, S. Cheng, and J. Timonen, "Measuring guided waves in long bones: modeling and experiments in free and immersed plates, Ultrasound in Medicine and Biology **32,** 709–719 (2006).
36. M. G. Vavva, V. C. Protopappas, L. N. Gergidis, A. Charalambopoulos, D. I. Fotiadis, and D. Polyzos, "Velocity dispersion of guided waves propagating in a free gradient elastic plate: application to cortical bone," Journal of the Acoustical Society of America **125** (2009).
37. A. Papacharalampopoulos, M. G. Vavva, V. C. Protopappas, and D. I. Fotiadis, "A numerical study on the propagation of Rayleigh and guided waves in cortical bone according to Mindlin's Form II gradient elastic theory", Journal of the Acoustical Society of America **130**, 1060–1070 (2011).
38. M. Ben-Amoz, "A dynamic theory for composite materials," Journ. Appl. Math. Phys. **27**, 83–99 (1976).
39. G. M. Luo, J. J. Kaufman, A. Chiabrera, B. Bianco, J. H. Kinney, D. Haupt, J. T. Ryaby, and R. S. Siffert, "Computational methods for ultrasonic bone assessment," Ultrasound in Medicine and Biology **25**, 823–830 (1999).
40. A. Hosokawa, "Simulation of ultrasound propagation through bovine cancellous bone using elastic and biot's finite-difference time-domain methods," Journal of the Acoustical Society of America **118**, 1782–1789 (2005).
41. A. Hosokawa and T. Otani, "Ultrasonic wave propagation in bovine cancellous bone," Journal of the Acoustical Society of America **101**, 558–562 (1997).
42. A. Hosokawa, "Ultrasonic pulse waves in cancellous bone analyzed by finite-difference time domain methods," Ultrasonics **44** (Suppl 1), E227–E231 (2006).
43. A. Hosokawa, "Numerical simulations of change in trabecular structure due to bone remodeling under ultrasound propagation," Journal of Mechanics in Medicine and Biology **13**, (2013).
44. V.- H. Nguyen, S. Naili, and Sansalone, "Simulation of ultrasonic wave propagation in anisotropic cancellous bone immersed in fluid," Wave Motion **47,** 117–129 (2010).
45. K. A. Wear, "The dependencies of phase velocity and dispersion on cancellous thickness and spacing in cancellous bone-mimicking phantoms", Journal of the Acoustical Society of America **118,** 1186–1192 (2005).
46. G. Haiat, A. Lhemery, F. Renaud, F. Padilla, P. Laugier, and S. Naili, "Velocity dispersion in trabecular bone: influence of multiple scattering and of absorption," J Acoust Soc Am **124**(6), 4047–4058 (2008).
47. R. B. Yang, and A. K. Mal, "Multiple-scattering of elastic waves in a fiber-reinforced composite," Journal of the Mechanics and Physics of Solids **42**, 1945–1968 (1994).
48. G. Haïat and S. Naili, "Independent scattering model and velocity dispersion in trabecular bone: comparison with a multiple scattering model," Biomechanics and Modeling in Mechanobiology **10**, 95–108 (2011).
49. E. Bossy, F. Padilla, F. Peyrin and P. Laugier, "Three-dimensional simulation of ultrasound propagation through trabecular bone structures measured by synchrotron microtomography", Phys. Med. Biol. **50,** 5545–5556 (2005).
50. P. Moilanen, M. Talmant, P. H. F. Nicholson, S. L. Cheng, J. Timonen, and P. Laugier, "Ultrasonically determined thickness of long cortical bones: Three-dimensional simulations of in vitro experiments," The Journal of the Acoustical Society of America **122**(4), 2439–2445 (2007).
51. P. Moilanen, M. Talmant, V. Kilappa, P. Nicholson, S. L. Cheng, J. Timonen, and P. Laugier, "Modeling the impact of soft tissue on axial transmission measurements of ultrasonic guided waves in human radius", Journal of the Acoustical Society of America **124**, 2364–2373 (2008).
52. J. Chen, L. Cheng, Z. Su, and L. Qin, "Modeling elastic waves in coupled media: Estimate of soft tissue influence and application to quantitative ultrasound", Ultrasonics **53**, 350–362 (2013).

53. E. Bossy, F. Padilla, F. Peyrin, and P. Laugier, "Three-dimensional simulation of ultrasound propagation through trabecular bone structures measured by synchrotron microtomography," Physics in Medicine and Biology **50**, 5545–5556 (2005).
54. E. Bossy, P. Laugier, F. Peyrin, and F. Padilla, "Attenuation in trabecular bone: a comparison between numerical simulation and experimental results in human femur," Journal of the Acoustical Society of America **122**, 2469–2475 (2007).
55. Y. Nagatani, K. Mizuno, T. Saeki, M. Matsukawa, T. Sakaguchi, and H. Hosoi, "Numerical and experimental study on the wave attenuation in bone – fdtd simulation of ultrasound propagation in cancellous bone," Ultrasonics **48**, 607–612 (2008).
56. A. Hosokawa, "Development of a numerical cancellous bone model for finite-difference time-domain simulations of ultrasound propagation," IEEE Transactions on Ultrasonics Ferroelectrics and Frequency Control **55**, 1219–1233 (2008).
57. G. Haiat, F. Padilla, F. Peyrin, and P. Laugier, "Fast wave ultrasonic propagation in trabecular bone: numerical study of the influence of porosity and structural anisotropy," The Journal of the Acoustical Society of America **123**(3), 1694–1705 (2008).
58. G. Haiat, F. Padilla, M. Svrcekova, Y. Chevalier, D. Pahr, F. Peyrin, P. Laugier, and P. Zysset, "Relationship between ultrasonic parameters and apparent trabecular bone elastic modulus: a numerical approach," Journal of Biomechanics **42**(13), 2033–2039 (2009).
59. M. G. Vavva, V. C. Protopappas, D. I. Fotiadis, and K. N. Malizos "Ultrasound velocity measurements on healing bones using the external fixation pins: a two-dimensional simulation study," Journal of the Serbian Society for Computational Mechanics **2**(2), 1–15 (2008).
60. S. P. Dodd, J. L. Cunningham, A. W. Miles, S. Gheduzzi, and V.F. Humphrey, "An in vitro study of ultrasound signal loss across simple fractures in cortical bone mimics and bovine cortical bone samples," Bone (2006).
61. S. P. Dodd, A. W. Miles, S. Gheduzzi, V. F. Humphrey, and J. L. Cunningham, "Modelling the effects of different fracture geometries and healing stages on ultrasound signal loss across a long bone fracture," Computer Methods in Biomechanics and Biomedical Engineering **10**, 371–375 (2007).
62. S. P. Dodd, J. L. Cunningham, A. W. Miles, S. Gheduzzi, and V. F. Humphrey, "Ultrasound transmission loss across transverse and oblique bone fractures: an in vitro study," Ultrasound in Medicine&Biology **34**(3), 454–462 (2008).
63. V. T. Potsika, V. C. Protopappas, M. G. Vavva, K. Raum, D. Rohrbach, D. Polyzos, and D. I. Fotiadis, "Two-dimensional simulations of wave propagation in healing long bones based on scanning acoustic microscopy images," IEEE International Ultrasonics Symposium, Dresden (2012).
64. V. T. Potsika, V. C. Protopappas, M. G. Vavva, K. Raum, D. Rohrbach, D. Polyzos, and D. I. Fotiadis, "An iterative effective medium approximation for wave dispersion and attenuation estimations in the healing of long bones," 5th European Symposium on Ultrasonic Characterization of Bone, Granada (2013).
65. B. Preininger, S. Checa, F.L. Molnar, P. Fratzl, G.N. Duda, and K. Raum, "Spatial-temporal mapping of bone structural and elastic properties in a sheep model following osteotomy," Ultrasound in Medicine & Biology **37**, 474–483 (2011).
66. D. G. Aggelis, S. V. Tsinopoulos, and D. Polyzos, "An iterative effective medium approximation for wave dispersion and attenuation predictions in particulate composites, suspensions and emulsions," Journal of the Acoustical Society of America **9**, 3443–3452 (2004).
67. V. C. Protopappas, I. C. Kourtis, L. C. Kourtis, K. N. Malizos, C. V. Massalas, and D. I. Fotiadis, "Three-dimensional finite element modeling of guided ultrasound wave propagation in intact and healing long bones", Journal of the Acoustical Society of America **121**, 3907–3921 (2007).

Printed in the United States
By Bookmasters